D0207544

Basics of Physical Stratigraphy and Sedimentology

Basics of Physical Stratigraphy and Sedimentology

William J. Fritz
Department of Geology, Georgia State University, Atlanta

Johnnie N. Moore
Department of Geology, University of Montana, Missoula

John Wiley & Sons, Inc.
New York Chichester Brisbane Toronto Singapore

Library of Congress Cataloging in Publication Data:

Fritz, William J.
Basics of physical stratigraphy and sedimentology.

Bibliography: p. 337
Includes index.
1. Geology, Stratigraphic. 2. Sedimentology.
I. Moore, Johnnie N. II. Title.
QE651.F76 1988 551.7 87-29458
ISBN 0-471-80235-2

Printed in the United States of America

10 9 8 7 6 5 4 3 2 1

To those who taught us
and to those whom we have taught
—we hope the cycle continues.

Preface

We have written this book for students taking their first course in stratigraphy/sedimentology at the university level. Because students in these courses invariably have a variety of backgrounds, we have covered a wide range of material and emphasized basic physical principles. This mixture is intentional. We have presented the concepts of correlation and processes because an understanding of these are integral to understanding facies models and rock sequences. We intend this book to be a precursor to books in which facies models and modern environments are well documented. If used for a quarter- or semester-long course, the instructor might wish to use *Facies Models* (Walker, 1984) as a supplement to provide reading for the various facies. A course following our book might use any of the numerous treatments of modern environments and facies listed in the appropriate chapters. Some may wish we had included this material here. We have not for two reasons. First, this material is adequately addressed in the previously mentioned works; we can think of little to add. Second, we believe that many courses prematurely teach modern environments before students are well-grounded in the basics of correlation, the history of stratigraphy, classification, and fluid dynamics. We believe that this process-oriented background is vital to the correct understanding of modern environments and to interpreting rock sequences.

We intend this book for students with no geology background other than introductory physical and historical geology. To make for easier reading we have referred to few references in the text. Instead, we have lumped important references and outside reading at either the start or end of each chapter; complete citations for these are given in the bibliography at the end of the book. By using these lists it should be possible to find the original readings for the material that we present in each chapter. Many of these references can be assigned as outside reading if the instructor wishes to present material not included in our work. Boldface terms are included in a glossary at the end of the book.

The systematic study of sedimentary rocks began with formulation of the principles of correlation by Smith, Hutton, Lyell, and others in the late eighteenth and early nineteenth centuries. Since those efforts, the discipline of stratigraphy has changed dramatically. Work done during the first hundred years emphasized correlation of strata throughout the world. During the latter stages of this "classical" phase of stratigraphy, various stratigraphers heralded a "new stratigraphy" that not only described in detail the textures of sedimentary rocks and sediment but also used specific grain properties to interpret the processes and environment of deposition. During the early evolution of stratigraphy, only Henry Clifton Sorby made advances in understanding the affects of fluid flow on sedimentary structures and realized the importance of "hydrodynamics" in interpreting sedimentary rocks. By the 1950s and 1960s the science of stratigraphy had advanced again, splitting off the disciplines of sedimentology and sedimentation with the modeling of modern carbonate and terrigenous depositional environments. These major changes in direction from description, classification, and correlation to interpretation by comparison with modern environments were accompanied by experimentation in flumes that led to the flow regime concept. This powerful interpretive model allowed detailed observations of sedimentary structures to be used directly in interpreting paleoflow.

Study of diagenetic processes paralleled these advances in interpreting depositional environments and processes. Thus, an integrated approach can now be used in understanding the formation of sedimentary rock sequences that includes all processes from production of the original sediment by weathering to final lithification and diagenesis after burial. This exciting evolution in the science of soft-rock geology has molded stratigraphy into an integrated, complex system of description and interpretation based on modern equivalents and experimentation that long ago left the realm of "correlation."

Although stratigraphy has advanced toward a complete integration of process sedimentology and a stratigraphic interpretation of ancient deposits, textbooks in the field have not followed suit. Most textbooks on soft-rock geology follow one of two approaches: (1) correlation and principles (typified by Krumbein and Sloss, 1961) and (2) sedimentology (e.g., Friedman and Sanders, 1978). Neither of these two approaches satisfies the needs of a modern undergraduate (sophomore–junior) stratigraphy/sedimentology course that emphasizes interpretation of sedimentary rocks. Such a course must integrate both approaches to describe, classify, and interpret sedimentary sequences. We believe that this text fills this need and differs in content, scope, and purpose from any of the current texts available.

This text may come the closest to that of Dunbar and Rodgers (1957), who discussed the various "principles of stratigraphy" as they were practiced at the time. In this book, we discuss the principles of interpreting layered rocks in Chapters 1 and 2 by tracing the historical development of the science, by discussing the principles of naming rock units as dictated by the North American Stratigraphic Code (included in Appendix A), and by introducing the naming and classification of sedimentary rocks. This material is followed by a description of sedimentary structures in Chapter 3. Chapters 4–8 discuss the principles of sediment transport and deposition through unidirectional flowing water and wind, bidirectional currents, and sediment flows. In Chapter 9, we conclude with the principles of sedimentary facies analyses and an introduction to facies models.

In any project as large as the writing of a text, numerous people have helped with the project. We especially want to thank Don Deneck for his early guidance, inspiration, and encouragement in writing and Cliff Mills, Editor, Kevin Murphy, Designer, Safra Nimrod, Photo Editor, and Pam Pelton, Production Supervisor, at John Wiley & Sons for their hard work and assistance in the details of manuscript preparation. All or portions of the manuscript were reviewed by Scott Brande, University of Alabama at Birmingham, H. Edward Clifton, U.S. Geological Survey, William J. Frazier, Columbus College, Clemens A. Nelson, University of California, Los Angeles, Dave Rubin, U.S. Geological Survey, Kenneth J. Terrell, Georgia State University, and numerous students in our introductory sedimentology and stratigraphy courses at Georgia State University and the University of Montana. We thank these people for their time, effort, and suggestions in helping us to improve both style and content. Welcome drafting assistance was provided by John Cuplin, University of Montana, and Frank Drago, Georgia State University. Bonnie Johnson Fritz provided innumerable suggestions to improve the readability.

We appreciate all this help and criticism, but of course take full responsibility for the content of this textbook. We have not tried to include discussions (detailed or general) about every aspect of stratigraphy and sedimentology. Instead, we have taken a quote from Anatole France literally and offer a book that builds a framework of the important principles of soft-rock geology.

> *Do not try to satisfy your vanity by teaching a great many things. Awaken people's curiosity. It is enough to open minds; do not overload them. Put there just a spark. If there is some good inflammable stuff, it will catch fire.*
>
> *Anatole France,* **The Earth Speaks, 1983**

Atlanta and Missoula — **Bill Fritz**

March 1987 — **Johnnie Moore**

Contents

Chapter 1

Development and Practice of Basic Principles of Stratigraphy

HISTORY AND DEFINITION OF STRATIGRAPHY AND SEDIMENTOLOGY

Until well into the twentieth century the major goal of stratigraphers was to determine the chronological relationship of various sedimentary sequences. They used basic stratigraphic principles developed in the seventeenth and eighteenth centuries to describe and name stratigraphic units and to establish correlations over broad regions. Because of the success of these stratigraphers, we now know the approximate age of nearly every stratigraphic sequence throughout the world and have a fair idea about what units correlate with others across and between continents. In other words, a worldwide stratigraphic framework has been established. From its inception, **stratigraphy** has been equated with Historical Geology, and stratigraphers have strived to fill in the sequence of events of the evolving earth. To understand fully the history and definition of stratigraphy and its offshoot **sedimentology,** it is important to go back in time, unravel the roots of the science, and trace its early development. A look backward often makes it easier to understand the origin of modern thinking and of current controversies in the discipline that have evolved from earlier times. This is especially true for stratigraphic nomenclature, for which a detailed examination of previous work on the unit or use of a proposed term is mandatory.

Many detailed questions remain unanswered; therefore, the needs of doing stratigraphy have changed as more specialized disciplines have been embraced. The goal of modern stratigraphers is to establish details, such as the age, depositional environment, fossil content, and magnetic similarities of stratigraphic packages around the world, as well as to refine the global stratigraphic framework. This endeavor requires different tools from those used to establish local or regional stratigraphic relationships. Moreover, interpreting the details of sedimentary packages out of context of the broad stratigraphic framework, as was done in the past, causes great duress (and rightly so) to those stratigraphers dedicated to establishing earth history. Thus, one must learn the history of geology, its heroes and its villains, to acquire a broader view of the science and thereby understand the modern practice of stratigraphy and sedimentology. Figure 1.1 gives the names of most of the people mentioned in the text and can be used to gain a perspective on the times when they worked.

It is difficult to trace the roots of geology back into ancient times. Certainly many cultures several thousand years ago knew of minerals, operated mines, and used other resources of the earth. However, we are not sure whether they understood the nature of these resources or even cared to investigate their formation. During the Greek period, Halicarnassus (?484–425 B.C.) studied earthquakes and equated those events with faults and movement of the ground. Eratosthenes (276–195 B.C.) studied the earth and arrived at a fairly close estimate (within 20%) of its diameter. As is often the case in science, these accurate observations were

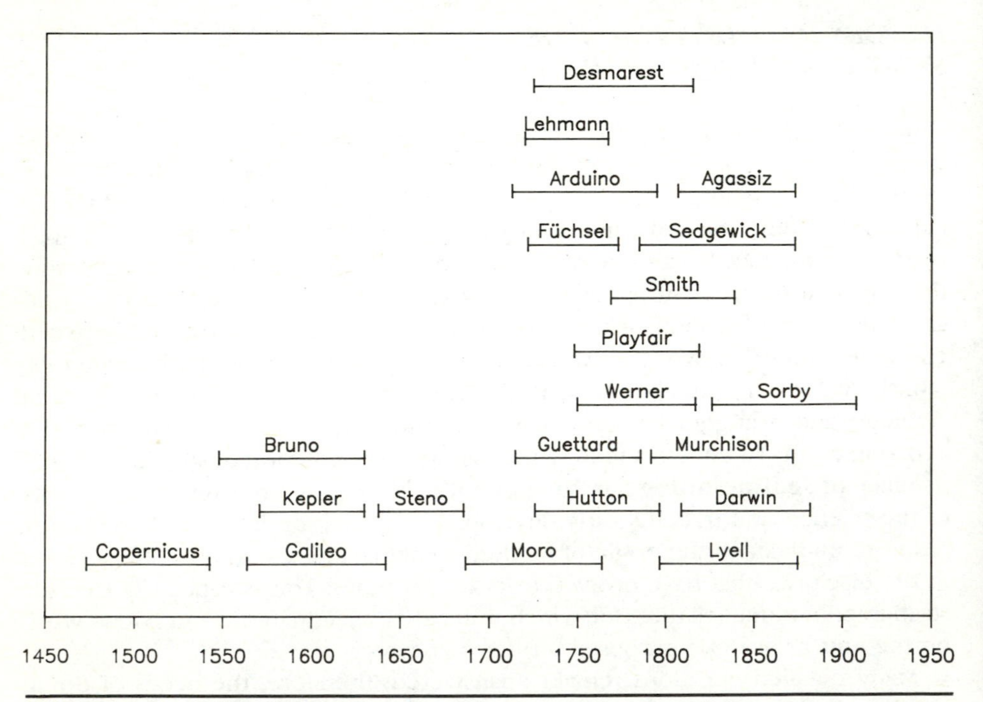

FIGURE 1.1 Table illustrating the life spans of some of the major scientists responsible for the development of the principles of geology.

forgotten and had to be "rediscovered" centuries later. For example, 1800 years after Eratosthenes, most Europeans still thought the earth to be flat!

Even though few great scientific advances were made by the Romans, Pliny the Elder (A.D. 23–A.D. 79) was described by his contemporaries as a "tower of wisdom." He was a tireless worker who spent long hours in the service of his emperor and also was a voluminous reader. Out of his travels and reading, he wrote the encyclopedia *Natural History,* which remained in use for over 1500 years. The eruption of Mount Vesuvius in A.D. 79 was just the kind of natural phenomenon that Pliny, who then was a fleet commander, could not resist investigating. Against the urging of his crew, he sailed his ship into the bay near the ruined Pompeii and was overcome and killed by gases from the eruption. Pliny's nephew, Pliny the Younger, witnessed the entire eruption from a distance and later, in a report describing his uncle's death, detailed an extremely accurate account and interpretation of the Vesuvius eruption. This type of eruption that produces a tall column is still called a **plinian eruption** in his honor.

Over 1700 years after the advances of Eratosthenes and other Greek scholars, most of the scientific community in Europe still believed in an earth-centered universe with the sun and planets revolving around the earth. This explanation, the result of a dogmatic literalist interpretation of scripture, was imposed by the Roman Catholic Church, as well as many protestant denominations, and persisted for centuries. Eventually and inevitably, the theological walls began to crumble. Some of the earliest opponents of such dogmatic theology were sixteenth-century astronomers, who published their views at great personal risk. An early attempt to break with tradition is found in the writings and books of Copernicus (1473–1543), who envisioned a sun-centered universe. However, his writings and books were condemned and banned by the church. Galileo (1564–1642), a proponent of the Copernican view of the universe, was tried and persecuted for his beliefs, and Giordanno Bruno (1548–1600) was burned at the stake during the Inquisition for placing Genesis on a level with the Greek myths (Fenton and Fenton, 1952). Kepler (1571–1630) continued, despite these grim examples, to champion the views of a sun-centered universe.

Even though these great astronomers were in an excellent position to expand their views to a study of the earth, scientific interest had not yet focused on the earth sciences, which were, in the Middle Ages, in more of a shambles than cosmology. Sharks' teeth called *glossopterae* (tonguestones), found along the Italian coast, and Stone Age axes and fossil belemnites called *cerauniae* (thunder wedges) were thought to have fallen from the heavens and were believed to have killed people in their falling (Fig. 1.2). Ironically, one of the first scientists to oppose the church in defense of geology was a young medical doctor from Denmark, Niels Stenson (1638–1686), who arrived in Florence, Italy, at the invitation of the Grand Duke Ferdinand II. Soon after his arrival he was brought the carcass of a shark that had been washed ashore. His dissection of the beast (Fig. 1.3) brought him to the correct conclusion that *glossopterae* were nothing

FIGURE 1.2 ***Glossopterae*** **and** ***ceranuiae*** **were thought to fall from the sky and kill people. This was the popular view of the earth sciences when Steno developed his principles in 1667. From F. D. Adams, 1954,** ***The Birth and Development of the Geological Sciences,*** **Dover, New York, Fig. 22, p. 119. Reprinted by permission.**

more than ancient sharks' teeth, a discovery that prompted him to delve into earth science to determine how one solid, such as a shark's tooth or a fossil clam shell, could be contained in another solid layer of rock. In his *Dissertation of a Solid Naturally Contained within a Solid,* first published in an early form around 1667, Stenson, or Nicoli Stenonis (Steno) as the Italians knew him, outlined what were to become known as Steno's three laws of stratigraphy (Fig. 1.4): that layered rocks were deposited in horizontal beds or **strata,** that these once lay laterally continuous across the surface of the earth, and that they were deposited in chronological succession such that the oldest bed lay at the bottom and the youngest at the top. These principles, termed the law of **original horizontality,** the law of **lateral continuity,** and the law of **superposition,** respectively, were crucial in the development of a correct view of geology. Stenson's medical and anatomical background served him well in his studies of fossil organisms. He boldly proclaimed that even though no living marine organism like the fossil had ever been found, their detailed anatomy shows that they are the remains of organisms rather than the inorganic freaks of nature. Thus, without explicitly stating the principle, Stenson used the present as a key to help interpret the past.

Unfortunately, Stenson could not bear the pressure from the church and the accusation of heresy that his views promoted. Before the publication of his work in 1667, he abjured his native Lutheran Protestantism, converted to Catholicism, and wrote no more on geology. Thus, even though Stenson is best known for his contribution in geology, his entire career in the earth sciences took place before he was 30 and lasted for just over a year.

Even though Stenson had correctly laid the groundwork for later advances in stratigraphy, he misinterpreted many other types of rocks and geologic processes. For example, he was apparently influenced by Descartes (1596–1650) and held that mountains resulted from collapse of subterranean caverns and that volcanoes were not constructional and did not produce new rock. This view was challenged in the early 1700s by Abbé Anton-Lazzaro Moro (1687–1764), who collected accounts of volcanic eruptions that formed islands. Moro was so convinced that

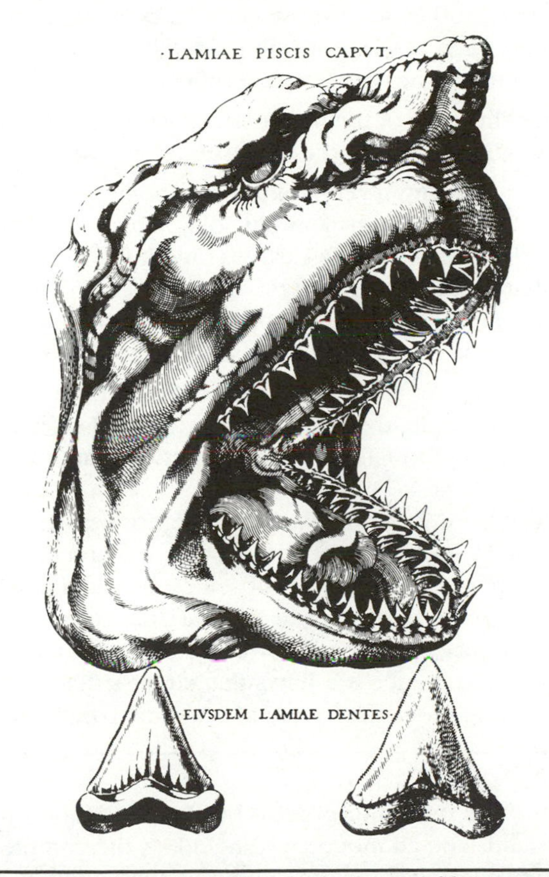

FIGURE 1.3 Drawing of shark published in Steno (1667). The dissection of such a beast allowed Steno to conclude that *glossopterae* were actually fossil shark's teeth. From H. Faul and C. Faul, 1983, *It Began with a Stone,* Wiley, New York, Fig. 4.2, p. 36. Copyright © 1983. Reprinted by permission of John Wiley & Sons, Inc.

FIGURE 1.4 Steno's (1667b) statement of the basic principles of stratigraphy. From *The Prodromus of Nicolaus Steno's Dissertation Concerning a Solid Body Enclosed by Process of Nature within a Solid,* University of Michigan, Humanistic Studies, Vol. XI, Part 2, 1916, translated from the Latin by J. G. Winter, pp. 229–230, Macmillan, New York, 1916.

Concerning the position of strata, the following can be considered as certain:

1. At the time when a given stratum was being formed, there was beneath it another substance which prevented the further descent of the comminuted matter; and so at the time when the lowest stratum was being formed either another solid substance was beneath it, or if some fluid existed there, then it was not only of a different character from the upper fluid, but also heavier than the solid sediment of the upper fluid.
2. At the time when one of the upper strata was being formed, the lower stratum had already gained the consistency of a solid.
3. At the time when any given stratum was being formed it was either encompassed on its sides by another solid substance, or it covered the entire spherical surface of the earth. Hence it follows that in whatever place the bared sides of the strata are seen, either a continuation of the same strata must be sought, or another solid substance must be found which kept the matter of the strata from dispersion.
4. At the time when any given stratum was being formed, all the matter resting upon it was fluid, and, therefore, at the time when the lowest stratum was being formed, none of the upper strata existed.

As regards form, it is certain that at the time when any given stratum was being produced its lower surface, as also its lateral surfaces, corresponded to the surfaces of the lower substance and lateral substances, but that the upper surface was parallel to the horizon, so far as possible; and that all strata, therefore, except the lowest, were bounded by two planes parallel to the horizon. Hence it follows that strata either perpendicular to the horizon or inclined toward it, were at one time parallel to the horizon.

volcanic eruptions were the major process of island formation that he extended his model to all islands and all mountains, including the complex assemblage of rocks on Britain. He became so engrossed by his ideas that even limestone was ascribed a volcanic origin to explain away the reef deposits found on some of his "volcanic" islands. Such erroneous interpretations were based on data collected from books and often accompanied by tremendous speculation. Moro knew nothing of the concepts of precise observation and classification and depended more on scripture to explain volcanic eruptions. Two French geologists working during the latter years of Moro's life developed the procedures needed to interpret volcanic rocks precisely and accurately.

In 1780, sixteen years after Moro died, Jean-Étienne Guettard (1715–1786) presented a monograph of volcanoes found in south-central France, based on field study and comparisons to modern volcanoes and volcanic rock. Even though Guettard accurately mapped these volcanic centers, he still misunderstood the nature of the formation of basalt and argued that it was a chemical precipitate from seawater. Fellow French geologist Nicholas Desmarest (1725–1815), unimpressed by the lack of detail in the initial study by Guettard, continued study on the volcanic centers for 30 years and composed a precisely surveyed map of cones, flow, and other volcanic features. Even though his study was completed much earlier and presented to the Academy of Sciences in Paris in 1765, it was not formally published until 1774 and his final views until 1806. In this study he determined the detailed eruptive history of the region and interpreted correctly that basalt was formed from hot molten lava flows. He based his work on careful description and interpretation of rocks in the field and not on previously established models or the word of the established religious doctrine.

Desmarest developed what Brenner (1980) has more recently termed a process response model. Desmarest described the details of volcanic rock and features, compared them to deposits known to have formed from recently active volcanoes, and only then devised an explanation. This process approach was a major philosophical change in the direction of science in the eighteenth century and was unfortunately submerged beneath the "Geology by Dictum" (Fenton and Fenton, 1952) presented by Abraham Gottlob Werner (Fig. 1.5) later in the century, thus obscuring years of astute scientific progress conducted at great risk to brave men and leaving the field of geology stagnant for decades.

Werner (1750–1817) was an extremely influential leader of scientific thought in the late eighteenth century, a popular teacher, and a charismatic personality. His attainment of such prominence seems difficult to understand. Werner had seldom traveled from his native area of Germany, did not like to write, and hated to read. Yet he dared to teach a total view of the origin of the earth and the geology of every major continent (Fenton and Fenton, 1952). Werner wrote little; most of his ideas are recorded by his students. However, he did publish a short paper that sets down the basic premise of his ideas. Werner's unscientific approach almost jumps out of his writing in numerous phrases such as "we know . . . we are persuaded . . . we are certain . . . we are convinced . . . ."

Werner modified an older scheme developed by others, such as Giovanni Arduino (1714–1795), Johann Gottlob Lehmann (1719–1767), and George Christian Füchsel (1722–1773), and divided all rocks into four series. He interpreted the first and oldest "Primitive" or "Primary" rocks that formed as chemical precipitates from seawater of the "original sea." These included granite, slate, basalt, marble, and schist and were described as devoid of fossils. The next oldest were

FIGURE 1.5 Portrait of Abraham Gottlob Werner. From F. D. Adams, 1954, *The Birth and Development of the Geological Sciences,* Dover, New York, Plate IV. Reprinted by permission.

the *Floetz* or stratified formations, including limestone, sandstone, conglomerate, coal, chalk, and gypsum. Werner interpreted these both as chemical precipitates and as reworked chunks of the "Primary" rocks. Volcanic rocks, Werner argued, represented only a very minor and recent part of the earth and were formed from the burning of limestone and coal from the *Floetz* rocks. Last were the washed deposits, which included alluvium from water of the shrinking "original sea." Eventually, Werner did allow for a fifth division, the "Transition Series" between the "Primary" and *Floetz* series.

Werner's teaching, called *geognosy,* was defined as the "science which treats of the solid body of the earth as a whole and of the different occurrences of minerals and rocks of which it is composed and of the origin of these and their relations to one another" (Hallam, 1983). Werner also took credit for "discovering" that geological formations cover the entire globe in an onionlike fashion.

Eventually, Werner and his followers such as Robert Jameson (1808) became known as the Neptunists because of their views that most rocks formed from chemical precipitation from seawater.

It is certain that Werner's ideas, although a throwback to outdated ideas popular long before his time that were being refuted while he was still an infant, caught on because his original sea from which most of the geologic column was precipitated was easily accounted for in the Biblical account of Noah's flood and fit a 6000-year chronology. In fact, the church actively supported Werner from the pulpit and denounced as heresy those views opposing him. The church's support seems a little surprising because nowhere in Werner's writing does he mention God, Creation, or Moses (Faul and Faul, 1983) and he apparently did not interject any personal religious beliefs into this teaching.

The short chronology for the earth was popularized by John Lightfoot, who in 1642 argued that, according to his study of Genesis, the earth was created at 9:00 A.M. on the 17th of September. Two years later he even specified the year of creation: 3928 B.C. In 1658, Archbishop Ussher, Primate of Ireland, refined the original date of creation to the 23rd of October, 4004 B.C. Archbishop Lloyd added this interpretation as a marginal comment in the great edition of the English Bible in 1701 (Dunbar and Waage, 1969). Following Lloyd's comment, this view was held as divine dogma by the church. Thus, Werner's views of a global geology, consistent with scripture, appealed to a mass public in spite of an overwhelming body of scientific data to the contrary that was known even at the time.

In contrast to the global geology of Werner, James Hutton (1726–1797), a Scottish geologist from Edinburgh (Fig. 1.6), began to develop the concept that geological processes operating today were similar to those that produced much of the geologic column. Hutton spent much of his time describing breaks in the rock record (Figs. 1.7 and 1.8), later called unconformities, and observing modern processes. He argued that vast amounts of time were needed to produce the field relationships that he observed around him.

Furthermore, Hutton saw evidence and argued convincingly that many unconformities and geological contacts were the result of the intrusion of magma. This interpretation was quite controversial in light of Werner's theory that basalts and lavas were chemical precipitates. Hutton recognized two main types of intrusive contacts. At Arthurs Seat, near Edinburgh, Scotland, he recognized sills of basaltic rock that had been intruded into the surrounding sediment. By applying the law of inclusion he argued that the magma must be intrusive and younger than the sediments because it contained blocks of the surrounding sediment included in the basalt. In addition, some of the older rock was baked, leading Hutton to conclude that the intrusive rock was not only molten, but very hot. In another locality, Hutton found evidence of the intrusive nature of granite. In outcrops along a local river, Hutton found granite in contact with schist. Because inclusions or xenoliths of schist were contained within the granite, he argued that the granite was younger and was intruded in a hot molten state. Again, his

FIGURE 1.6 Portrait of James Hutton. From C. L. Fenton and M. A. Fenton, 1952, *Giants of Geology*. Copyright © 1952 by C. L. Fenton and M. A. Fenton. Reprinted by permission of Doubleday & Company, Inc.

observations were at odds with the views of Werner, who held that all granite was part of the Primitive Series that were the original part of the earth's crust.

Hutton, and his spokesman John Playfair (1748–1819), thus established the Volcanist or Plutonist point of view in opposition to that of Werner and the Neptunists. After many decades of bitter argument, the debate was partially resolved to the satisfaction of most scientists by the overwhelming field evidence that basalts originated as lava flows (discussed in Chapter 2) and that much granite is intrusive and younger than the *Floetz* rocks. However, even in light of this field evidence of the magmatic origin of basalt flows, the Neptunists continued to argue for a geologic column that originated as chemical precipitates in the original sea.

Even today this idea is still held in one form or another by a large segment of the population in the United States, who attempt to fit geology into a 6000-year history. For example, some "Scientific" Creationists still argue that gypsum deposits in Southern California formed as a chemical precipitate in the waning stages of Noah's flood, that coal originated as floating mats in the flood, that many limestone reef rocks are debris piled up as windrows by its waves, and that recent volcanism is caused by coal and limestone burning in underground fires. These ideas are still prevalent in the 1980s and illustrate that there is always danger of yielding to nonscientific interpretation of scientific phenomena. To maintain a

FIGURE 1.7 Photograph of the angular unconformity between Silurian turbidites (lower) and overlying conglomerate of the Devonian Old Red Sandstone at Siccar Point, Scotland. This was one of the localities at which Hutton recognized the importance of breaks in the rock record. (Photograph by W. J. Fritz, 1984).

scientific perspective, geologists need to understand the origins of scientific thought and be aware of the pitfalls that have impeded the progress of their field.

As the eighteenth century closed, process-oriented thinking was established by James Hutton (1726–1797) in his *Theory of the Earth with Proofs and Illustrations,* published in 1795, which is the basis for modern geologic interpretation. In his work, Hutton equated modern processes with those involved in the formation of rocks and realized the vast time required for the accumulation of the geologic column. Of special interest to Hutton were breaks in the stratigraphic record called unconformities that seemed to call for enormous amounts of missing time. Hutton, and later followers, have classified these unconformities into **angular unconformities, nonconformities, disconformities,** and **paraconformities.** Hutton and Playfair also argued that the "present is the key to the past" and refined this phrase into the concept of uniformitarianism. This view, that processes and events operating today can be used to interpret the geologic column, was promoted by Charles Lyell (1797–1875) in his textbooks on geology.

However, as in most scientific controversies, it is probably not fair to view one side as completely wrong and the other as right. For example, even though the Neptunist views seem almost ridiculous today, they did attempt to organize the various rock strata into an organized geologic column. In some respects, this was a better approach than that of the Plutonists, who developed no such model of stratigraphy.

FIGURE 1.8 Sketch of the Siccar Point unconformity by Sir James Hall, who acted as an illustrator for Hutton. From G. Y. Craig, D. B. McIntyre, and C. D. Watterson, 1978, *James Hutton's Theory of the Earth: The Lost Drawings,* Scottish Academic Press, Edinburgh, Fig. 42, p. 61. Reprinted by permission of the Scottish Academic Press courtesy of Sir John Clark. Copyright © by Sir John Clark.

The concepts presented by early geologists such as Guettard and Desmarest, James Hutton, and John Playfair established that particular processes account for particular characteristics of geological structures. Such insights provided high drama even into the mid-1800s when Hugh Miller (1860), as a young quarryman, described his work in *The Old Red Sandstone:*

> *The gunpowder had loosened a large mass in one of the inferior strata, and our first employment, on resuming our labors, was to raise it from its bed. I assisted the other workmen in placing it on edge, and was much struck by the appearance of the platform on which it had rested. The entire surface was ridged and furrowed like a bank of sand that had been left by the tide an hour before. I could trace every bend and curvature, every cross hollow and counter ridge of the corresponding phenomena; for the resemblance was no half resemblance—it was the thing itself; and I had observed it a hundred and a hundred times, when sailing my little schooner in the shallows left by the ebb tide.*

Unfortunately, both then and now, many people ascribed such phenomena to Noah's flood. Such mythical explanations do little to advance understanding of earth history. That is why the scientific method of observation requires that stratigraphers and sedimentologists avoid the easy route of comparison with established speculation and instead should always describe the characteristics of a sedimentary rock, determine processes and mechanisms that formed those characteristics, and only then build a model describing the conditions during and after sedimentation. This method is especially important in deciphering the rock record, because there are a number of ancient depositional systems that do not exist today in precisely the same form and, therefore, encourage naive interpretations such as those described above.

The raging controversies between the Neptunists and Volcanists failed to impress others of that time who realized that theology in no way invalidated the principles of Steno for practical application. William Smith (1769–1839), a canal builder in England, became interested in the different layers of rock that his workers had exposed. A keen observer of geology, Smith kept detailed notes as he rode over 10,000 miles each year on horseback and soon recognized an order to the various strata that could be predicted using Steno's three principles. Furthermore, he observed that various layers of rocks contained unique assemblages of fossils, which, together with other characteristics, allowed him to recognize a certain formation and then predict the next layer that his workers would encounter. From these observations, Smith formulated the concept of biotic succession, established the value of fossils in stratigraphic analysis, and presented this information in detailed and accurate geologic maps. Smith's analysis of fossil succession was later championed by Charles Darwin (1809–1882) and used as part of his proof for the theory of organic evolution.

TOWARD THE DEVELOPMENT OF A GEOLOGIC COLUMN

The founders of geology long recognized that there was an order to the rock record and attempted to organize the various layers within the crust into a systematic column. As early as the mid-1700s, Giovanni Arduino (1713–1795), Johann Gottlob Lehmann (1719–1767), and George Christian Füchsel (1722–1773), working in Italy and Germany, developed a three-part "geological column." These views were later expanded by Werner into his five-part subdivision of the crust. Thus, even though Werner held views on the origin of rocks that seem almost ludicrous today, he did recognize that there was an order to the column and a need for a "standard column" that could be applied on a continental scale. In this respect, Werner actually had a superior stratigraphy to that of Hutton and the Plutonists, who proposed no stratigraphy whatsoever, and furthermore saw no need for one (Hallam, 1983). What appeared to finally topple the views of the Neptunists was a better stratigraphy based largely on fossils that was developed by Smith in England and Georges Cuvier (1769–1832) in work on the Paris basin. It is interesting to note that Cuvier did acknowledge an indebtedness to Werner for having developed the idea of a standard stratigraphy.

By the early 1800s, geologists applying and combining the concepts of Steno, Hutton, Smith, Cuvier, and others recognized that there was a complex order in the rock record that could be represented as a geologic column. Work on this column lasted until nearly the end of the nineteenth century and produced the standard geologic column in use today (Fig. 1.9). The Paleozoic (meaning early life) part of the column was developed by geologists working in England and Wales. Roderick Impey Murchison (1792–1872) and Adam Sedgewick (1785–1873), working both separately and jointly, established the Cambrian, Silurian, and Devonian systems, taking their terms from Latin names for Wales and its early tribes. Other divisions soon followed. The Ordovician, also named for an early tribe of Wales, was proposed as a solution to a bitter controversy between Sedgewick and Murchison over rocks intermediate in character and position between Cambrian and Silurian. The Carboniferous was named for rocks containing abundant coal measures in central England. The Permian was later described by Murchison on a trip through Russia and named for rocks in the province of Perm.

The description of the Mesozoic (meaning middle life) was accomplished mostly by German geologists, who named the Triassic for a threefold division of red sandstones and shales in central Germany, the Jurassic for outcrops in the Jura mountains in Switzerland, and the Cretaceous, Latin for chalk, for extensive deposits of this type in Belgium.

Work on the Cenozoic (recent life) proceeded at a slower pace. Many of the Cenozoic terms still in use (Fig. 1.9) stem from outdated concepts of Lyell, who held that he could derive the age of a rock by determining how many of the fossils contained in the rock represented living species. This procedure may have worked for marine fossils in Italy, where he made his observations, but was useless for other areas and for other fossil types. Today the Cenozoic is subdivided by various **biozones** of different types of fossils and by detailed work on lithology. Unfortunately, these divisions are not as easy to establish because the Cenozoic strata contain many continental deposits that change lithologic character abruptly and cannot be correlated for long distances.

Subdivision of the Precambrian has proceeded even more slowly; even today new subdivisions are constantly being proposed based on new knowledge of radiometric dates or on more detailed descriptions. Because Precambrian rocks contain few fossils useful for correlation, it is difficult to arrange them into a coherent stratigraphic order without detailed descriptions, chemical analyses, and radiometric dating techniques that were unavailable to the early geologists. It was probably the development of the radiometric dating technique by Boltwood in 1905 and a more thorough application by Holmes in 1911 that finally allowed for more detailed subdivisions into Archean and Proterozoic eons. Recent work

FIGURE 1.9 Standard geologic column. From A. R. Palmer (comp.), 1983, *Geology,* 11, 504. Reprinted by permission.

DNAG

DECADE OF NORTH AMERICAN GEOLOGY

1983 GEOLOGIC TIME SCALE

GEOLOGICAL SOCIETY OF AMERICA

CENOZOIC

Period		Epoch		Age	Picks (Ma)
QUATERNARY		HOLOCENE			0.01
		PLEISTOCENE		CALABRIAN	1.6
TERTIARY	NEOGENE	PLIOCENE	L	PIACENZIAN	3.4
			E	ZANCLEAN	5.3
		MIOCENE	L	MESSINIAN	6.5
				TORTONIAN	11.2
			M	SERRAVALLIAN	15.1
				LANGHIAN	16.6
			E	BURDIGALIAN	21.8
				AQUITANIAN	23.7
	PALEOGENE	OLIGOCENE	L	CHATTIAN	30.0
			E	RUPELIAN	36.6
		EOCENE	L	PRIABONIAN	40.0
			M	BARTONIAN	43.6
				LUTETIAN	52.0
			E	YPRESIAN	57.8
		PALEOCENE	L	THANETIAN (SELANDIAN)	60.6
				UNNAMED (SELANDIAN)	63.6
			E	DANIAN	66.4

MESOZOIC

Period	Epoch	Age	Picks (Ma)	Uncert (m.y.)
CRETACEOUS			66.4	
	LATE	MAASTRICHTIAN	74.5	4
		CAMPANIAN	84.0	4.5
		SANTONIAN	87.5	
		CONIACIAN	88.5	
		TURONIAN	91	2.5
		CENOMANIAN	97.5	2.5
	EARLY	ALBIAN	113	4
		APTIAN	119	9
	NEOCOMIAN	BARREMIAN	124	9
		HAUTERIVIAN	131	8
		VALANGINIAN	138	5
		BERRIASIAN	144	5
JURASSIC	LATE	TITHONIAN	152	12
		KIMMERIDGIAN	156	6
		OXFORDIAN	163	15
	MIDDLE	CALLOVIAN	169	15
		BATHONIAN	176	34
		BAJOCIAN	183	34
		AALENIAN	187	34
	EARLY	TOARCIAN	193	28
		PLIENSBACHIAN	198	32
		SINEMURIAN	204	18
		HETTANGIAN	208	18
TRIASSIC	LATE	NORIAN	225	8
		CARNIAN	230	22
	MIDDLE	LADINIAN	235	10
		ANISIAN	240	22
	EARLY	SCYTHIAN	245	20

Magnetic polarity: RAPID POLARITY CHANGES

PALEOZOIC

Period		Epoch	Age	Picks (Ma)	Uncert (m.y.)
PERMIAN				245	20
		LATE	TATARIAN	253	20
			KAZANIAN	258	24
			UFIMIAN	263	22
			KUNGURIAN	268	12
			ARTINSKIAN		
		EARLY	SAKMARIAN		
			ASSELIAN	286	12
CARBONIFEROUS	PENNSYLVANIAN	LATE	GZELIAN / KASIMOVIAN (S.)	296	10
			MOSCOVIAN (W.)		
			BASHKIRIAN	315	20
	MISSISSIPPIAN	EARLY	SERPUKHOVIAN (N.)	320	
				333	22
			VISEAN	352	8
			TOURNAISIAN	360	10
DEVONIAN		LATE	FAMMENIAN	367	12
			FRASNIAN	374	18
		MIDDLE	GIVETIAN	380	18
			EIFELIAN	387	28
		EARLY	EMSIAN	394	22
			SIEGENIAN	401	18
			GEDINNIAN	408	12
SILURIAN		LATE	PRIDOLIAN	414	12
			LUDLOVIAN	421	12
		EARLY	WENLOCKIAN	428	8
			LLANDOVERIAN	438	12
ORDOVICIAN		LATE	ASHGILLIAN	448	12
			CARADOCIAN	458	16
		MIDDLE	LLANDEILAN	468	16
			LLANVIRNIAN	478	16
		EARLY	ARENIGIAN	488	20
			TREMADOCIAN	505	32
CAMBRIAN		LATE	TREMPEALEAUAN / FRANCONIAN / DRESBACHIAN	523	36
		MIDDLE		540	28
		EARLY		570	

PRECAMBRIAN

Eon	Era	Bdy. Ages (Ma)
PROTEROZOIC		570
	LATE	900
	MIDDLE	1600
	EARLY	2500
ARCHEAN	LATE	3000
	MIDDLE	3400
	EARLY	3800?

on microfossils, trace fossils, and impressions of soft-bodied metazoa has shown that there is a more complex biotic succession to life in the Precambrian than was long suspected and that there is indeed a rich history of Precambrian life. This succession is proving useful for establishing new subdivisions in these ancient rocks.

> *. . . Some drill and bore*
> *the solid earth, and from the strata there*
> *Extract a register, by which we learn*
> *That he who made it, and reveal'd its date*
> *To Moses, was mistaken in its age.*
>
> ***William Cowper, 1785, from "The Task"***

PRINCIPLES OF STRATIGRAPHY

The early concepts of geology outlined in the preceding history comprise what are now called the classic principles of stratigraphy. Probably foremost of these are the concepts of Steno. The first, that of **original horizontality,** states that regardless of the attitude in which we find sedimentary strata today, they were originally deposited in horizontal layers. Of course, even though this is a valid principle, it has been applied in areas where it is not strictly true. For example, in nonmarine, coarse-grained deposits, deposition can often take place on slopes of up to 30 degrees. Even in marine sections layers often accumulate at angles of a degree or so. These apparent exceptions do not invalidate the original concepts of Steno. Instead they allow us to restate the principle of original horizontality to include all sediments deposited at less than their original angle of repose. Because this angle of repose is nearly horizontal for most sedimentary strata, geologists can interpret folding and the tectonic history from the study of layers that have been deformed from their original horizontality.

Because sediments are deposited as horizontal layers, they are useful for determining relative time. The law of **superposition** holds that in a vertical stack of layered rocks the bottom ones were deposited first and are older than those higher in the sequence. Even though it is possible to imagine exceptions to this law, such as deposition in a cave eroded into and underneath older sediments, these exceptions have never been found except in very local, easy to interpret cases. There is a danger in overapplying this concept, however. Although superposition must always work for a vertical section in one area or outcrop, the bottom beds in a sequence are not necessarily older than the top beds correlated into a different region. The law also applies only to continuous sequences that have not been disturbed by faulting. For example, in thrust terrains and subduction complexes, one can often find a sequence from young rocks at the base to older ones on top. However, close examination reveals that these have been tectonically emplaced and do not represent conditions of deposition.

Johanes Walther (1893–1894) developed the law of facies, which argues that, in modern environments, different sediment types accumulate beside each other. However, as these facies or environments migrate through time they stack up on top of each other. Thus, in a continuous vertical sequence with no major unconformities, all rocks and environments represented must have existed at the same time, spread out across the ancient countryside (Fig. 1.10). This concept means that the lower rocks in a vertical sequence might be the same age, or younger, than rocks at the top of the same sequence but several hundred miles

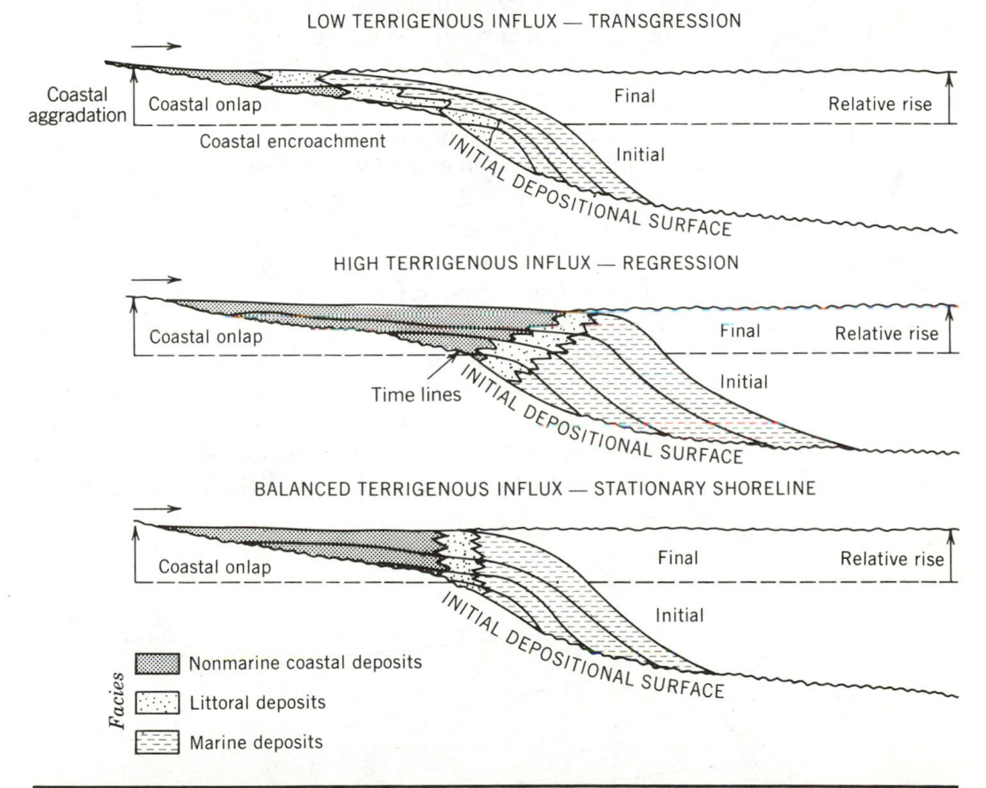

FIGURE 1.10 Drawing to illustrate facies migration resulting from different amounts of sediment supply during a rise in sea level. Note that lithologic units in all cases cross time lines. From P. R. Vail, R. M. Mitchum, Jr., and S. Thompson, III, 1977, Seismic stratigraphy and global changes of sea level. Part 3. Relative changes of sea level from coastal onlap. In C. E. Peyton (Ed.), *Seismic Stratigraphy—Application to Hydrocarbon Exploration,* American Association of Petroleum Geologists Memoir 26, Fig. 3, p. 66. Reprinted by permission of the American Association of Petroleum Geologists.

away. Thus, the law of superposition should only be applied to strata in a given area. A corollary of Walther's law of facies is that only environments that occur in lateral contact with each other in modern settings, or ones that contacted each other in the past, can come to lie on top of each other. This concept gives rise to facies models that relate depositional processes and the resulting sediment type in a variety of modern environments. These facies models will be presented in Chapter 9.

Another major premise of stratigraphy is that of lateral continuity. This concept states that strata were deposited in continuous layers across the depositional basin and may be correlated over distance and across valleys that have cut them (Fig. 1.11). Although this concept is crucial in understanding a regional stratigraphy, it must be limited to a given depositional basin. Indiscriminate application of this law gave rise to early theories of "onion skin" geology that presented the earth as a sphere with concentric shells of sedimentary rock that circled the globe as the skin of an onion (Fig. 1.12). We now know that sedimentary layers pinch out and end at the margin of the depositional basin and also grade into other rock types as the result of facies changes from different time-equivalent depositional environments. A strict insistence on "layer-cake" stratigraphy can also lead to erroneous interpretations. Rock units must be viewed as layers that change thickness, often abruptly in continental environments, end at basin margins, and grade laterally into other lithologic types.

Another basic principle of stratigraphy is that of cross-cutting relationships and its corollary the law of inclusions. These concepts hold that anything that cuts a layer of sedimentary rock is younger than the rock layer. Thus a dike that injects sedimentary strata is younger than the strata. Likewise an inclusion, such as a clast or shell in a conglomerate, must be older and must have existed before the layer was deposited. This may well be one of the most basic of the laws of

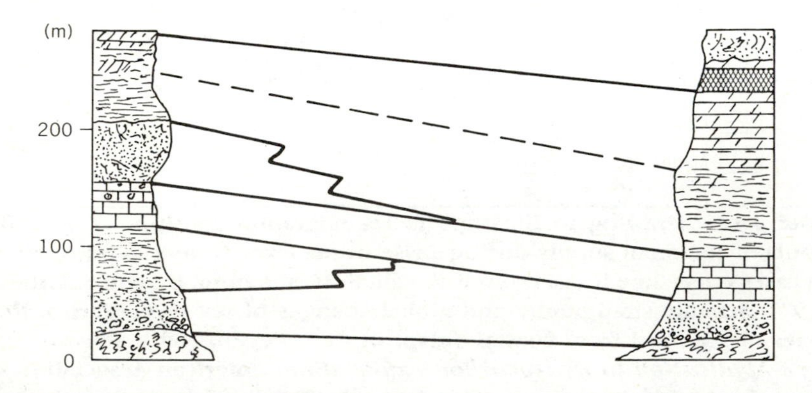

FIGURE 1.11 Correlation of lithologic units between two measured sections.

FIGURE 1.12 Drawing of the earth and the universe with a series of "onion skin" layers. From F. D. Adams, 1938, *The Birth and Development of the Geological Sciences,* Dover, New York, Figs. 2 and 3, p. 58. Reprinted by permission.

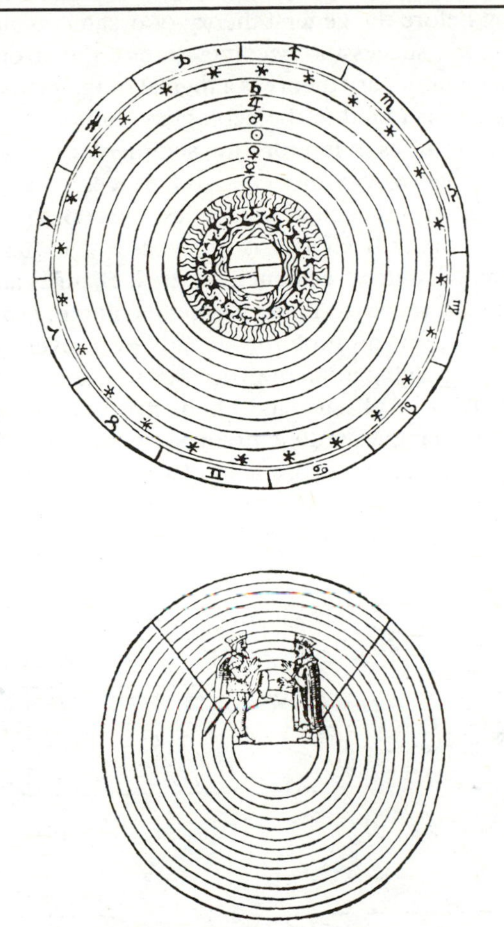

stratigraphy, as it was the one that intrigued Niles Stenson when he investigated the inclusion of a solid within a solid.

The principles just mentioned allowed geologists in the early nineteenth century to develop a basic framework for the succession of rock types on a continental or global scale. The developing relative time scale allowed scientists such as William Smith and Charles Darwin to realize that there is a definite biotic succession to the types of plant and animal fossils found in sedimentary rocks. Remember that this succession was put to practical use by Smith in his canal building enterprise and was interpreted by Darwin and others as having resulted from organic evolution over long periods of time. It is interesting to note that some modern-day "Scientific" Creationists charge that the time scale and geologic column with its different types of life-forms were invented by geologists as a proof of organic evolution. However, this cannot possibly be the case because biotic succession was established long before the general theory of organic evolution gained wide acceptance. Thus, biotic succession becomes one of the strongest cases for organic evolution; the evolutionary theory is a logical outgrowth of and explanation for the sequence of fossils found in the geologic column.

One of the major premises in interpreting sedimentary rocks is that sedimentologists can study modern environments and processes and use information gained in this way to interpret ancient sequences. This does not mean that everything in the past has *always* been as it is today, as a strict uniformitarian interpretation would demand. Indeed, evidence for great changes and vast differences from today can be found recorded in the geologic column, especially in the very oldest rocks of the Archean and Proterozoic. However, even these vast changes can be interpreted by applying modern processes to the past. The basic premise is that fundamental laws of physics and chemistry have remained unchanged throughout time. For example, sedimentologists who study Precambrian sedimentary strata that accumulated before land plants can apply a process-oriented approach to the interpretation of sedimentary structures (such as those described in Chapters 3 and 4) to unravel their unique history.

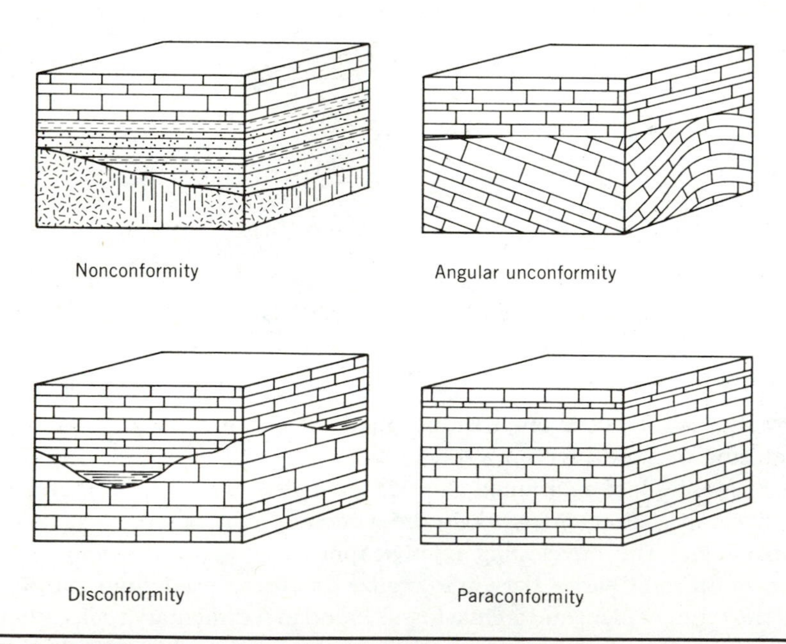

FIGURE 1.13 Drawing of the four basic types of unconformities. From C. O. Dunbar and J. Rodgers, 1957, *Principles of Stratigraphy,* Wiley, New York, Fig. 57, p. 117. Copyright © 1957. Reprinted by permission of John Wiley & Sons, Inc.

To apply the basic principles of stratigraphy just described, one must be able to determine the original stratigraphic position or tops of beds. This is easy in areas of undisturbed, flat-lying layers. However, it is not always easy to apply laws such as superposition in areas where rock layers have been folded and deformed. In a sequence of vertical beds, for example, superposition cannot be applied without knowing the original stratigraphic top of the sequence. Tops of beds are best interpreted from the regional stratigraphic sequence as correlated from areas of little deformation or from an analysis of the sedimentary structures described in Chapter 3.

Another important observation in attempting to establish a regional correlation of rock units is that of completeness or incompleteness of the stratigraphic record. One way to approach this is to ask how much time is missing from a given rock record. Hutton, late in the eighteenth century, recognized the importance of missing time in his description of breaks in the rock record (unconformities). **Angular unconformities** result from the uplift and folding of older strata. These are then planed (leveled) by erosion and new units deposited on top of them (Figs. 1.7, 1.13, 1.14, 1.15, and 1.16), resulting in an angular contact between the two sequences of rocks of different ages. Hutton first recognized the importance of the missing time in such a contact at Siccar Point in southern Scotland (Figs. 1.7 and 1.8). An elegant description of Hutton's impression of the unconformity is provided by Playfair, who was with Hutton at the time of discovery. Playfair's description makes fascinating and informative reading even today.

FIGURE 1.14 Photograph of an angular unconformity between Silurian and Devonian rocks near Siccar Point, Scotland. (Photograph by W. J. Fritz, 1984).

FIGURE 1.15 Angular unconformity at Flint Creek Canyon, near Phillipsburg, southwestern Montana, between Proterozoic rocks of the Belt Supergroup to the right and Middle Cambrian rocks to the left. Even though the difference in the angle of contact is less than 20 degrees, this unconformity represents a hiatus of around two hundred million years. (Photograph by W. J. Fritz, 1984).

On us who saw these phenomena for the first time, the impression made will not easily be forgotten. The palpable evidence presented to us, of one of the most extraordinary and important facts in the natural history of the earth, gave a reality and substance to those theoretical speculations, which, however probable, had never till now been directly authenticated by the testimony of the senses. We often said to ourselves, what clearer evidence could we have had of the different formation of these rocks, and of the long interval which separated their formation, had we actually seen them emerging from the bosom of the deep? We felt ourselves nec-

FIGURE 1.16 Angular unconformity between older tilted Cretaceous marine sediments of the Hudspeth Formation and flat-lying Eocene volcanic sediments of the Clarno Formation in central Oregon south of the town of Mitchell. (Photograph by Dave Alt, 1976).

essarily carried back to the time when the shistus on which we stood was yet at the bottom of the sea, and when the sandstone before us was only beginning to be deposited, in the shape of sand or mud, from the waters of a superincombent ocean. An epocha still more remote presented itself, when even the most ancient of these rocks instead of standing upright in vertical beds, lay in horizontal planes at the bottom of the sea, and was not yet disturbed by that immeasurable force which was burst asunder the solid pavement of the globe. Revolutions still more remote appeared in the distance of this extraordinary perspective. The mind seemed to grow giddy by looking so far into the abyss of time; and while we listened with earnestness and admiration to the philosopher who was now unfolding to us the order and series of these wonderful events, we became sensible how much farther reason may sometimes go than imagination can venture to follow.

John Playfair, 1805

In studying angular unconformities, care must be taken not to use the angle of the contact as an indication of the amount of time missing from the sequence.

FIGURE 1.17 Drawing illustrating that the angle of contact in an angular unconformity depends on the position within the fold and not on the amount of missing time. From C. O. Dunbar and J. Rodgers, 1957, *Principles of Stratigraphy,* Wiley, New York, Fig. 64, p. 123. Copyright © 1957. Reprinted by permission of John Wiley & Sons, Inc.

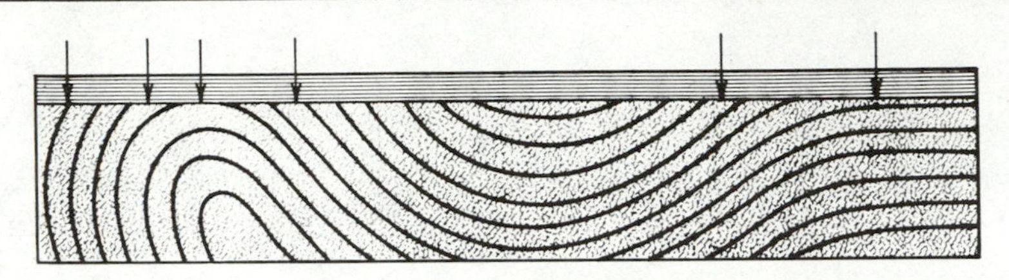

Figure 1.17 shows how all angles of contact can exist from deposition on top of a series of eroded asymmetrical folds. In this case, nearly the same amount of time is missing at the contact. Another error in recognizing angular unconformities in the field is to confuse large-scale sweeping cross beds for an unconformity resulting from tectonic uplift, erosion, and new deposition (Fig. 1.18).

Disconformities occur where an upper and lower sequence are parallel with each other but the contact is an old erosional surface with significant local relief (Figs. 1.19 and 1.20). **Paraconformities** also have two parallel sequences of rock in contact with each other (Fig. 1.21). However, in this case no erosional surface is obvious and the missing time must be inferred from other types of dating such as missing fossils or from radiometric techniques. To avoid confusion between the concepts of missing time and erosional contacts, the entire use of the term unconformity or unconformable contact (first introduced into English from German by Blackwell in 1815) is restricted to apply only to the actual rock surface or interface between vertically adjacent strata. The amount of time not represented and inferred by the surface is termed a **hiatus.** Thus, a hiatus is an intangible concept and must be inferred from field evidence.

Nonconformities, unlike the other types, exist where two very different types of rocks, such as plutonic and sedimentary, are in contact with each other. One type exists where stratified rock lies on top of old crystalline rocks that formed deep within the earth's crust (Fig. 1.22). The four basic types of unconformities found in the field can often grade into one another. Figure 1.23 illustrates how uplift along a mountain front can produce a nonconformity that grades into an angular unconformity that in turn becomes a disconformity, a paraconformity, and finally a continuous sequence out in the basin away from active uplift. The term **diastem** has often been used to describe a local base level above which sediment cannot accumulate. Thus, any diastem represents a short hiatus and missing time.

Another type of stratigraphic contact recognized by Hutton is an intrusive contact. In this case, plutonic or high-level intrusive rocks, such as dikes and sills,

FIGURE 1.18 Horizontal sediments of the Carmel Formation resting on horizontal but cross-bedded Navajo Sandstone produce a field relationship that looks much like a major angular unconformity but is in reality a normal contact. From C. O. Dunbar and J. Rodgers, 1957, *Principles of Stratigraphy,* Wiley, New York, Fig. 65, p. 124. Copyright © 1957. Reprinted by permission of John Wiley & Sons, Inc. (Photograph by Carew McFall).

are in contact with stratified sediments. Such contacts can be interpreted from Steno's law of crosscutting relationships, which states that intrusions cutting a sediment are younger than the beds they cut.

The four basic types of unconformities, with the possible exception of paraconformities, produce field relationships that are easy to recognize. From the existence of these great unconformities, it is easy to infer an incomplete stratigraphic record with vast amounts of time missing from the record. Such in-

FIGURE 1.19 Photograph of a disconformity between Lexington Limestone and Tyrone Limestone in a roadcut on Kentucky Route No. 33. Head of hammer rests on contact between bioclastic calarenite and calcirudite of Curdsville Limestone Member of Lexington above and calcilutite of Tyrone below. Erosional contact is inferred from fragments of Tyrone Limestone in basal bed of Curdsville Limestone Member. (Photograph by E. R. Cressman, U.S. Geological Survey, ER2)

terpretations of missing time are not as obvious in apparently continuous stratigraphic packages. Even today two schools of thought exist about such sections. The first holds that, with the exception of the great unconformities, the sedimentary strata record a nearly complete record of earth's history with little missing time. The other school argues that each bedding surface, no matter how small, represents missing time. Thus, more time may be represented in the bedding surfaces than in the rocks themselves.

Such arguments regarding the completeness of the stratigraphic record strike at the very core of our interpretation of evolution and the need for transitional fossils. Paleontologists and sedimentologists who operate under the assumption of nearly continuous deposition and a complete record of earth's history often must interpret very rapid rates of evolution to explain the sudden appearance of new taxa in the fossil record. On the other hand, sedimentologists and paleontologists who allow for much missing time assume that considerable change could occur during the long time intervals not represented in the rock record. They argue that only the odd sedimentary layer becomes preserved. Most strata are deposited only to be reworked by the next event and thus are only rarely preserved for long-term inclusion in the record. The truth probably lies somewhere in between and must be assessed according to rock type. Sandstone and conglomerate beds often represent deposition in storms or other high-energy episodic point events and thus their bedding surfaces represent great amounts of missing time. Many shales, on the other hand, accumulate slowly from almost continuous deposition and may therefore represent a nearly continuous rock record. The thickness of a bed also has little to do with accumulation rates without considering processes and rock types. For example, storms episodically deposit a great amount of material in a short time, whereas continuous deposition of centuries often forms only thin layers. The concept of correlating missing time with bedding surfaces (Dott, 1983) is illustrated in Fig. 1.24 and 1.25 and Table 1.1. The estimates in Table 1.1 are useful in attempting to analyze the completeness of the stratigraphic record.

In his work on seismic stratigraphy, P. R. Vail and colleagues at Exxon have attempted to interpret missing time from unconformities located by seismic re-

FIGURE 1.20 Disconformable contact, about two-thirds of the way up the cliff, between the lower Mississippian Madison Limestone and the overlying Amsden Formation, Bighorn Canyon, Montana. (Photograph by W. J. Fritz, 1984).

FIGURE 1.21 Presumed paraconformity between the Middle Silurian Louisville Limestone (lower) and Middle Devonian Jefferson Limestone (upper) in Beargrass Quarry, Louisville, Kentucky. From C. O. Dunbar and J. Rodgers, 1957, *Principles of Stratigraphy*, Wiley, New York, Fig. 61, p. 121. Copyright © 1957. Reprinted by permission of John Wiley & Sons, Inc. (Photograph by Charles Schuchert).

TABLE 1.1
Time Required for Deposition of Various Types of Sedimentary Beds

Duration	Event
Seconds	Beach lamination
Minutes	One hummocky bed
Hours	Flash flood or turbidite deposit
Weeks	Scablands flood
1 year	Seasonal varve
10^3 years	One centimeter of pelagic sediment
10^7 years	Large clastic wedge
10^8 years	Cratonic transgressive sequence

Source: From Dott (1983), Table 1, p. 8. Reprinted by permission of the Society of Economic Paleontologists and Mineralogists, Tulsa, Okla.

flection (Fig. 1.26). These profiles are generated by passing seismic energy into the ground by means of explosions or vibrations produced by a large truck. Microphones placed over a broad area record the returning signals reflected back from layers within the earth. The seismic stratigrapher must then correlate these reflections with the known stratigraphy of the basin and must also infer missing time from the unconformities.

STRATIGRAPHIC CODE

Geologists working with the development of the standard geologic column in the late nineteenth century soon realized that it was very important to separate

FIGURE 1.22 Nonconformity. Middle Ordovician Trenton Limestone overlaps Precambrian granite. Montmoreney Falls, east of Quebec City. From C. O. Dunbar and J. Rodgers, 1957, *Principles of Stratigraphy*, Wiley, New York, Fig. 58, p. 118. Copyright © 1957. Reprinted by permission of John Wiley & Sons, Inc. (Photograph by Charles Schuchert).

FIGURE 1.23 Variable expression of an unconformity at the base of the Pennsylvanian system in southeastern Wyoming from uplift into the depositional basin. From C. O. Dunbar and J. Rodgers, 1957, *Principles of Stratigraphy,* Wiley, New York, Fig. 64, p. 123. Copyright © 1957. Reprinted by permission of John Wiley & Sons, Inc.

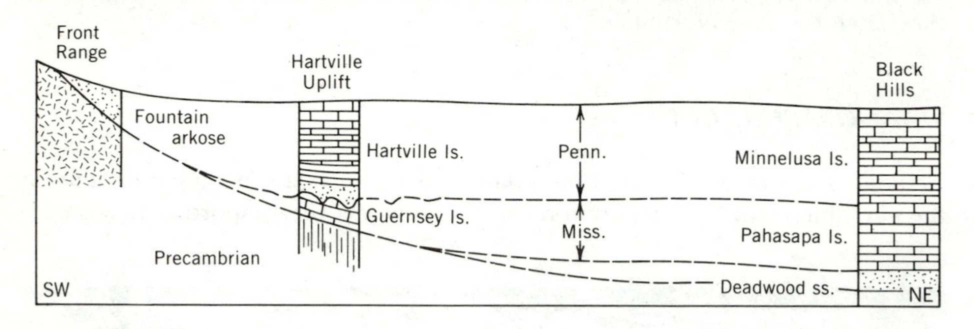

the concept of geologic time and subdivisions of time from a classification of rocks deposited during those periods of time. This distinction led to the development of **geologic time units** to refer to various **periods** of geologic time and **time-rock units** as the tangible **systems** of rocks formed during those periods. Geologic time units are divided, in order of decreasing intervals of time, into eons, eras, periods, epochs, and ages. Time-rock units deposited during these

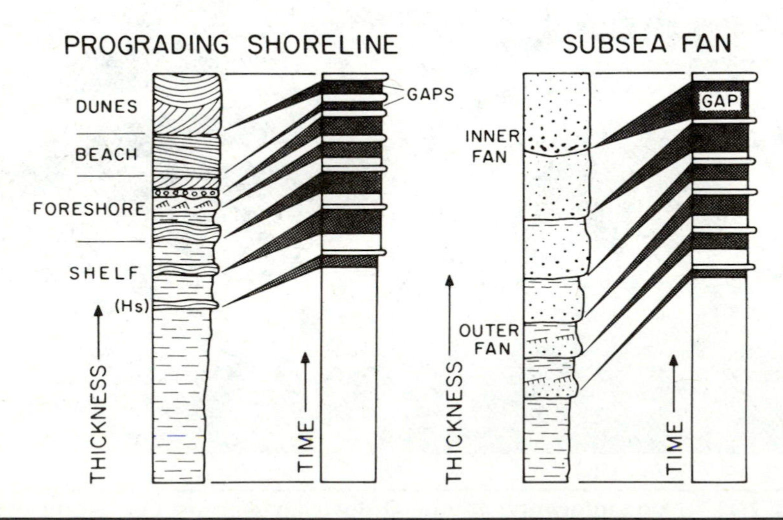

FIGURE 1.24 Diagram illustrating the concept of completeness of the stratigraphic record by a correlation of lithologic breaks and time. From R. H. Dott, Jr., 1983, *Journal of Sedimentary Petrology,* 53, Fig. 3, p. 8. Reprinted by permission of SEPM, Tulsa, Okla.

FIGURE 1.25 Three-dimensional conceptual graph illustrating relationships among volume of material deposited, frequency of event, and degree of postdepositional modification. Minimum preservation potential occurs at the lower right corner and increases along the "preservation vector" to a maximum at the upper right. From R. H. Dott, Jr., 1983, *Journal of Sedimentary Petrology,* 53, Fig. 20, p. 21. Reprinted by permission of SEPM, Tulsa, Okla.

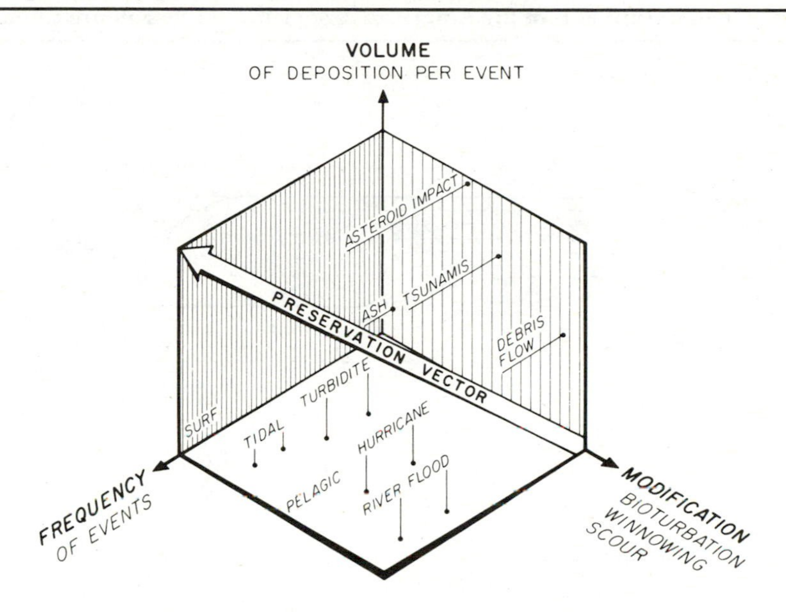

times are broken into erathem, system, series, and stage. The correlation between the rocks deposited during a specified interval of time are given in Table 1.2. Both the time-rock and the time units take their names from the worldwide standard geologic column (Fig. 1.9). Thus it is crucial to follow the name taken from the column by a time term or a time-rock term. For example, the term Cambrian period refers to all time between about 570 and 505 million years ago

TABLE 1.2
Correlation of Terms Used for Geologic Time and Time-Rock Units

Time-Rock Units	Time Units
Eonothem	Eon
Erathem	Era
System	Period
Series	Epoch
Stage	Age

FIGURE 1.26 Diagram illustrating the correlation of unconformities with missing time as interpreted from a seismic profile. From R. M. Mitchum, Jr., P. R. Vail, and S. Thompson, III, 1977, Seismic stratigraphy and global changes of sea level. Part 2. The depositional sequence as a basic unit for stratigraphic analysis. In C. E. Payton (Ed.), *Seismic Stratigraphy—Applications to Hydrocarbon Exploration,* American Association of Petroleum Geologists Memoir 26, Fig. 1, p. 54. Reprinted with permission of the American Association of Petroleum Geologists.

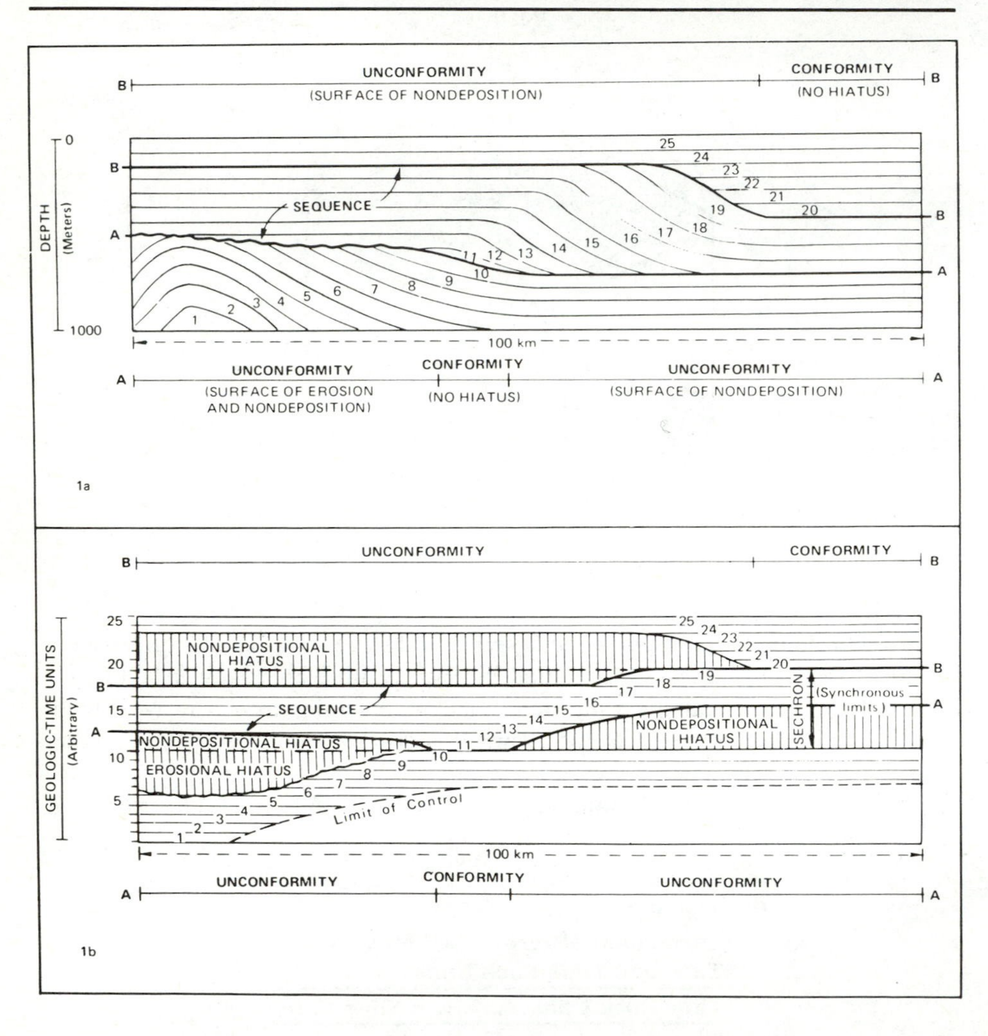

(Ma) and the term Cambrian System refers to all rocks deposited during that period of time.

Another important concept in the subdivision and classification of rocks is that of lithologic units that are independent of time. Rock units are identified by lithologic character with no consideration for age. The standard subdivision of a rock unit is a **formation.** Related formations can be lumped into **groups** and **supergroups,** or subdivided into **members** and **beds** (Table 1.3).

By the early 1900s, it became apparent to government surveys, editors of scientific journals, and all geologists who wished to maintain accuracy in terms and communication that there was a need to formalize the definitions of the major subdivisions of rocks and time. This need led to the publication in 1933 of a stratigraphic code (Committee on Stratigraphic Nomenclature, 1933) that was widely used in North America and, eventually, in one form or another on a worldwide basis by the International Subcommission on Stratigraphic Classification (ISSC, 1976). The most recent (1983) edition of the North American code continues to guide the practice of subdividing and naming rock units.

Because any student of stratigraphy should learn how these units and terms are formalized and named, we have included the entire text of the 1983 code in

TABLE 1.3
Classification of Stratigraphic Units, Followed by the Basic Unit, as Presented in the 1986 North American Stratigraphic Code (Boldface Units Are Discussed in the Text)

Unit	Basic Unit
I. Material Units	
A. **Lithostratigraphic**	**Formation**
B. Lithodemic	Lithodeme
C. Magnetopolarity	Polarity Zone
D. **Biostratigraphic**	**Biozone**
E. Pedostratigraphic	Geosol
F. Allostratigraphic	Alloformation
G. **Unconformity-Bounded Stratigraphic Unit**	**Synthem**
II. Units Related to Geologic Age	
A. Time Units	
1. **Geochronologic**	**Period**
2. Polarity-Chronologic	Polarity Chron
3. Diachronic	Episode
B. Material Units Deposited during Specified Time Spans	
1. **Chronostratigraphic**	**System**
2. Polarity-Chronostratigraphic	Polarity Chronozone

Source: From North American Commission on Stratigraphic Nomenclature (1983), Table 1, p. 848. Reprinted by permission of the American Association of Petroleum Geologists, Tulsa, Okla. Addition of unconformity-bounded stratigraphic unit from ISSC (1987).

Appendix A. In the following paragraphs, we will summarize and explain the most commonly used of these subdivisions. However, for a complete understanding of all terms and methods of classification you should gain a working knowledge of the actual code. It is important to note that the stratigraphic code is a changing system of classification that attempts to formalize the terms in use in the science. As new methods of subdividing rock or time units are discovered, new units are included in the code. Likewise, old terms are deleted from subsequent editions.

The philosophy of the North American Stratigraphic Code is to simplify and standardize stratigraphic classification so that stratigraphers around the world can communicate easily with one another. A major goal of the stratigraphic classification is to allow for correlation of stratigraphic sections. This type of correlation is to establish the relationships between sections based on such criteria as age, fossil content, magnetic properties, and other considerations. To this end, the 1983 code has established eleven categories of stratigraphic units (Appendix A, Table 2). These units (Table 1.3) are divided into material categories based on physical description and categories that express geologic age. The categories of geologic age can also be subdivided into temporal (nonmaterial) units and material categories used to define time spans and rocks that were deposited during a specified portion of geologic time. Table 1.3 gives the various categories of units followed by the name of the basic working subdivision for each. Four of these units, lithostratigraphic, biostratigraphic, chronostratigraphic, and geochronologic, have been widely used by stratigraphers and will be discussed in the text. The additional units should be studied directly from the code in Appendix A and are based on various subdivisions of magnetic properties of rocks, unconformities, and fossil soils.

Lithostratigraphic units, or simply rock units, are probably the most fundamental category to the science of stratigraphy. Units in this category are recognized solely by their unique lithologic character as compared to surrounding units. Lithostratigraphic units can be composed of only one basic rock type or some combination of rock types that in some way makes it recognizable from adjacent units (Fig. 1.27). Contacts between two lithostratigraphic units may be either sharp or gradational. If the gradational zone is large enough, it may, at the discretion of the stratigrapher, be designated as a third unit. When establishing a new lithostratigraphic unit, it is necessary to designate a **type section** that is representative of the unit. The type section should be picked not only for its typical lithologic character but for its accessibility as well. This type section, called the Unit Stratotype by the International Subcommission on Stratigraphic Classification, and the upper and lower contacts (Boundary Stratotype) should be formally designated for the unit (Fig. 1.28).

The basic subdivision of a lithologic unit is the formation. Similar formations are lumped into groups and similar groups into supergroups. Formations may be subdivided into members and members into beds. Figure 1.28, taken from the

FIGURE 1.27 Examples of various lithostratigraphic boundaries and classification. From North American Commission on Stratigraphic Nomenclature, 1983, North American Stratigraphic Code, *American Association of Petroleum Geologists Bulletin,* 67, Fig. 2, p. 857. Reprinted by permission of the American Association of Petroleum Geologists.

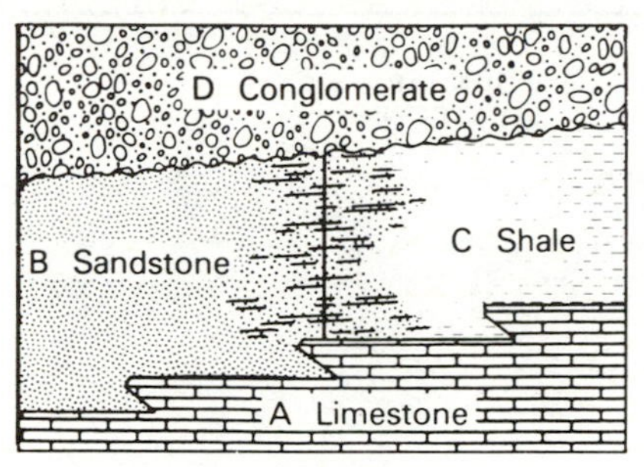

A.--Boundaries at sharp lithologic contacts and in laterally gradational sequence.

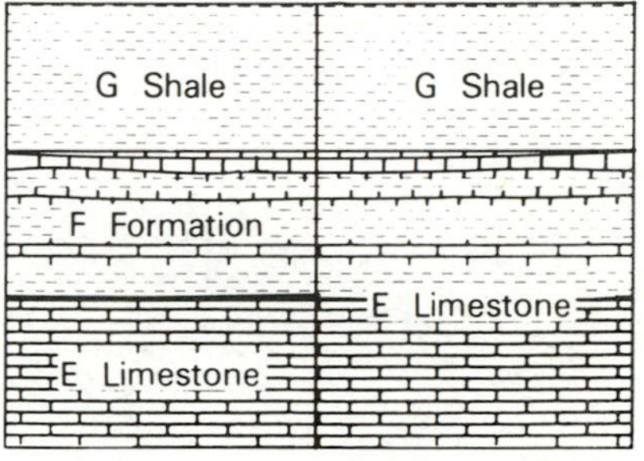

B.--Alternative boundaries in a vertically gradational or interlayered sequence.

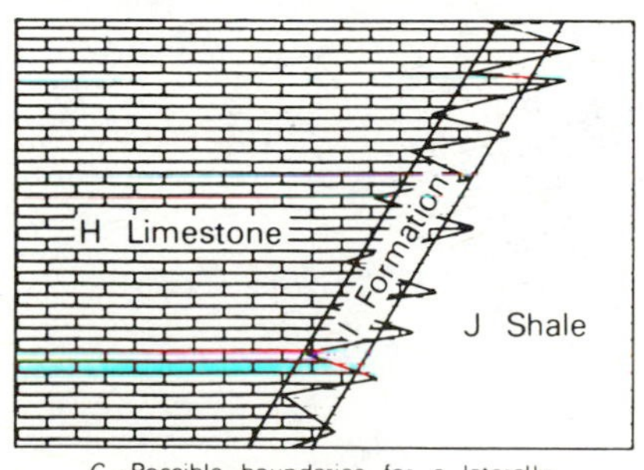

C.--Possible boundaries for a laterally intertonguing sequence.

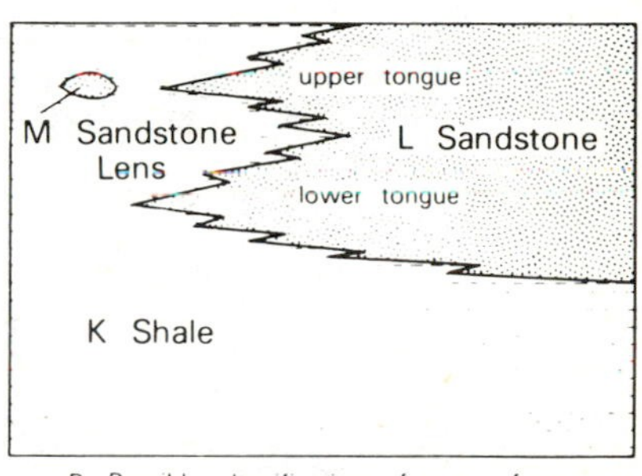

D. Possible classification of parts of an intertonguing sequence.

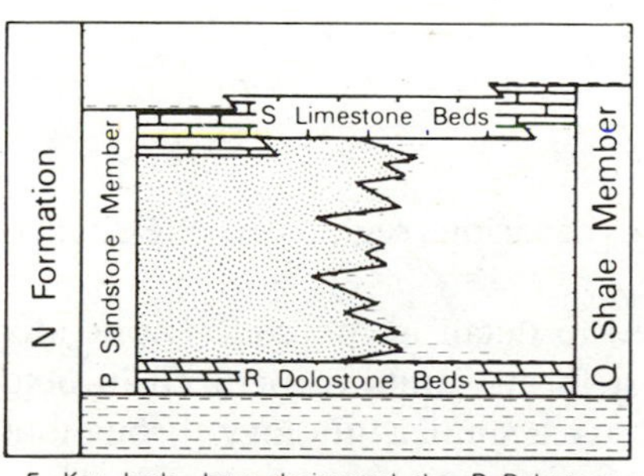

E.--Key beds, here designated the R Dolostone Beds and the S Limestone Beds, are used as boundaries to distinguish the Q Shale Member from the other parts of the N Formation. A lateral change in composition between the key beds requires that another name, P Sandstone Member, be applied. The key beds are part of each member.

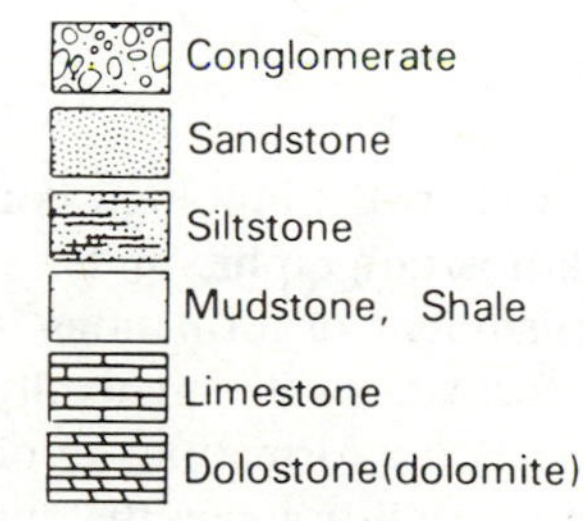

FIGURE 1.28 Illustration of the placement of Unit Stratotype and Boundary Stratotype for (*a*) lithostratigraphic unit and (*b*) biostratigraphic unit. From International Subcommission on Stratigraphic Classification of IUGS Commission on Stratigraphy, H. D. Hedberg (Ed.), 1978, *International Stratigraphic Guide*, Wiley, New York, Fig. 2, p. 25. Copyright © 1978. Reprinted by permission of John Wiley & Sons, Inc.

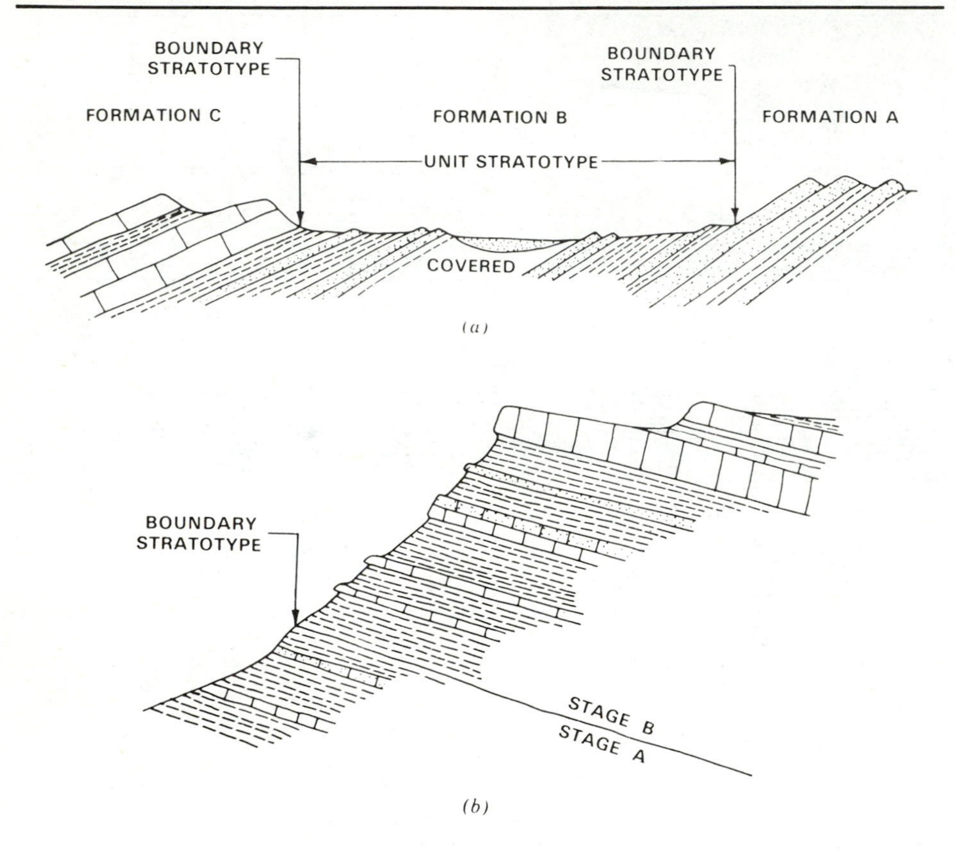

code, illustrates how the lithologic complexities of various regions may be divided into formal lithostratigraphic units.

The establishment of formations is covered in detail in the code (Appendix A). In brief, formations are generally the first formal units recognized in field work. To be called a formation, a unit must have a similar lithology throughout or a combination of lithologies that makes it unique from surrounding units. Thus, a formation must also have upper and lower contacts that may be sharp or gradational. Because formations are field units, they must be mappable, that is, large enough to be shown on a map of the scale used in the region where they are recognized.

Formations that have been designated as formal units are named by a geographic term followed by either a single term that designates the rock type or by the word "formation." Thus, the Rome Formation in northwest Georgia is names for outcrops near Rome, Georgia, and the Three Forks Shale for a formation composed dominantly of shale near Three Forks, Montana. Note that the names of formations are formal names and both parts must be capitalized.

Biostratigraphic units are divisions of rocks based on the fossil content. The only criterion for establishment of a **biozone** is that it contain a unique type or assemblage of fossils as compared to the surrounding rocks. Biozones are established independent of lithologic type or age considerations; thus, the boundaries may or may not be the same as other units. Because biostratigraphic units are based on fossils, they are established with regard to the abundance or occurrence of fossil taxa. Interval biozones are defined as those strata between the lowest and highest stratigraphic occurrence of one or two specified fossil taxa. If this is based on the occurrence of a single taxon, it is called a **taxon range zone.** Often more refined biostratigraphic units can be established based on the simultaneous occurrence in the ranges of more than two taxa. Such **assemblage zones** provide for detailed units and precise correlations. **Abundance zones** are a third type of biozones that are established based on the time of greatest abundance of one or more taxa.

Names of biostratigraphic units are based on the scientific name for the fossil, or most abundant fossil, used to identify the zone. Thus, a unit of strata identified by Miocene fossil clams might be formally named the *Echinophoria apta* biozone, taking its name from the scientific name for that clam.

Chronostratigraphic or **time-stratigraphic units** are defined as the rocks that formed during a specified interval of geologic time. These units traditionally serve as the reference for all the rocks forming during that portion of time. It is important to note that the boundaries of such units are defined as being synchronous, that is, as being the same age everywhere. The basic rank of chronostratigraphic units is the familiar **systems** included in the standard geologic column, which, for the most part, were established during the nineteenth century.

Geochronologic or geologic time units are intangible portions of geologic time often recognized by worldwide patterns in sedimentary rock deposition as seen in chronostratigraphic units. The basic rank is the **period,** which corresponds to the system of the chronostratigraphic units. Periods are grouped into eras and eons and subdivided into epochs and ages.

Recently the International Subcommission on Stratigraphic Classification (1987) has defined a new type of unconformity-bounded stratigraphic unit. These units are recognized as a body of rock bounded at both upper and lower contact by major unconformities (angular unconformity, disconformity, etc.) of regional or interregional extent (Fig. 1.29). These units may contain mixtures of any rock type, fossils, and ages. They are recognized solely on the basis of the unconformities and not on any other criteria. **Synthems,** the basic unconformity-bounded

FIGURE 1.29 Relationship of unconformity-bounded units and (*a*) lithostratigraphic units, (*b*) biostratigraphic units, and (*c*) chronostratigraphic units. From International Commission on Stratigraphic Classification (Amos Salvador, Chairman), 1987, *Geological Society for America Bulletin*, 98, Fig. 1, p. 235.

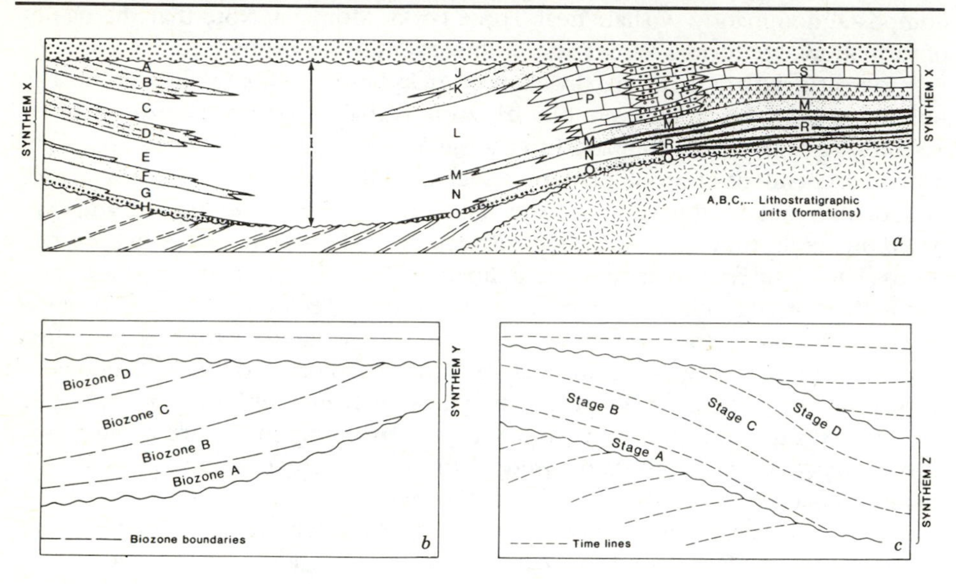

unit, were defined to formalize major unconformity-bounded units that stratigraphers such as Sloss (1963) identified as continent wide transgressive and regressive sequences (Fig. 1.30). More recently, others have proposed that these result from worldwide changes in sea level and therefore may be used to correlate between continents. The next Subcommission meeting will determine if this unit will be adopted and included in the North American Stratigraphic Code.

Other units are recognized by the code. One category of unit is identified on the basis of the remnant magnetism in rocks and is called a **magnetopolarity unit. Pedostratigraphic units** are established on the basis of fossil soils and **allostratigraphic units** on the basis of bounding unconformities. Lithodemic units are used for unstratified, high-grade metamorphic and plutonic rocks. **Polarity-chronologic, diachronic,** and **geochronometric** units are other examples of time units that may help name and classify the stratigraphic record.

It should be apparent from the stratigraphic code that there are many ways to classify, subdivide, and correlate stratified rocks. Any sequence of strata may be classified by all or most of these units to produce units with very different boundaries. Figure 1.31 illustrates a sequence of strata in west Texas that has been correlated on the basis of time and lithology. Note that these produce very different boundaries. When establishing units it is always mandatory to identify

FIGURE 1.30 Unconformity-bounded synthems in North America produced by continental transgressions and regressions. Dark areas represent large hiatuses and light and stippled areas represent deposition. Notice that the synthems thin against the presumed highlands of the cratonic interior. Modified from L. L. Sloss, 1963, *Geological Society of America Bulletin,* 74, Fig. 6, p. 110.

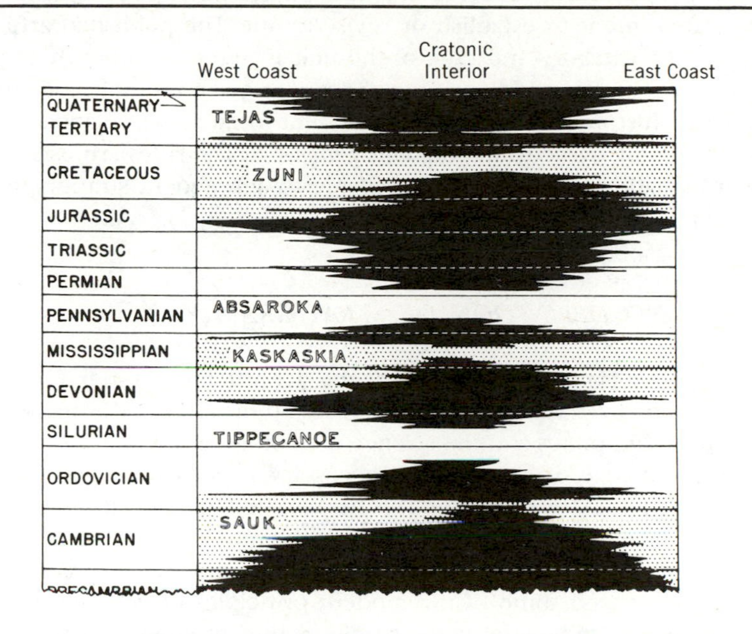

FIGURE 1.31 Correlation of Permian age lithologic units and time lines in the Guadeloupe Mountains–Delaware Basin area of Texas and New Mexico. From F. Meissner, 1972, Cyclic sedimentation in Middle Permian strata of the Permian Basin, West Texas and New Mexico. In J. C. Elam and S. Chuber (Eds.), *Cyclic Sedimentation in the Permian Basin,* 2nd ed., West Texas Geological Society, Fig. 8, p. 215. Reprinted by permission of the West Texas Geological Society.

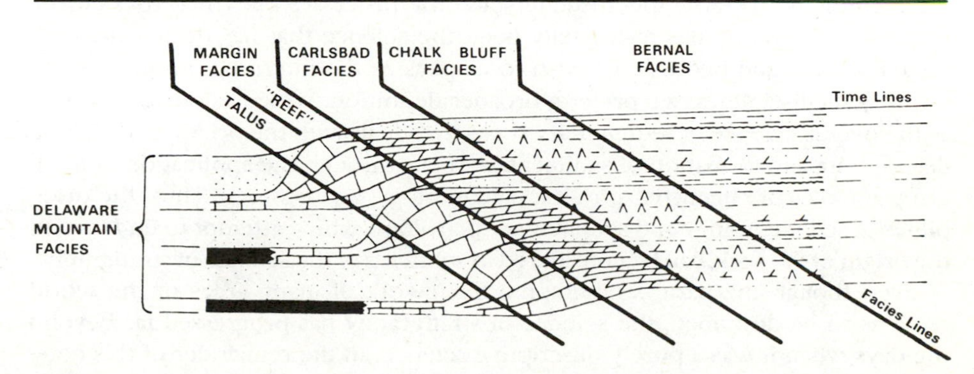

the type and category of units. Formations should never be established on the basis of fossils, as this would be a biostratigraphic rather than a lithostratigraphic unit.

The stratigraphic code details the prescribed steps to be followed to establish or revise any stratigraphic unit. This includes publication in a recognized journal with the stated intent to establish or revise a unit. The published article must clearly state the category and rank of the unit in question, how the name was chosen, the specification of a stratotype, description of the unit, a definition of its boundaries, historical background of the unit in question, regional shape and aspects of the unit, age correlation, and genesis. These rules are established so that the information will be readily available to subsequent stratigraphers who wish to examine and study the new unit.

SUMMARY AND DEFINITION OF STRATIGRAPHY

It should be clear from this chapter that the definition of stratigraphy and sedimentology has evolved over the years. Initially no distinction was made between these soft-rock disciplines and the science of geology as a whole. However, with increasing knowledge, it has become necessary to subdivide geology into its various subdisciplines. Even though this has allowed for specialization and great advances in the field, it has, at times, clouded the interrelationship of the various disciplines. For example, most stratigraphers must be able to practice paleontology and understand some of the modern principles of ecology to correctly interpret and establish biostratigraphic units. Sedimentologists must also be able to fathom the fields of physics and particle motion and use low-temperature geochemistry to unravel the origin and diagenetic history of sedimentary rocks. Magnetic stratigraphers draw heavily on concepts from the fields of physics and geophysics.

Some authors have tried to make a distinction between stratigraphy as a purely descriptive science that attempts to name and correlate rocks units and sedimentology that attempts to understand the processes by which rocks form. Because stratigraphy has historically been the science that has dealt with sedimentary rocks and because we wish to emphasize the interrelationships among the various disciplines, we prefer a broader definition. We envision stratigraphy as the overall science of sedimentary rocks that deals with the origin, correlation, description, and formation of sedimentary rocks through the application of modern principles of chemistry, physics, paleontology, and ecology. Thus, the disciplines of sedimentology and sedimentary petrology, which attempt to understand the origin of the rock, can be considered as subfields of the science of stratigraphy.

Even though the basic stratigraphic framework of many areas of the world remains to be described, the science of stratigraphy has progressed far beyond the days when it was a purely descriptive science. In the remainder of this book we will introduce you to a processes-oriented approach to the formation of sedimentary rocks that attempts to understand the origins of a given sequence. Even in areas where much descriptive work remains, a processes approach will allow for more accurate and detailed descriptions.

EPILOGUE

> *But, as there is not in human observation proper means for measuring the waste of land upon the globe, it is hence inferred, that we cannot estimate the duration of what we see at present, nor calculate the period at which it had begun; so that, with respect to human observation, this world has neither a beginning nor an end.*
>
> ***James Hutton, 1785***

To read the story recorded in sedimentary rocks we must look at the texture, structure, and relationships within and between individual beds, using detailed schemes for description and classification. It is important that these classifications are basically descriptive so that we do not bias our interpretation. In addition, they must also contain enough detail to allow comparisons between widely separated outcrops and encompass the vast variety of sedimentary textures and structures. The purpose of applying these classifications is to interpret the nature of sedimentary packages. Those interpretations should be based on a comparison to experimental and observational data, that is, physical, chemical, and biological mechanisms and processes of sediment formation, transport, deposition, and lithification.

To present you with the important constructs of interpreting sedimentary rocks, we have organized this textbook into nine chapters. The chapter following these introductory comments presents a summary of sedimentary rock classifications. We follow that discussion with a descriptive atlas of sedimentary structures that should allow you to name and describe most sedimentary structures found in all types of strata. The next six chapters present the major concepts of physical processes of sediment transport and deposition. These concepts and schemes will, it is hoped, prepare you for the rigors of interpreting sedimentary packages. As you read these chapters remember that many sedimentary situations that do not exist today were quite common throughout earth history, but nearly all the processes transporting and depositing sediment are either active today or can be modeled theoretically or reproduced in laboratory experiments. This uniformity of processes throughout time, rather than a uniformitarian view that things in the past have always been what they are today, is exactly the kind of **actualism** that Hutton advocated.

OUTSIDE READING

Selected Works by Early Geologists
Boltwood (1907); Holmes (1911); Hutton (1788, 1795, 1899 see also White, 1970); Jameson (1808, 1976); Lyell (1842); Miller (1860); Playfair (1805); Steno (1667a, b, 1916, 1969); Walther (1893–1894); Werner (1774, 1786).

History of Geology
Adams (1938); Conkin and Conkin (1984); Craig *et al.* (1978); Faul and Faul (1983); Fenton and Fenton (1952); Geikie (1905); Hallam (1983); Schneer (1969); White (1970).

Principles of Stratigraphy
Ager (1981, 1984); Berggren and Van Couvering (1984); Berry (1968); Boggs (1986); Committee on Stratigraphic Nomenclature (1933); Dott and Batten (1981); Dunbar and Rodgers (1957); Dunbar and Waage (1969); Eicher (1976); Friedman and Sanders (1978); Harbaugh (1968); International Subcommssion on Stratigraphic Classification (1976, 1987); Krumbein and Sloss (1963); Mathews (1984); Miall (1984); Middleton (1973); North American Commission on Stratigraphic Nomenclature (1983); Owen (1978, 1987); Palmer (1983); Shea (1982); Weller (1960).

Specific Examples
Berg and Wolverton (1985); Chang (1975); Childs (1983); Cloud and Glaessner (1982); Cohee *et al.* (1978); Cubitt and Reyment (1982); Donovan (1966); Dott (1983); Harland *et al.* (1982); Kauffman and Hazel (1977); Kennett (1980); Meissner (1972); Mitchum *et al.* (1977); Neidell (1979); Odin (1982); Poag and Ward (1987); Saddler (1981); Sloss (1963); Tarling (1983); Vail and Mitchum (1977); Vail *et al.* (1977a, b); Visher (1984).

Chapter 2

Classification of Sedimentary Particles and Rocks

INTRODUCTION TO ROCK CLASSIFICATION

One problem encountered by any student of sediments and sedimentary rocks is the seemingly confusing menu of classification systems. These classifications give many different names for the same rocks or may have similar sounding names for very different rocks. This apparent terminological jumble has evolved over the years partly as a result of independent work on classification of major sedimentary rock types by different specialists and partly because different classification schemes often have very different purposes. Field terms applied at the outcrop often require the observation of different characteristics and use of a different classification system than would be appropriate for examining a thin section under a microscope. These different nomenclatural systems contain valuable information and are necessary because they allow geologists to observe the same phenomena from different perspectives. Unfortunately, they often confuse anyone learning the system for the first time. To ease this initial difficulty, we hope to combine and present these terms and classification systems in a way that will allow you to communicate with other geologists and to comprehend major concepts in sedimentology at an introductory level. With this foundation, you should be able to handle the "terminological jumble" on your own as you advance in the discipline and read the scientific literature.

Controversies over the origin and naming of rocks are not new in geology. Many of the early seventeenth- and eighteenth-century geologists argued over the origin and classification of rocks. Remember that the Neptunists (from Chapter

1) proposed that most rocks were deposited as chemical precipitates from water. This idea may seem ridiculous today, but it gained widespread acceptance in light of the theological beliefs of the time that called for a 6000-year-old earth and the formation of most rocks during a worldwide deluge. Another group, the Plutonists, argued that many of these same rocks formed from the action of fire or volcanism. Werner's work, starting around 1775, supposed a nonuniformitarian earth and, as such, was championed by conservative elements in the church. A specific test of Werner's ideas centered around the origin of basalts interbedded with marine sediments around his native Freiberg, Germany. Because the interbedded sediments contained fossils that were presumed to be from the flood, Werner also classified the basalts, as he did all rocks, as marine chemical precipitates. The Plutonists, having observed volcanic eruptions and volcanic islands, correctly interpreted these as volcanic lava flows. However, they also pushed their views to the extreme by in turn proposing that many sedimentary rocks were the result of volcanic activity. This history serves as a reminder not to let biases and preconceived belief systems interfere with basic descriptions and observations. The controversy also suggests that there may be a danger of overgeneralization in classification or, worse, of applying one classification scheme to all rocks. As it turns out, the Plutonists were correct in their interpretation of basalt, but both groups were right in that both water-laid and volcanic rocks are included in the geologic record. It is now well established that the four basic rock types (intrusive, extrusive, metamorphic, and sedimentary) form by very different processes and that one, all-encompassing origin must be discounted.

One of the most important reasons to classify anything, whether plants, animals, birds, cars, rock formations, or sedimentary rocks, is communication. Thus, a "good" name should be clear, concise, short, and as unambiguous as possible. A name should be neither so specific that each individual specimen needs a new name nor so general that important differences are missed. In attempting to develop such a classification for sedimentary rocks, two methods have been tried. The first, a descriptive one, attempts to describe the appearance of the rock in as few words as possible without implying anything about its origins. Conversely, genetic classification attempts to include in the rock name information about origins of the rock. A genetic classification works well if rocks formed in one way look significantly different from one formed by other processes; in such a case, the genetic name adds much information about the rock. Problems develop when two rocks that look alike were formed in different ways. Many very common and widely used names for sedimentary rocks have been used in both a descriptive and a genetic sense by different authors, making it difficult to communicate without knowing both systems. We believe sedimentary particles are best interpreted by studying both the genesis as well as the physical characteristics of the particles. The same is true for the resulting rocks. In this chapter we present and compare various descriptive and genetic classifications that have been used widely in the sedimentary literature and present summarized classifications that cover both compositions and textures of most sedimentary rocks.

Many classifications group the common sedimentary rocks into four basic categories: conglomerates, sandstones, carbonates (limestones and dolostones), and mudstones (shales). This simple subdivision is based on grain composition and grain size. In addition to these names, other rock types exist that do not fit readily into this scheme. These include mixtures of carbonate and terrigenous grains, coal, siliceous rocks, evaporites, iron-bearing sediments, phosphatic sediments, and rocks altered from their original textures. In this chapter, we present classifications for sedimentary rocks in the following order: conglomerates and sandstones, then mudstones and carbonates, and, finally, the common remaining rock types such as chert, phosphorites, coal, and other organically derived carbonaceous rocks. A few explanations about classification are needed before we discuss the details of each rock type.

Although any solid can potentially become a sedimentary particle, there are various origins of grains found in sediments and sedimentary rocks. Sedimentary grains formed from the breakdown of other solids are called **clasts.** Grains derived from weathering of previous rocks form **terrigenous clastic sediments,** and those generated by volcanic eruptions, form the **volcaniclastic sediments.** The mechanical, chemical, and biological breakdown of skeletal parts produces **bioclastic sediment.** Unbroken biological particles, the direct organic precipitation of minerals by organisms, and inorganic precipitation of mineral grains also produce sediment. All these sedimentary grains can be transported and deposited by surface processes to form sedimentary rocks.

Minerals are also generated after deposition of sediment by direct inorganic precipitation; these grains are generally termed authigenic minerals. Sedimentary rocks also form from the direct precipitation of minerals by biological and inorganic chemical processes to form rocks without the transportation of particles.

The grain size of a sedimentary rock made of transported particles must be considered independent of the composition or genesis of the grains. On the Wentworth grain-size scale, sediment is commonly grouped into three main size categories: (1) gravel (grains larger than 2 mm); (2) sand (grains 63 μm–2 mm); and (3) mud (grains less than 63 μm) (Fig. 2.1). Many mixtures of these categories of grain sizes and compositions exist in nature, making tidy groupings impossible. Various names, listed on triangular classification diagrams, have been given for mixed sediments of sand, silt, and clay (Fig. 2.2). These classifications are based on the relative percentage of a particular size fraction. A rock composed of more than 90% sand is termed a sandstone, more than 90% clay a claystone, and more than 90% silt a siltstone. Mixtures of these size fractions require combining terms, such as sandy mudstone. When using these classifications, remember that mixtures occur somewhat often; borderline cases should be handled by lumping or combining terms rather than inventing a new name.

Sediments composed of terrigenous clastics in addition to nonterrigenous carbonate and/or silica require another classification. These mixed sediments are very common in the rock record (Fig. 2.3) but have only recently been described and classified in detail (Mount, 1984). Mixed sediments, like all sedimentary

Size in Meters	Class Boundary in Millimeters	Size Classes			Phi (ϕ) Units
		Gravel	Boulders	very large	
	2048				−11
		Gravel	Boulders	large	
1	1024				−10
		Gravel	Boulders	medium	
	512				−9
		Gravel	Boulders	small	
	256				−8
		Gravel	Cobbles	large	
10^{-1}	128				−7
		Gravel	Cobbles	small	
	64				−6
		Gravel	Pebbles	very coarse	
	32				−5
		Gravel	Pebbles	coarse	
	16				−4
		Gravel	Pebbles	medium	
10^{-2}	8				−3
		Gravel	Pebbles	fine	
	4				−2
		Gravel	Grit	very fine	
	2				−1
		Sand		very coarse	
10^{-3}	1				0
		Sand		coarse	
	1/2 (500μm)				1
		Sand		medium	
	1/4 (250μm)				2
		Sand		fine	
10^{-4}	1/8 (125μm)				3
		Sand		very fine	
	1/16 (63μm)				4
		Mud	Silt		
	1/32 (31μm)				5
	1/64 (16μm)				6
10^{-5}	1/128 (8μm)				7
	1/256 (4μm)				8
		Mud	Clay		
	1/512 (2μm)				9
10^{-6}					

FIGURE 2.1 Table of names for sedimentary particles based on Wentworth's classification of grain size.

FIGURE 2.2 Triangular diagram showing one method of classification of sediments based on the relative percentages of sand, silt, and clay.

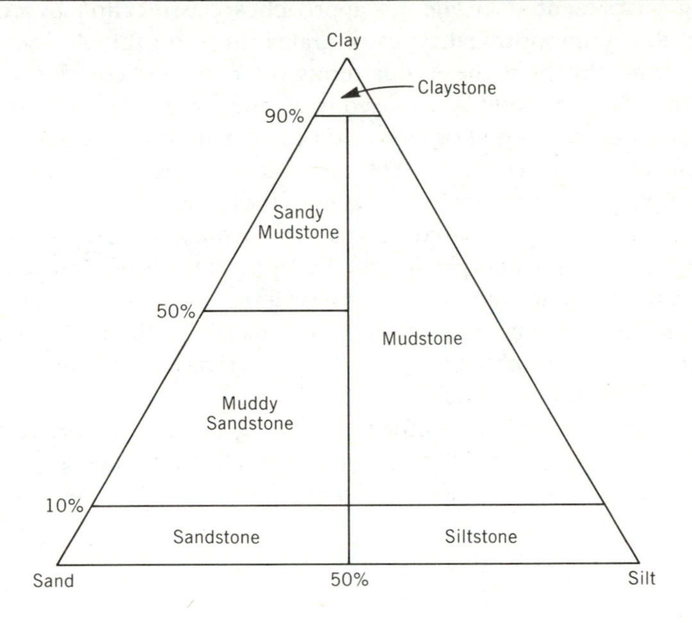

FIGURE 2.3 Mixed sediment of calcareous clam shells and terrigenous mud, Miocene Calvert Cliffs Formation, Maryland. (Photograph by W. J. Fritz, 1975).

rocks, can be classified by considering either the size of the grains or the grain composition.

The diagram presented in Fig. 2.4 approaches classification by considering mixtures of four compositional types of grains. In using this diagram, it is imperative to remember that these four terms refer to the "composition" of the grains and that the term sand is also used in a compositional sense. For example, a sedimentary rock composed of sand-sized carbonate bioclasts would be called a sandstone in a textural classification (Fig. 2.1) and a limestone in a compositional classification (Fig. 2.4). We emphasize that both these names for the same sediment are "correct." Each could be used depending on the objectives of a particular study. A sedimentologist interpreting the hydrodynamics and depositional environment of a rock would probably want to use a textural sandstone classification, whereas someone studying the same rock to understand the entire depositional history might wish to emphasize the lack of terrigenous sediment input and choose the compositional classification.

Any composition of sediment other than terrigenous sand, terrigenous mud, carbonate, and nonterrigenous silica can also be included in such a compositional classification (Fig. 2.4). For example, when classifying a rock with a high per-

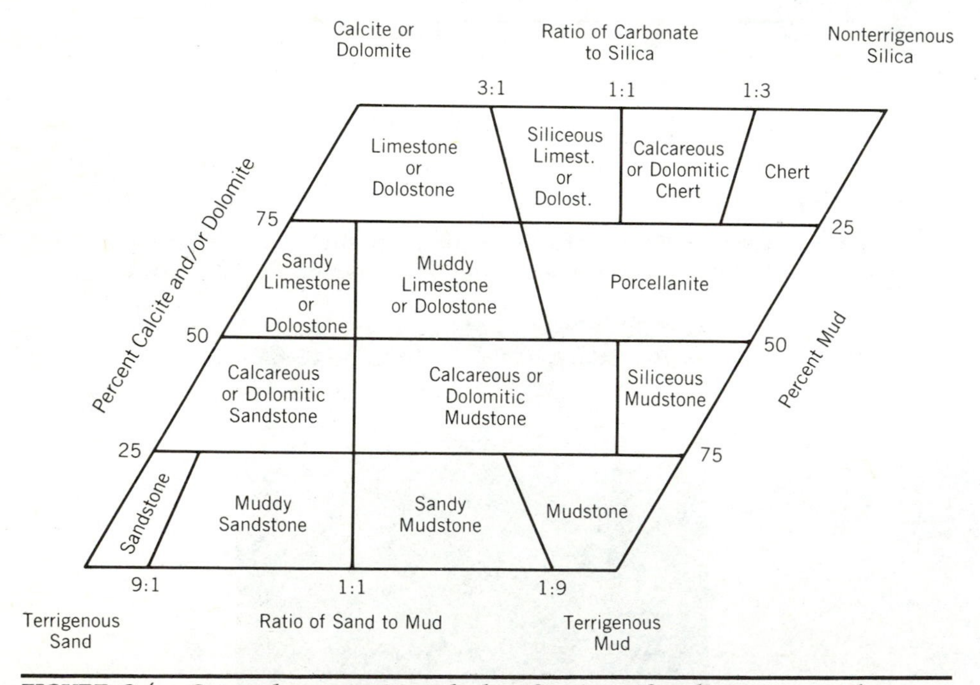

FIGURE 2.4 General compositional classification of sedimentary rocks composed of terrigenous sand, terrigenous mud, carbonate (calcite or dolomite), and nonterrigenous silica.

centage of phosphate grains, it might be best to replace "nonterrigenous silica" with "phosphate." The resulting rock would then be called a **phosphorite** or phosphatic mudstone, depending on the amount of phosphate.

The general compositional classification parallelogram (Fig. 2.4) names most rocks based on the relative percentages of terrigenous sand, carbonate, nonterrigenous silica, and mud. Once these percentages have been identified by examining the rock, they can be plotted on the diagram and a name chosen from the appropriate field of the diagram.

Diagenetic processes that cement and transform sediment into rock are also important in interpreting sedimentary processes. Cement type can also be included in the rock name by modifying the primary name with the cement type. Thus, a "calcite-cemented sandstone" is composed of terrigenous clastic sand derived from erosion of existing rocks and cemented by calcite. Terrigenous sediments are variously cemented and the cement type should be stated in all cases.

To expand this brief introduction, let us now look at classification schemes for the major groups of rocks.

TERRIGENOUS CLASTIC ROCKS

Terrigenous clastic rocks include those sedimentary rocks composed of clasts produced by physical, chemical, and biological weathering processes that break down preexisting rocks. The sediment thus produced is then physically transported and deposited. Terrigenous clastic rocks comprise a very large group of rocks spanning a great range of grain sizes from conglomerate to mudstone. Variations within this range form a continuum based on grain size and texture of the rock. Thus, terrigenous clastic rocks are classified according to grain size, grain composition, and grain textures such as rounding, sorting, shape, and size.

Probably the most important texture to consider first in classifying terrigenous clastic rocks is the size of the clasts that make up the rock. Such a classification is a very good first-order subdivision that helps a sedimentologist interpret the dynamics of the processes that transported the grains. Various categories have been proposed for assigning names to the sizes of particles based on diameter (Fig. 2.1). One problem encountered when first attempting to use a table of size categories is that several different classifications have been used. These include the diameter of the particles in inches, diameter in millimeters, size in sieve mesh size based on the opening of a screen that a sediment will pass through (see Chapter 4), and **Phi units** (negative log base 2 of the diameter in millimeters). Phi units are a logarithmic translation of the Wentworth particle size scale that takes into account the large range in grain diameters that exist in clastic rocks (Krumbein, 1934). Phi units for the major size classes are given in Fig. 2.1. The use of such a system allows the easy presentation of grain-size data in a quantitative manner. In this system, the higher the Phi number, the finer grained are the particles. Thus, units with a negative number are in the gravel to boulder range and numbers greater than four indicate silt- and clay-sized particles.

$$\phi = -\log_2 (D)$$

where: ϕ is the Phi number
D is the grain diameter in mm

Because it is impossible to classify by absolute size without producing infinite numbers of terms, names are given to certain ranges of particle size (Fig. 2.1). Even though these divisions may at first seem arbitrary, they are based on accepted usage and somewhat on hydrodynamic principles. Particles less than 4 μm in diameter are termed **clay,** in a textural sense. The word clay has been used in

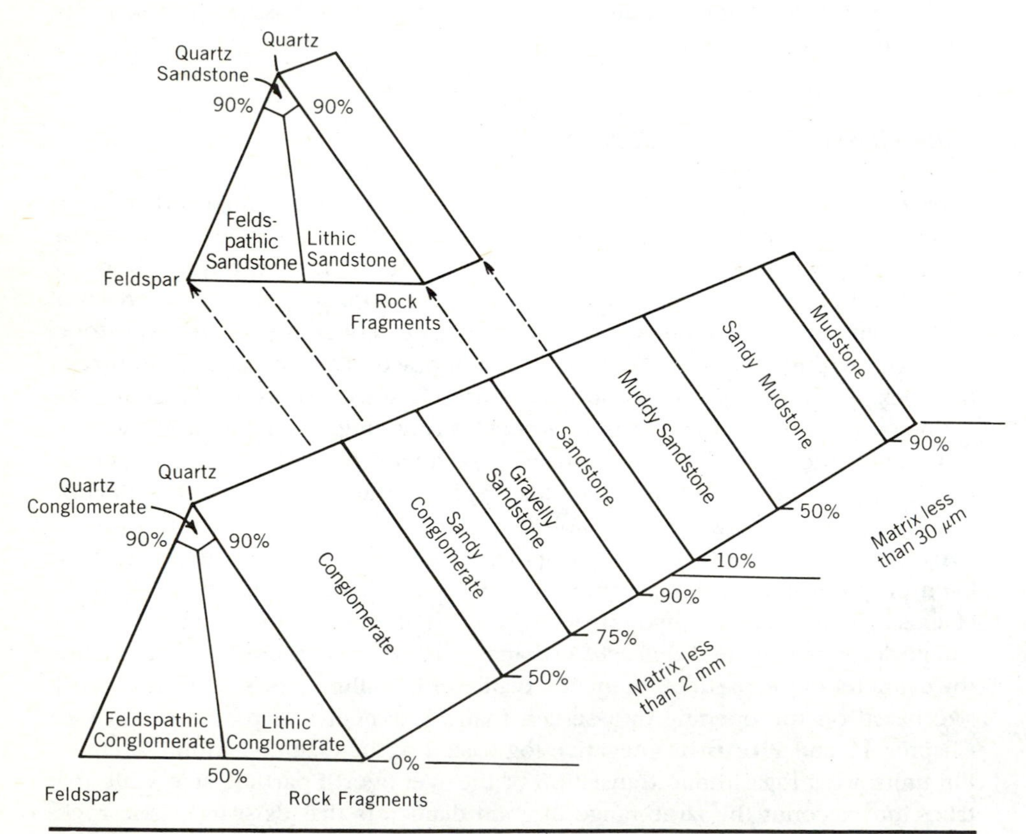

FIGURE 2.5 Classification of sedimentary rocks with clay to boulders composed of quartz, feldspar, and lithic rock fragments.

FIGURE 2.6 Mathematical method of calculating roundness of a sedimentary grain.

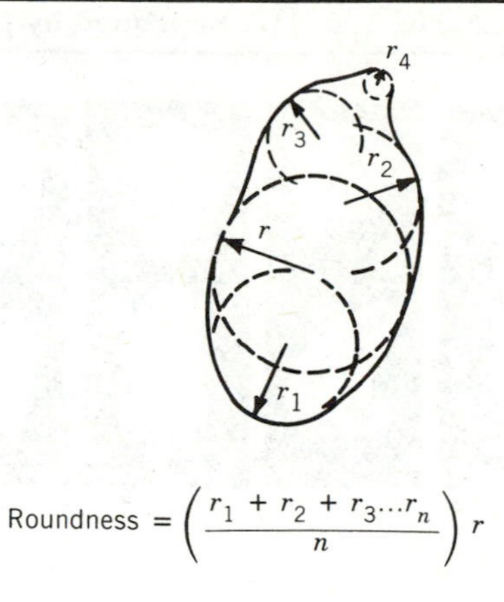

$$\text{Roundness} = \left(\frac{r_1 + r_2 + r_3 \ldots r_n}{n}\right) r$$

at least two different ways. To some, the word implies a **clay mineral,** and to others it refers only to the size of the particle. Thus, in a textural sense, a particle of quartz less than 4 μm in diameter should be considered a clay-sized particle, but not a clay mineral. The argument is largely academic because few particles other than clay minerals are as small as clay-sized particles. Quartz, for example, rarely degrades to grains smaller than silt size. Nevertheless, to avoid confusion, we will use the term clay in a purely textural sense and the word clay mineral when referring to the composition of a particle.

Most sedimentary rocks composed of grains of sand size and larger are named by a combination of mineralogical composition and grain textures. The classification by mineralogy of the grains is fairly straightforward. (Because just about anything can become a sedimentary particle, names are based on the most abundant type of mineral.) Common types included in most triangular classification diagrams (Fig. 2.5) are quartz, feldspar, and rock fragments of any preexisting rock type. Although they are useful, triangular classification diagrams that name rocks based on only three characteristics do not work as well for sedimentary rocks as they do for igneous rocks, because it is often difficult to reduce sedimentary rocks to only three important constituents.

Sandstones and conglomerates are often classified by various textures of their sedimentary grains such as grain size, shape (roundness or angularity and sphericity), and **surface texture.** Roundness in a sedimentary particle is defined as the absence of sharp corners. Mathematically it can be calculated by the tedious method shown in Fig. 2.6. You can see by examining the grain in Fig. 2.6 that if a grain has more sharp corners, it will have a lower numerical roundness. A grain with smooth edges will have a numerical roundness near 1. Most grains fall within the range from 0.1 to 0.9 "roundness units." The simplest method, used by most

FIGURE 2.7 Figure for estimating roundness of sedimentary grains based on comparisons with grains of calculated roundness. Based on M. C. Powers, 1953, *Sedimentary Petrology*, 23, Fig. 1, p. 118. Reprinted by permission of SEPM.

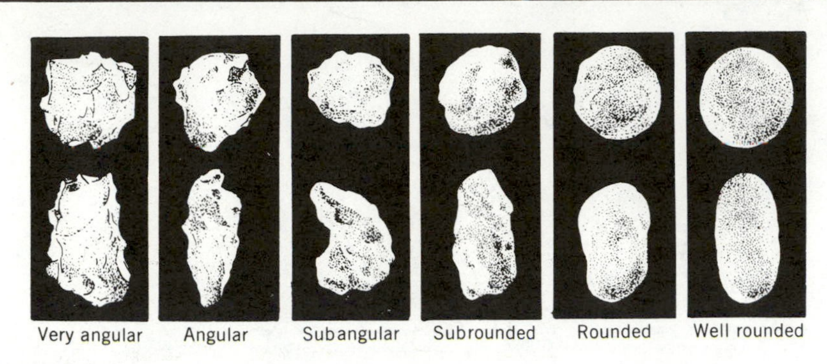

FIGURE 2.8 Medium-grained sand of quartz, feldspar, and various heavy minerals. Note the variety of grain shapes and surface textures. Try to plot individual grains on Figs. 2.7 and 2.8. (Photograph by William M. Johnson, Soil Survey SCS, Berkeley, CA.).

sedimentologists, to determine the roundness of grains is by visual comparison of the sample with drawings of grains of varying degrees of roundness (Fig. 2.7). This method is adequate for most field studies, can be reproduced easily, and is surprisingly accurate with a little practice. For example, practice determining the roundness and shape of the sand grains shown in Fig. 2.8.

The form or shape property of sphericity of sedimentary particles is considered independent of the degree of rounding or angularity and is defined by how close the grain approximates a sphere (Wadell, 1932). Names such as sphere, disk, blade, and roller are given to four basic shapes (Fig. 2.9). Another classification uses different names for the same shapes of grains: compact (equant), platey (oblate), bladed, and elongate (prolate), respectively. Each of these shape names can be used for particles ranging from very angular to well rounded.

Another very useful textural feature of sedimentary rocks is the degree of **sorting,** a measure of the dispersion of the grain-size distribution of a sediment. Even though sorting can be determined by mathematical analysis, most field-based sorting determinations are made by comparison to a drawing of previously determined sorting indices (Fig. 2.10).

Now that we have seen the diversity of rock types included as terrigenous

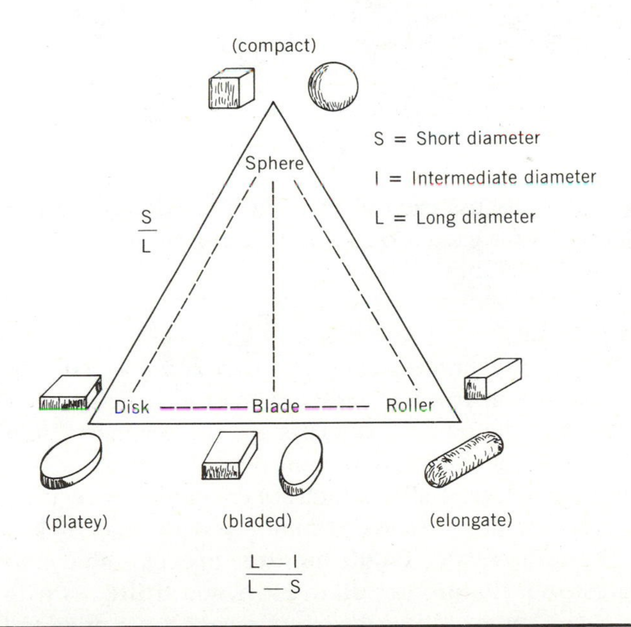

FIGURE 2.9 Triangular diagram illustrating the names given to various shapes of sedimentary particles. Grains are plotted on the diagram using various calculations of the long (L), intermediate (I), and short (S) diameters of the grain. Based on a method developed by E. D. Sneed and R. L. Folk, 1958, *Journal of Geology,* 66, 114–150.

FIGURE 2.10 Figure illustrating sorting indices for various mixtures of sedimentary particles. Note that no scale is implied in these drawings; the large clasts in 5–7 could be smaller than the grains in 1. From R. R. Compton, 1985, *Geology in the Field,* Wiley, New York, Fig. 4.1, p. 50. Copyright © 1985, John Wiley & Sons, Inc. Reprinted by permission of John Wiley & Sons, Inc.

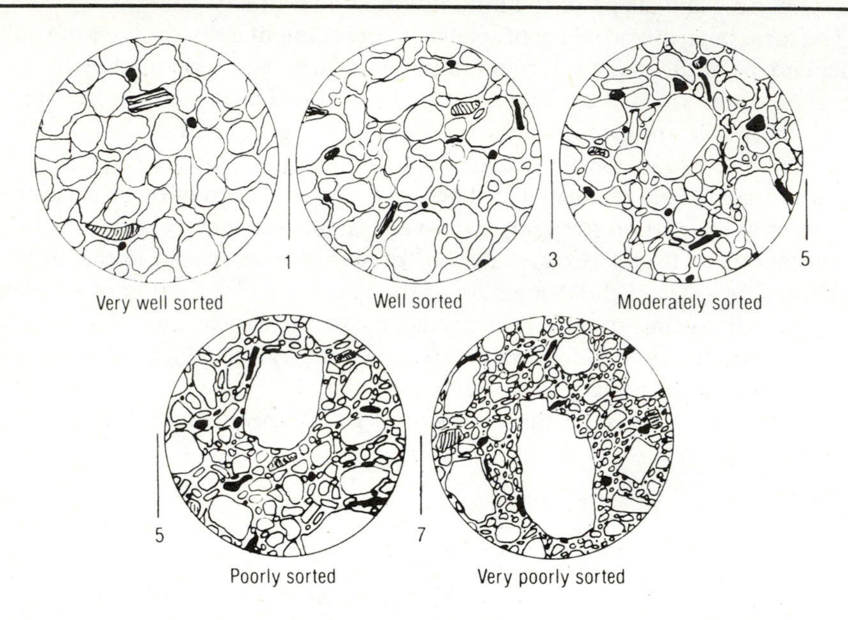

clastics, we will next discuss each of them in more detail and present specific detailed classifications for each. Conglomerates will be presented first, followed by sandstones and mudstones.

Conglomerates/Rudites

Conglomerates, often called rudites (Fig. 2.11), are composed of large framework clasts and smaller-sized matrix that fits between the large grains. The majority (greater than 25%) of the framework clasts are coarser than 2 mm in diameter and range upward to any size that can be transported (Fig. 2.5). Clast sizes include pebbles, cobbles, and boulders, all of which are encompassed by the term "gravel" (Figs. 2.1 and 2.12). Names for some common sedimentary rocks composed of gravel include **conglomerate, fanglomerate, breccia, sharpstone, roundstone, puddingstone, diamictite, diamicton,** and **tillite.** As with other types of sedimentary rocks, some of these terms are purely descriptive and others have strong genetic implications. Also, the terms are not equivalent names but illustrate the range of types included in the conglomerate/rudite category.

The term conglomerate (Figs. 2.1 and 2.5) is reserved for rocks with greater than 25% clasts that are larger than sand size. Rocks with less than 25% of these

FIGURE 2.11 Cobble conglomerate (Miocene) in Tapanga Canyon, west of Santa Monica, California. (Photograph by D. W. Hyndman).

FIGURE 2.12 Gravel conglomerate, Pleistocene Powell Terrace, Cody, Wyoming. (Photograph by W. J. Fritz, 1984).

FIGURE 2.13 Matrix supported conglomerate from a lahar of the Minoan eruption of Santorini, Greece. Large angular clasts are in a matrix of ash and pumice. (Photograph courtesy of M. E. Tucker, University of Newcastle-upon-Tyne, England).

larger conglomerate-sized clasts are given a sandstone or mudstone name and modified by the words gravelly, pebbly, cobbly, or bouldery; rocks containing between 25 and 50% clasts (Fig. 2.13) are modified by the words sandy or muddy (Fig. 2.14). The name **breccia** is reserved for a conglomerate composed of very angular clasts (Fig. 2.15). Some geologists argue that it might be better to discard this term completely as a sedimentary rock name because of the confusion among fault breccia, volcanic breccia, and a sedimentary breccia. If these terms are used with the proper prefix, however, there is no problem. To illustrate the potential confusion that could exist, when a volcanologist uses the word breccia alone, he or she is likely referring to a nonsedimentary breccia that forms from the cracking of cooling blocks of lava at the top of a flow. Other synonymous terms for sedimentary breccias are angular conglomerate or sharpstone conglomerate.

Unlike these descriptive terms, some conglomerate names imply a certain genesis or depositional environment. Tillite, for example, refers to a very poorly sorted conglomerate deposited directly by glaciers. Care should be taken in using names that imply specific depositional environments to be sure that such interpretations are warranted. Often it is best to use a descriptive name until sufficient

FIGURE 2.14 Classification of conglomerates based on percentages of gravel, sand, and mud.

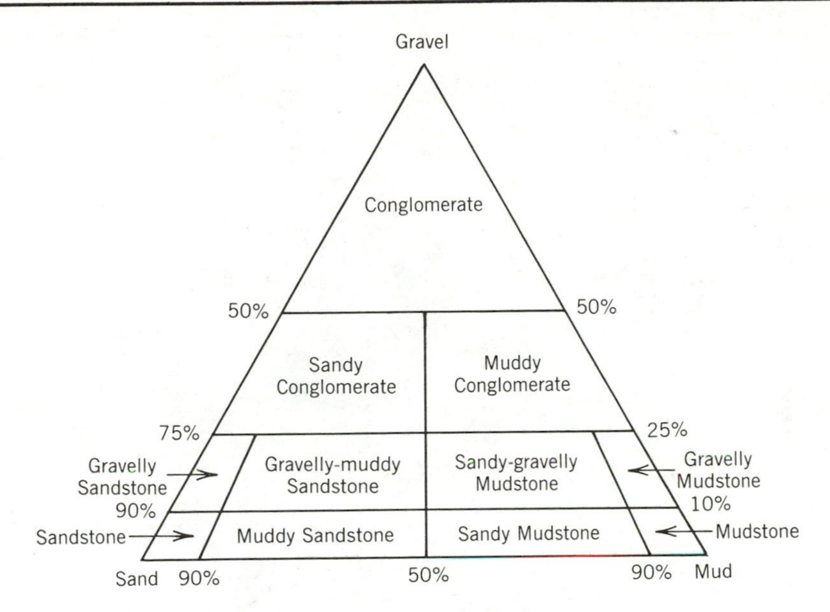

evidence is collected from facies relationships, sedimentary structures, and fossils to determine the processes that deposited the conglomerate.

Conglomerates are composed of both matrix and clasts. In this case, matrix is defined as the smaller-sized grains that fit between the large clasts. If the large clasts "float" in the fine-grained matrix mostly without touching, the rock is **matrix-supported** (Fig. 2.16). For example, diamictons are matrix-supported conglomerates of undetermined origin but often ascribed to mudflows. By contrast, rocks with the large clasts all touching at some point and with either no matrix or matrix filling only the void pore spaces in the rock are **grain-supported** (Fig. 2.17).

Another way to describe conglomerates is by the composition of the clasts (Fig. 2.5). Conglomerates can be divided into oligomictic or **monomictic** (clasts all of one composition) and **polymictic** (clasts of several or many different mineralogies or rock types). In describing a conglomerate it is important to describe both the matrix and the grain types separately. Thus, it is helpful to remember that a sedimentary rock is composed of particles, matrix, and cement.

Another important feature of conglomerates is the clast **fabric,** or orientation. Clast orientation can give information useful for deciphering the hydrodynamics of paleocurrents within the depositional system. Clasts in gravels deposited by particular fluvial and sediment gravity flow processes have a strong tendency to dip upcurrent, called **imbrication** (Fig.2.18), and cobbles on modern beaches

FIGURE 2.15 Sedimentary breccia (sharpstone conglomerate) with sand matrix. East slope of Pinyon Peak, Utah County, Utah. (Photograph by H. T. Morris, #30, U.S. Geological Survey, ca. 1952).

commonly dip in a seaward direction (Fig. 2.19). The long axes of bladed and roller-shaped grains can also become oriented parallel to flow, especially in laminar flows in glaciers and sediment flows. Flow perpendicular orientation occurs in bedload transported grains and is caused by shear in low-viscosity turbulent fluids. Oriented clasts in conglomerate deposited in other environments are not nearly as well documented. The fabrics of mudflow deposits, for example, range from a random clast orientation to strongly imbricated, similar to that found in stream gravels. Various terminologies used to describe the petrofabrics of these gravels are given in Fig. 2.20.

Now that we have seen how to classify conglomerates, sediments with clasts larger than 2 mm, we will move to classification of sandstones, terrigenous sedimentary rocks with grains from 2 mm to 63 μm.

Sandstones

According to Wickman (1954), sandstones make up only about 8% of all sedimentary rocks. However, probably about 85% of the studies on sedimentary rocks come from the information contained and preserved in sandstones, a fact that has led to the development of myriad sandstone classification schemes.

One of the most commonly used sandstone classifications (Dott, 1964; Pettijohn *et al.* 1972) combines composition, the relative percentage of quartz, feldspar, and rock fragments, with **matrix** percentage, here defined as sedimentary grains smaller than 30 μm (Fig. 2.5). This classification separates sandstones with less than 10% matrix from muddy sandstones (**wackes**), which contain between 10 and 50% matrix. Rocks in this latter group are of special interest because of its high matrix content; there are apparently few modern depositional environments with processes that deposit a mixture of silt, clay, and sand. In most environments, sandstones are much better sorted, with the sand being concentrated and the silt and clay carried away in suspension. It is hydrodynamically difficult to deposit both under the same flow conditions (see Chapter 4). Thus, such a rock implies special depositional conditions such as a turbidity flow, mudflow, or changes after deposition such as bioturbation or secondary alteration during diagenesis.

The term graywacke has often been used to describe rocks containing mixtures

FIGURE 2.16 Recent matrix-supported volcaniclastic conglomerate from the 1980 debris flow, Mount St. Helens, Washington. (Photograph by J. N. Moore, 1980).

FIGURE 2.17 Grain-supported conglomerate from the Eocene Lamar River Formation, Mount Hornaday, Yellowstone National Park, Wyoming. Gm unit shows crude horizontal bedding and imbrication; Gmu unit is normally graded with no grain fabric. (Photograph by W. J. Fritz, 1982).

of sand and mud, but there are a number of problems connected with its use. Graywacke is a term that was first used in the 1700s as a mining term for green-gray altered rock, its origin having nothing to do with sedimentology or sandstone classification. In fact, this word has been used to describe nine out of ten sandstones under various classifications. Moreover, many authors believe that a graywacke must have a green matrix of the mineral chlorite. Because chlorite is a metamorphic mineral, a graywacke with chlorite that formed after deposition of the original sediment would thus be a metamorphic, not a sedimentary rock. However, a graywacke with detrital chlorite grains that were eroded and transported from a low-grade metamorphic source terrain would, of course, be considered as a sedimentary rock. Furthermore, the word has been used to classify rocks ascribed to a turbidite origin. Confusion exists because not all turbidites deposit muddy sandstones and muddy sandstones may form from processes other than turbidity flow. An even worse problem with the term "graywacke" is that

FIGURE 2.18 Imbricated gravel from the alluvial/hyperconcentrated stream flow facies of the Eocene Lamar River Formation, Pebble Creek, Yellowstone National Park, Wyoming, showing paleoflow from left to right. (Photograph by W. J. Fritz, 1984).

FIGURE 2.19 Beach gravel with imbricated, seaward-dipping clasts on the northern Oregon coast. (Photograph by W. J. Fritz, 1983).

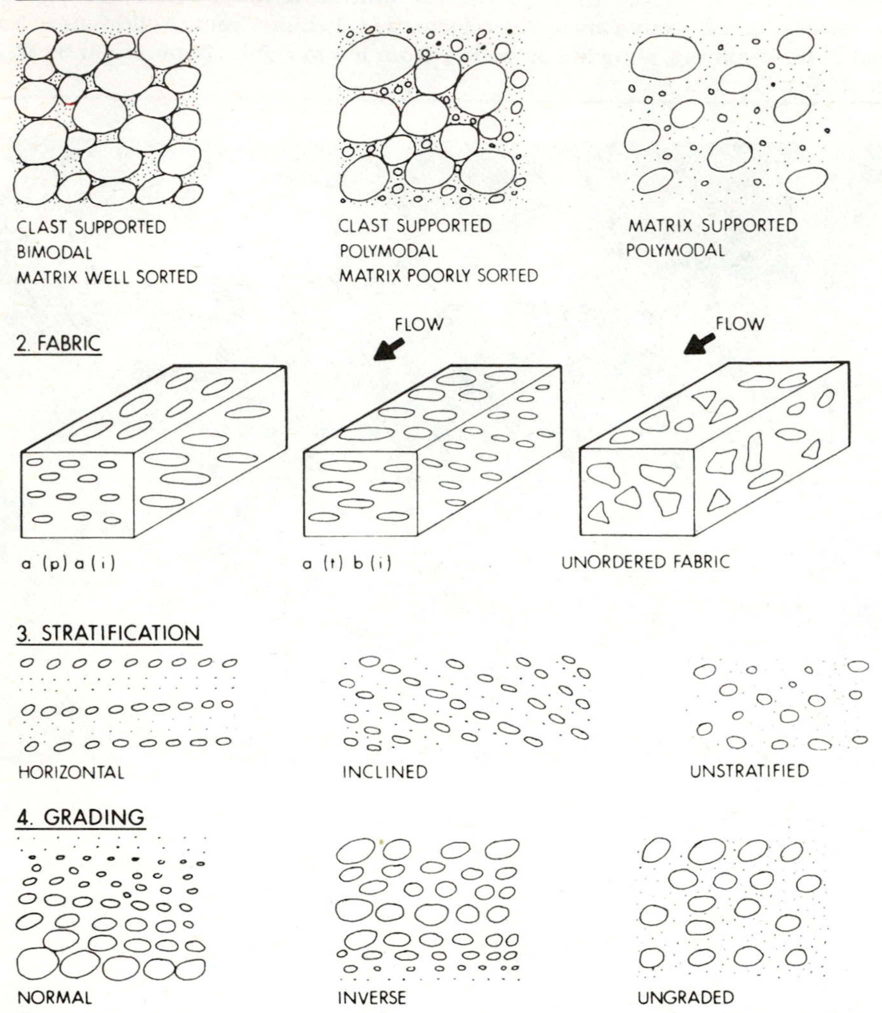

FIGURE 2.20 Diagram of terms for various features and fabrics of conglomerates. For fabric, the coding *a*(p) *a*(i) means that the *a* axis (long diameter) of the grain is parallel to the flow direction and imbricated. The coding *a*(t) *b*(i) means the *a* axis (long diameter) is transverse to the flow with the *b* axis or intermediate diameter imbricated. From J. C. Harms, J. B. Southard, D. R. Spearing, and R. G. Walker, 1975, *Depositional Environments as Interpreted from Primary Sedimentary Structures and Stratification Sequences,* Society of Economic Paleontologists and Mineralogists Short Course No. 2, Fig. 7-1, p. 134. Reprinted by permission of SEPM.

it stems from the erroneous concepts of Werner and Neptunism in the eighteenth century. Werner defined Greywacke or Grauwacke as an assemblage of sandstone, slate, quartzite, and schists that he put together into the Transition Group (Fenton and Fenton, 1952). Werner's student Jameson defined these rocks (Greywacke) as recording the "passage or transition of the earth from its chaotic to its habitable state." Probably the best view of this sandstone term is that of Rev. Adam Sedgewick, who, in the mid-1800s, discovered lava sheets among sediments of the greywacke and began to realize that the term "was merely a theoretical catchall for rocks which no one understood" (Fenton and Fenton, 1952).

Some sedimentologists argue that no wackes are forming today, and that they are encountered only at depth in drill cores or as outcrops and result from secondary alteration of a better-sorted sandstone. In this theory, unstable mineral grains are altered by heat, pressure, and ground water during burial to form various clay minerals. Often this happens at very low temperatures and pressure and is not considered metamorphism, but a diagenetic process. Feldspars are very susceptible to this type of alteration. Some argue that graywacke should be used in a purely descriptive sense for any sandstone with a clay matrix; however, there is confusion even here because many different percentages have been proposed. For all these reasons, graywacke is considered an invalid term by most modern sedimentologists and the term **wacke** or muddy sandstone is used to name sandstones with more than 10% clay matrix.

An important concept in understanding sandstones is that of **maturity**, the extent to which a sediment's properties approach the ultimate end product toward which they are being driven by the processes that affect it. Maturity can be measured in one of two ways. Compositionally mature sandstones are composed mostly of one mineral type. Because quartz is the most resistant of the major rock-forming minerals, the most compositionally mature sandstone is a quartz sandstone. However, quartz sandstones do contain small amounts of heavy minerals such as zircon, rutile, or tourmaline that are as or more resistant than quartz. These do not make up the entire rock because they are very scarce in parent igneous and metamorphic rocks of the earth's crust. Textural maturity, on the other hand, reflects the change in a sediment's texture as it is transported downstream from its source area. Factors considered in textural maturity include the percentage of clay-sized particles and the degree of rounding and sorting as shown in Fig. 2.21.

Now that we have discussed the components and textures of sandstones, we will present several detailed sandstone classifications. As we have indicated, these classifications will be based on both grain size and grain composition.

Quartz sandstones occupy the quartz apex of the triangular compositional classification of sandstones (Fig. 2.5). Other commonly used names for this rock type are **orthoquartzite, quartzite,** or **quartz arenite.** By definition these compositionally mature sandstones contain at least 90% quartz grains and a ratio of sand to mud of greater than 9:1. Although it is not implied in the definition, most

FIGURE 2.21 Textural maturity classification. From R. L. Folk, 1951, *Journal of Sedimentary Petrology,* 21, Fig. 1, p. 128. Reprinted by permission of SEPM.

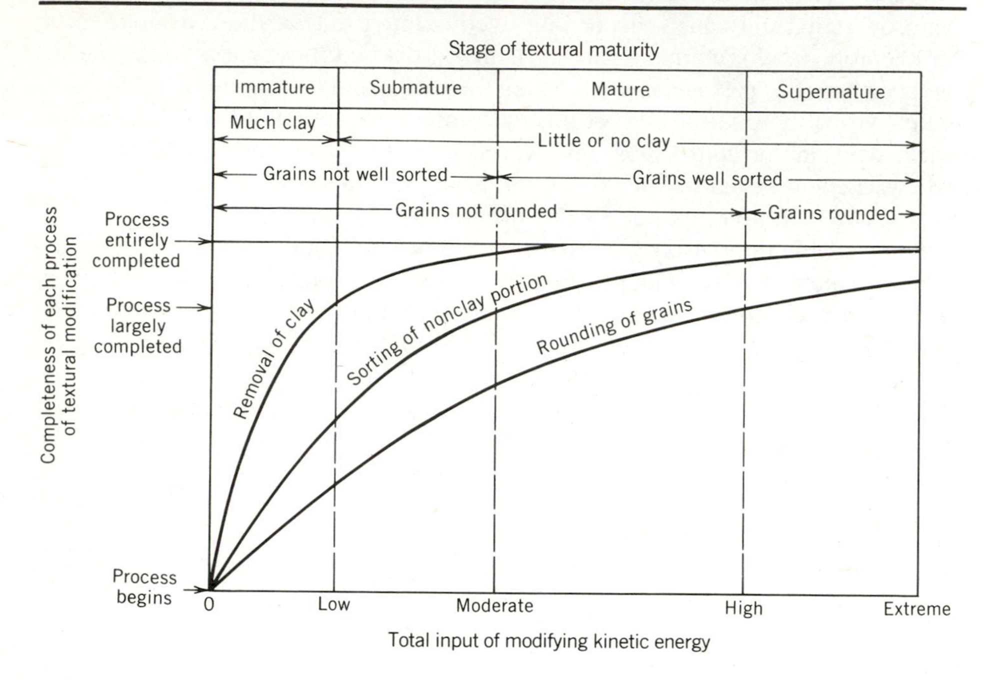

quartz sandstones, in addition to being compositionally mature, are also texturally mature with well rounded, highly spherical grains (Fig. 2.22). Any sediment that has been through sufficiently energetic processes and had enough time to become mostly quartz will also contain well-rounded, well-sorted, highly spherical grains.

The color of quartz sandstones can be extremely variable and is of little importance in any classification. Most quartz sandstones contain at least a small percentage of heavy minerals, many of which contain iron-bearing minerals (e.g., magnetite, illmenite, garnet). If this iron remains reduced, the color of the sandstone is almost snow-white, with a few dark flecks. If, however, the sand encounters oxidizing conditions, possibly from oxygenated groundwater flowing through the porous sand, some of the chemically unstable, heavy-mineral grains may become oxidized. Oxidized iron is red and even a small amount can stain the quartz sand various shades of brown and red. Thus, many red sandstones are in fact quartz sandstones because only a trace of iron stain can change the color of a sediment. Likewise, trace quantities of organic matter can stain a sandstone black, which implies chemically reducing conditions during deposition and diagenesis.

Quartz sandstones, like any sandstone, can form in two ways: from the weathering of an igneous or metamorphic source rock or from the reworking of an

FIGURE 2.22 Photomicrograph of a mature, well-rounded, well-sorted quartz sandstone showing highly spherical coarse grains. Ordovician St. Peter Sandstone, Twin City Brick Yard, St. Paul, Minnesota. Crossed nichols. (Photograph by F. J. Pettijohn, Johns Hopkins University, Dept. of Geology).

older sediment. Because it takes a lot of energy and/or time to remove all grains other than quartz, primary quartz sandstones (those derived directly from igneous and/or metamorphic source terranes) may be rare. No common depositional environment exists today that provides processes sufficiently energetic to directly concentrate 90% quartz from the 30–50% encountered in most igneous and metamorphic parent rocks in one erosional/transportational/depositional cycle. Many processes can concentrate up to 75% quartz. This sediment can then become lithified and, when uplifted, act as a source for a secondary sandstone. Thus, many quartz sandstones are second- or third-generation sandstones. Whatever the origin, quartz sandstones require an extreme amount of both chemical and mechanical weathering to remove all other mineral types and it is quite likely that they generally form by recycling older sediment.

Unlike the compositionally mature, mineralogically pure quartz sandstone, **feldspathic sandstones** contain a noticeable (greater that 10%) percentage of feldspar (Fig. 2.23) in addition to quartz. They often contain other minor components such as mica, rock fragments, and heavy minerals as well. The exact

FIGURE 2.23 Photomicrograph of a feldspathic sandstone showing angular, poorly rounded quartz (clear) and good multiple twining of the feldspar (mostly plagioclase) grains (clouded). Pennsylvanian Fountain Arkose, Colorado. (Photomicrograph by F. J. Pettijohn, Johns Hopkins University, Dept. of Geology).

percentage of feldspar in this category is controversial and, unfortunately, many genetic connotations are implied for sandstones containing feldspar. For example, Folk describes humid arkoses that form from intense chemical weathering in a tropical soil and arid arkoses that form from mechanical weathering and deposition on an alluvial fan. Some of these are probably warranted, but those terms should be kept separate from initial descriptions. In other words, the term "humid" refers to an unobserved environment and represents an interpretation based on the evidence of the features of the sandstone. Other names used in this category are **arkose, subarkose,** and **quartzo-feldspathic sandstone.**

Feldspathic sandstones provide unique information about the depositional system because feldspars break down relatively easily (compared to quartz). Two properties account for this characteristic. Mechanically, feldspars split apart easily along their two cleavages; chemically, they degrade as a result of the incongruent dissolution of potassium, sodium, and calcium during weathering. As feldspars are weathered, soluble ions go into solution and the insoluble silicate framework collapses to form clays that are carried away as part of the suspended load. Thus, it is nearly impossible to concentrate feldspar beyond the original percentage included in the parent rock from which it was weathered. Because feldspathic sandstones can form in several different climatic regimes, including arid and humid climates, clearly another control on sandstone composition is the source area.

Various rocks that make up source terranes also weather to provide a source for the sedimentary material that makes sandstones rich in those rock fragments. **Lithic sandstones** are characterized by sand grains composed of such identifiable rock fragments (Fig. 2.24). Examples of names that illustrate the range of different types of lithic sandstones include **subgraywacke, low-rank graywacke, lithic sandstone, lithic wacke, litharenite, sublitharenite,** or graywacke, depending on matrix content. Because there are so many possible ways in which different kinds of rocks might weather into sedimentary grains, each having unique weathering characteristics, it is crucial to differentiate various kinds of lithic sandstones. Although theoretically any rock can break down to sand-sized rock fragments that retain the characteristics of the original rock, only lithologies whose grain size is less than 2 mm can be considered prime candidates. Rocks with grains larger than 2 mm generally break into individual mineral crystals.

Extremes of composition in lithic sandstones vary from sandstones composed of nonresistant gypsum grains to a sandstone composed of extremely resistant chert grains that is compositionally as mature as a quartz sandstone. For these

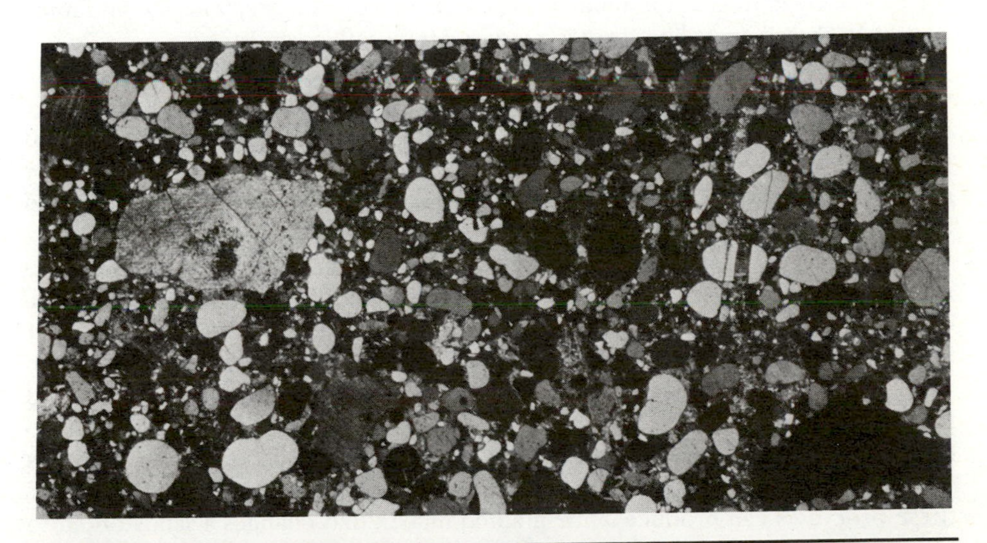

FIGURE 2.24 Photomicrograph of a lithic sandstone of the Slaven Chert, Nevada. The dark angular grains are shale and the light, poorly rounded grain to the left of center is limestone. Note the large, well-rounded quartz grains and poor sorting. (Photomicrograph by J. Gilluly, #230, U.S. Geological Survey, ca. 1953).

reasons, it is best to describe lithic sandstones by using the name of the dominant rock type as a modifier.

One common type of lithic sandstone contains grains of volcanic fragments. These might be primary volcanic ash grains or grains reworked from volcanic source rocks. Other names for this type of rock are volcanic sandstone, volcaniclastic sandstone, and tuffaceous sandstone. From a hydrodynamic viewpoint these sandstones are no different than any other terrigenous sandstone or quartz sandstone. Moving water or air acts on these grains just as it does on quartz grains; bedforms and other sedimentary structures form and are interpreted in the exact same way as with any other sandstone. The major difference between quartz and lithic sandstones is that the grains originated from nearby volcanic terranes and, therefore, require a particular tectonic setting. Thus, from a genetic standpoint, volcanic, or any lithic sandstone for that matter, gives unique information to help interpret the paleogeography and paleotectonics.

Metamorphic lithic sandstones, also called **phyllarenites,** are recognized and described by naming the type of metamorphic rock dominating the sand grains. Sandstones composed of fragments of low-grade metamorphic rocks commonly include slate, schist, phyllite, or argillite grains and greenstone fragments. These are normally very fine-grained sandstones because fine-grained metamorphic rocks do not resist weathering and transport; most break down to a small size very readily. Lithic sandstones derived from the weathering of high-grade metamorphic rocks tend to be very coarse-grained because of the large grain size in the parent rock. In fact, the tendency is for these large quartz, feldspar, and mica crystals to break apart and form a quartz or feldspathic sandstone.

Other types of lithic sandstones are also commonly found in stratigraphic sequences. Sedimentary lithic sandstones contain identifiable fragments of older sedimentary rocks such as chert, shale, sandstone, or limestone. These are named by the rock type of the lithic grain followed by the word "lithic sandstone." For example, a sandstone composed of grains of sedimentary chert might be called a "chert lithic sandstone."

Many diagenetic changes occur in sandstones after burial and during the cementation processes. One reason that sandstones undergo complex diagenetic changes is that they are often very porous and allow the migration of fluids that chemically alter the rock. These fluids can chemically alter unstable grains such as feldspar, iron-rich heavy minerals, and calcium-rich grains. Thus, lithic and feldspathic sandstones often exhibit more drastic diagenetic changes than do quartz sandstones. These diagenetic changes may alter lithic grains and feldspars into clays such as illite and chlorite and form secondary feldspar (Fig. 2.25). Thus, these processes can produce a fine-grained matrix that changes a sandstone into a wacke after deposition. In some cases, these diagentic clays are difficult to distinguish from **pseudomatrix** caused by the crushing and compaction of shale or slate lithoclasts.

Of economic importance is the fact that diagenetic changes can either destroy

FIGURE 2.25 Scanning electron micrograph of authigenic clay (kaolinite) between rounded sand grains. Rotliegend desert sandstone, Lower Permian, northern West Germany. From M. E. Tucker, 1981, *Sedimentary Petrology: An Introduction,* Halstead, New York, Fig. 2.48, p. 59. (Photograph by Dr. Dewey, University of Durham, England).

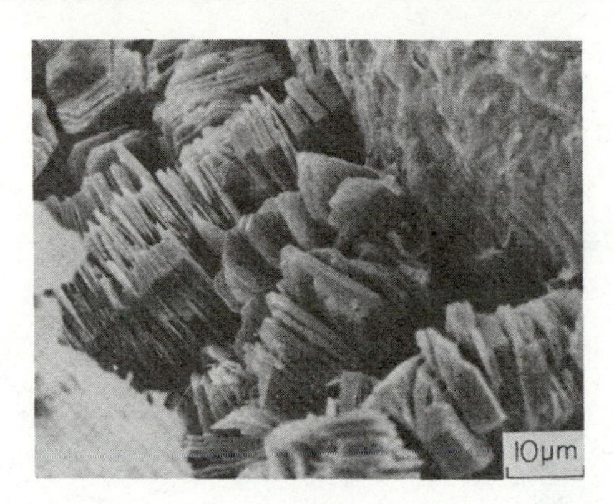

porosity in a rock by filling void spaces with cement or authigenic clays or increase porosity by dissolving unstable grains. Petrologists often use knowledge of these characteristics to determine the value of a sandstone as a reservoir rock for oil, natural gas, or water.

Mudrocks

Mudrocks, terrigenous clastic sedimentary rocks composed predominately of mud (Fig. 2.5), comprise over 80% of all sedimentary rocks. Other names used for various types of mudrocks include **mudstone, shale, claystone, siltstone, argillite,** and **siltite.** Mudrocks encompass all the terms and is most synonymous with mudstone. The term shale, when used alone, implies a mudrock with **fissility,** that is, a tendency to split apart along very thinly spaced bedding planes (Fig. 2.26). If necessary, shales can be further classified as clay shales or silt shales depending on the dominant grain size. The terms claystone or siltstone, when used alone, imply massive bedding (Fig. 2.27) and dominance of either clay- or silt-sized grains, respectively. The terms argillite and siltite are generally used for extremely well-indurated, slightly metamorphosed claystones and siltstones, respectively.

Like sandstones, mudrocks can originate from any source rock that can be broken down to silt and clay size, and the fragments are then transported mostly in suspension. The small grain size of mud partly determines the composition of the particles. Quartz, for example, resists mechanical breakage because it has no good cleavages and is almost never broken into particles smaller than silt. Clay

FIGURE 2.26 Photograph of a thinly laminated shale. (Photograph by W. H. Bradley, U.S. Geological Survey).

FIGURE 2.27 Outcrop of a massive fossiliferous claystone of Claiborne age, north of New Holland, Aiken County, South Carolina. (Photograph by G. E. Siple, #5, U.S. Geological Survey).

minerals make up the bulk of grains in mudrocks because they are among the few minerals that can exist in clay-sized particles.

Micas make up another important constituent of mudrocks and are often larger than the dominant silt- and clay-sized grains. This difference in grain size exists because sand-sized sheets of mica are carried in suspension (see Chapter 4) and deposited along with the suspended load of silt or clay. This occurs because the extremely platey flakes of mica have a very low settling velocity equivalent to that of much smaller spheres.

Organic particles make up a noticeable percentage of most mudrocks. Total organic content of a shale that may produce oil or natural gas when buried and heated varies from less than 1 to 10 or 15%. Organic components include bacteria, algae, and higher plant detritus in the form of kerogens, humus, asphalts, and other compound long-chain organic molecules.

Although the composition varies greatly, Shaw and Weaver (1965) report that the "average" mudrock is composed of approximately 61% clay minerals, 31% quartz and chert, 4.5% feldspar, 3.6% carbonates, 1% organic, and less than 0.5% iron oxides. Field determinations of these constituents are nearly impossible because of the small grain size, so X-ray diffraction, scanning electron microscopy, and chemical analyses are used to determine the composition.

Clay minerals make up the largest percentage of grains in almost all mudrocks. These form from the chemical alteration of feldspar, mica, and other clay minerals and are sometimes very useful in determining the weathering and diagenetic history of mudrocks because different climates, temperatures, rainfall, vegetation, and soil chemistry produce unique assemblages of clay minerals. Some of the more important clay minerals include illite, montmorillonite, and various members of the smectite group, kaolinite, chlorite, and gibbsite. Other types of clay minerals, such as mixed-layer illite-smectite and resulting rock types such as bentonite, form from the rapid weathering and hydration of unstable volcanic ash grains, or from heat and pressure during early diagenesis to low-temperature metamorphism.

As a result of the various processes of diagenesis, many changes occur in clay minerals. As previously mentioned, these include the breakdown of large unstable grains into clay minerals but also include alteration of the type of clay minerals present. For example, the smectite group clay minerals are changed to the mineral illite at about 360° Celsius.

Mudrocks do not exhibit the diversity of sedimentary structures and, therefore, bedforms found in sandstones because mud is carried mostly in suspension; they rarely contain sedimentary structures such as those found in rocks made of particles carried by traction or saltation. Because deposition from suspension dominates the deposition of mud, horizontal laminations are common in many mudrocks. In some environments, clay-sized particles would never come out of suspension except for the filtering of water by organisms or the flocculation of charged clay particles in salt water. Mudrocks vary from massive to horizontally laminated. Because finer-grained mudrocks are made largely of platey mica and clay minerals, the tendency is for claystones to exhibit a distinct fissility.

Many marine organisms live in environments conducive to deposition of mud, and mudrocks commonly exhibit bioturbation that ranges from scattered burrows to a complete churning of the sediment that destroys all traces of bedding, leaving it mottled in appearance.

One of the more significant nonbiogenic alterations of mudrocks is compaction. Compaction in mudrocks is generally significant because mud can contain up to 80% water when deposited. As this water is expelled during burial, it can disrupt the sediment and produce flame structures, soft-sediment folds, and clastic dikes (Chapter 3). The amount of compaction can often be determined by examining concretions that form during early, precompaction diagenesis. These concretions preserve the cemented and uncompacted structures and beds intact while the surrounding beds have become compacted. The degree of compaction is determined by comparing the thickness of the noncompacted beds with the corresponding altered ones (Fig. 2.28). Also, comparison of laminae within and without the concretion can allow interpretation of the timing of concretion formation.

A diagenetic change of considerable economic importance involves the alteration of the organic material contained in mudrocks into hydrocarbons of oil and natural gas. This change generally involves burial depths of 1 to 4 km and temperatures up to 200°C (Fig. 2.29). Most petroleum is formed from the breakdown of long-chain hydrocarbons originating from (in order of importance) bacteria, algae, zooplankton, and higher plant detritus. These organics are the source for the carbon in most black shales.

The color of many mudrocks is controlled by the amount and oxidation state of iron minerals as well as by the organic content. Even though iron constitutes only a trace percentage, the color of mudrocks reflects the ratio of oxidized to reduced sediments. The oxidation state of iron in a fine-grained rock does not necessarily reflect an oxidizing or reducing depositional environment but rather the diagenetic environment. Mudrocks deposited in various environments from Precambrian to Late Tertiary contain shales with mottled red and green patches, which record heterogeneous oxidation states within the sediment. In contrast to

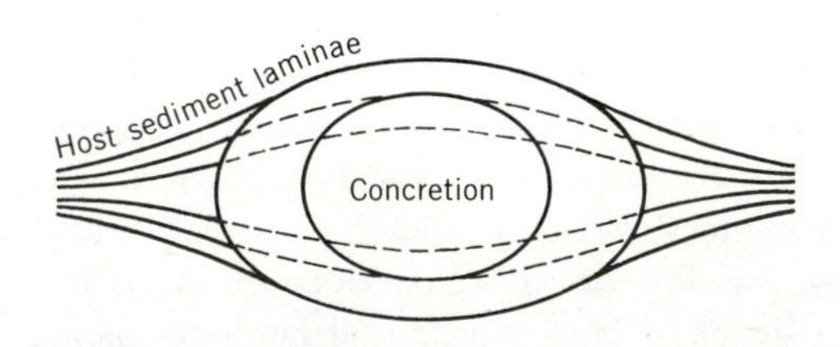

FIGURE 2.28 Drawing of a concretion showing early cementation and compaction. From R. Raiswell, 1971, *Sedimentology*, 17, Fig. 3, pp. 147–171. Reprinted by permission of Elsevier Science Publishers, B.V.

FIGURE 2.29 Thermal maturation of organic carbon as a function of burial depth. Modified from B. P. Tisote and D. H. Welte, 1978, *Petroleum Formation and Occurrence,* Springer-Verlag, New York, Fig. II.6.1, p. 185. Reprinted by permission of Springer-Verlag, Inc.

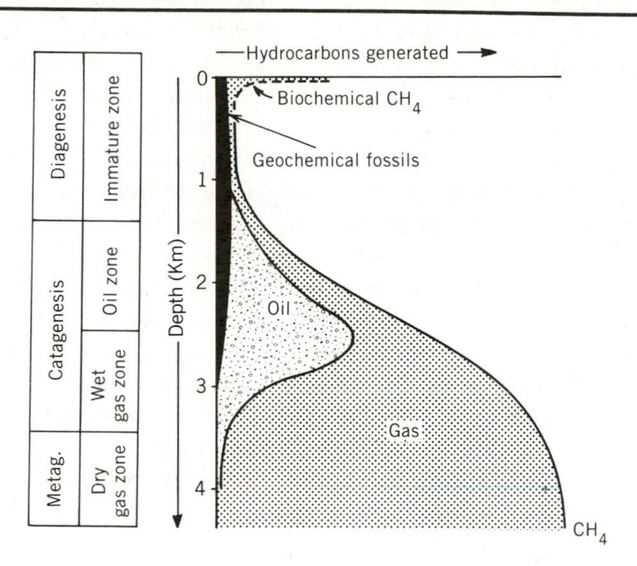

the mottled red and green colors of rocks with diagenetic oxidation/reduction spots, mudrocks with uniform red and green bands probably reflect changes in the organic content of the sediment and resulting redox conditions during diagenesis. Rock colors produced by different oxidation states of iron vary from greens to purple and red. Very black mudrocks are almost always the result of the organic carbon content often associated with very finely crystalline pyrite. The total organic carbon (TOC) content of these black shales can range anywhere from 1 to 15%, making them ideal as potential petroleum source rocks.

CARBONATES

As we have just discussed, terrigenous clastic rocks exhibit a great diversity of grain sizes and compositions. They can be classified either by emphasizing and describing the texture or by mineral composition of the grains. **Carbonates,** for the most part, could also be included in the previous terrigenous clastic classifications by considering only the physical textures of the grains. However, carbonates are unique because their major constituents were transported as part of the solution load and require no local source rocks; we will thus consider them under a separate classification.

Unlike the terrigenous clastics, the material that ultimately composes the carbonates is composed mostly of calcium and bicarbonate ions dissolved from the chemical weathering of rocks. These ions are then extracted from solution by organic or inorganic processes to make limestones and dolostones. For example, one very common way to extract ions from solution is in the form of small, coarse, sand-sized spheres of concentric layers, generally of calcium carbonate in the form of aragonite needles, called ooids. Ooids form when minerals dissolved in water supersaturated in calcium carbonate precipitate as aragonite needles around a small nucleus of some solid. These needles can agglutinate around the nucleus as rays tangent to the core, and so form a radial structure, or can form as concentric rings. Ooids can form in both marine and hypersaline lacustrine environments—modern examples include the Bahama Banks, the Great Salt Lake, the Persian Gulf, and Shark Bay, Australia.

Even though calcium carbonate ooids are most numerous, all the soluble ions are known to form "ooid"-type spheres. Examples include iron-rich hematite grains from the Silurian Red Mountain Formation near Birmingham, Alabama. These iron "ooids" have a flattened disk or flaxseed shape and are thought to have either been precipitated directly as iron or were altered from chamasite. Phosphate "ooids" in the Pennsylvanian Phosphoria Formation in the northwestern United States are another example of this phenomenon.

Organisms provide an especially important means of extracting ions from solution and turning them into sedimentary particles. This process usually involves the making of a hard exoskeleton or endoskeleton by the organism. These particles, either the whole tests of coccoliths, pteropods, and foraminiferans or

FIGURE 2.30 Bioclastic sandstone, upper foreshore, north end of Jekyll Island, Georgia. (Photograph by W. J. Fritz, 1985).

FIGURE 2.31 Classification of carbonate transported and authigenic grains based on grain size.

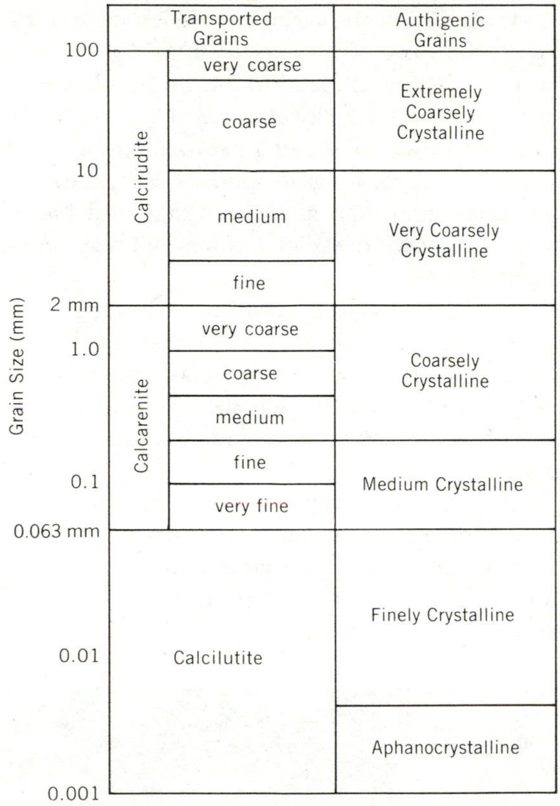

Grain Size (mm)	Transported Grains		Authigenic Grains
100	Calcirudite	very coarse	Extremely Coarsely Crystalline
		coarse	
10		medium	Very Coarsely Crystalline
		fine	
2 mm	Calcarenite	very coarse	Coarsely Crystalline
1.0		coarse	
		medium	
		fine	Medium Crystalline
0.1		very fine	
0.063 mm	Calcilutite		Finely Crystalline
0.01			
			Aphanocrystalline
0.001			

the broken shells and skeletons of marine organisms like mollusks, corals, brachiopods, and crinoids, can then be transported like any terrigenous clast. If the particles are large and mostly unbroken, they make a shell-clast sandstone called **coquina.** Calcium carbonate skeletons can also be bound directly into rock and are called **boundstones** in many organic **reefs.**

In addition to animals, many groups of marine algae also deposit calcium carbonate plated on their cell walls and, thus, can comprise bioclastic sediment. Plates of the algae *Halimeda* comprise a substantial fraction of modern carbonate sands on beaches in the Florida Keys, Bermuda, and the Persian Gulf. The green algae *Penicillus,* which lives in quite water, contains clay-sized grains of carbonate in the form of aragonite as part of the structure of the cell wall. When the organism dies and disintegrates, these carbonate needles form lime mud that is deposited in low-energy environments such as lagoons and bays.

Large skeletal hard parts can be broken down by mechanical abrasion and weathering after the death of the organisms. Even though the hard parts of any animal can potentially become a sedimentary particle, the most common, because of sheer numbers and volume, are the carbonate skeletons of marine invertebrates. These grains make up the bulk of **bioclastic limestones.** An example of this type of sediment is the "shell hash" sand found on beaches in recent carbonate environments, like those along the Florida coast (Fig. 2.30). These beach sediments are composed of a variety of skeletal parts, including the fragments of sea urchins (echinoderms), pelecypods, gastropods, corals, crinoids, coraline algae, and bryzoans. Microorganisms such as algae, fungi, and bacteria also play an important role in the breakdown of skeletal grains and may provide an important means of producing lime mud.

Because carbonates are deposited by organic or inorganic precipitation from solution, they are very different from terrigenous clastics. One exception is lithoclastic sedimentary rocks, which are made of clasts of carbonate grains eroded from a preexisting limestone. Even though they are carbonate in composition, these are actually terrigenous clastic rocks and should be classified as a lithic or rock fragment sandstone.

Carbonates include both limestones ($CaCO_3$: calcite or aragonite) and dolostones ($Ca\text{-}MgCO_3$: dolomite). Particular limestones can be considered a clastic rock because they are made of particles that are transported, like any clast. Most carbonate rocks are made of a mixture of particles that have been transported

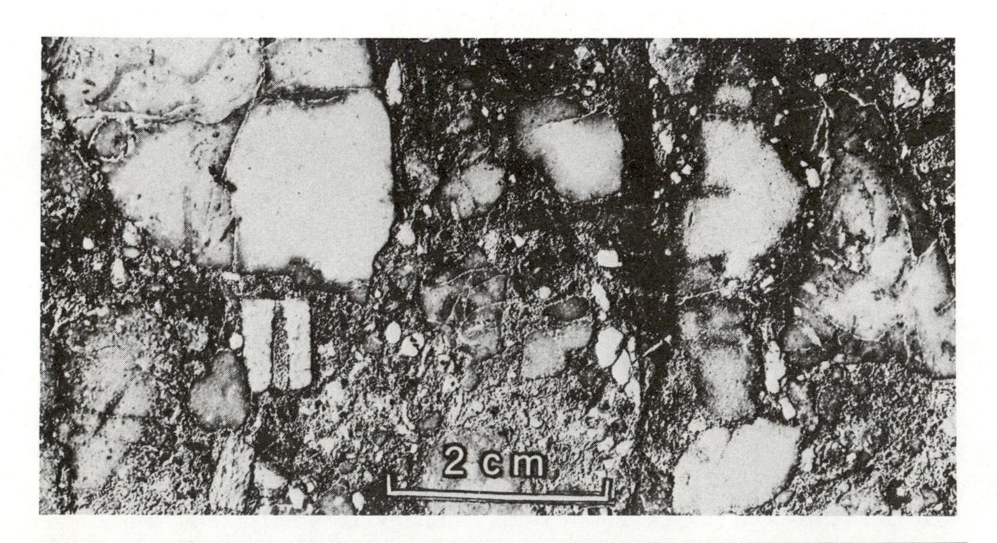

FIGURE 2.32 Core of calcirudite with large clasts of a preexisting reef limestone. From core of Horseshoe Atoll, Scurry County, Texas. (Photograph by P. T. Stafford, #3, U.S. Geological Survey, ca. 1952).

by physical processes and then cemented or recrystallized *in situ.* Therefore, it is possible to classify carbonates not only by looking at the transported grains, as in terrigenous clastic rocks, but by emphasizing the cement as well. One simple classification emphasizes that limestones are really nothing more than clastic rocks with carbonate grains. This system gives different names based on the size of the clasts (Fig. 2.31). Thus, a **calcirudite** is the carbonate equivalent of a terrigenous conglomerate with carbonate clasts larger than 2 mm (Fig. 2.32), a **calcarenite** is the equivalent of a sandstone with grains between 2 mm and 63 μm, and a **calcilutite** is a mudstone with grains smaller than 63 μm (Fig. 2.33).

As with terrigenous sandstones, grains, cement, and matrix are the important components of most limestone classifications. Grains, according to the classification of Folk (1959, 1962), can be broken into four major compositional types of allochemical constituents, namely, **bioclasts,** made of broken and whole skeletal parts ("bio"); **ooids** ("oo"); **intraclasts** ("intra"); and **fecal pellets** ("pel"). These grains are then held together by cement, generally **sparry calcite (sparite** or **spar),** and by a fine-grained matrix made of **lime mud (micrite),** the carbonate equivalent of clay. Occasionally, micrite can form authigenically to make cement, but because it most often forms as primary sediment, it is generally considered to be matrix for the purposes of classification.

The short terms in quotes are then used as prefixes to modify the rock name chosen from the cement type. As an example of how to use this classification, consider the name of a calcarenite composed of bioclastic fragments and cemented by sparry calcite. This rock would be called **biosparite** (Fig. 2.34), which

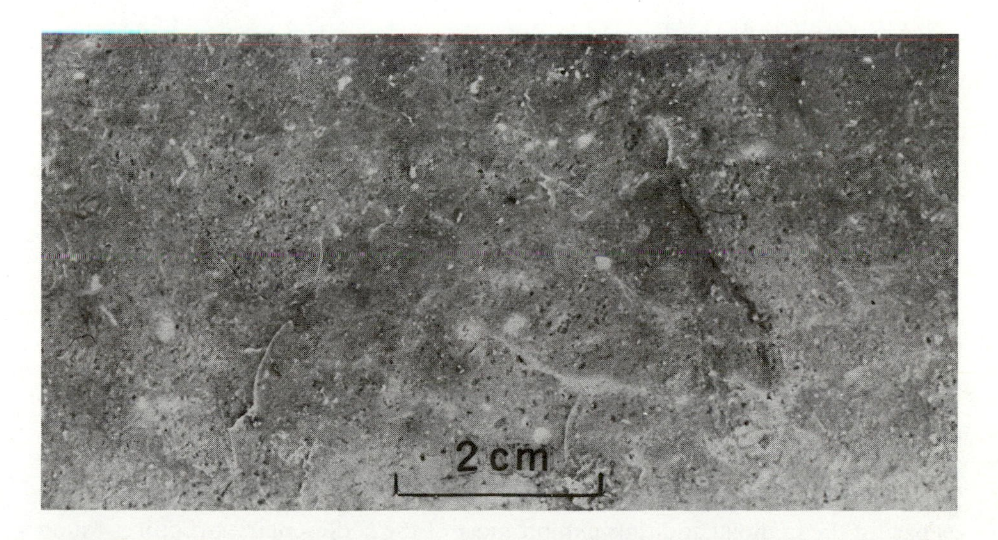

FIGURE 2.33 Core of calcilutite from Horseshoe Atoll, Scurry County, Texas. (Photograph by P. T. Stafford, #1, U.S. Geological Survey, ca 1952).

FIGURE 2.34 Photomicrograph of a biosparite with gastropod shells and ooids. Limestone facies of the Mississippian Pennington Formation (Chesterian) of Southern Tennessee. (Photomicrograph by W. J. Frazier).

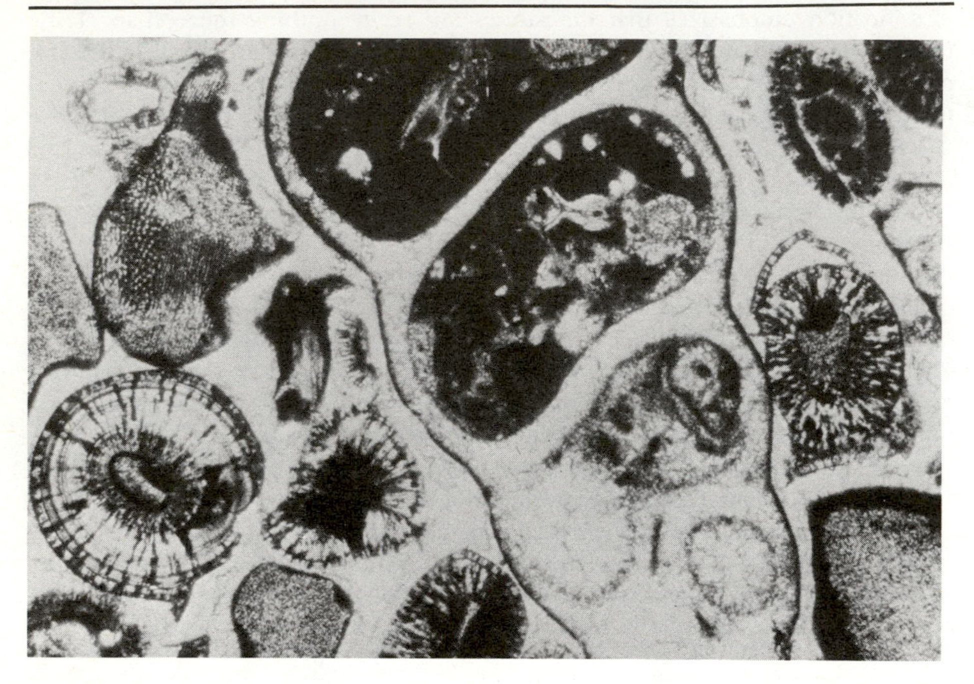

is equivalent to **bioclastic calcarenite, bioclastic limestone,** or **grainstone** and hydrodynamically equivalent to texturally mature terrigenous sandstone (see comparison of carbonate and terrigenous terms given in Fig. 2.35). In this system, the allochemical modifier used is that which is the most abundant. If, as is often the case, several allochemical constituents are present in the rock, several modifiers are used that represent the order of abundance. The least important constituent is placed first and the most abundant term is placed closest to the rock name. Thus a bio-oosparite contains less biological fragments than ooids. The classification that we illustrate in Fig. 2.35 contains terms used in both the Folk and Dunham systems and compares them with textures found in terrigenous sediments. Thus, this figure does not suggest a replacement classification for carbonates but should be viewed as a translation box for illustrating the similarities between the major terms used in carbonate rock classification.

As with sandstones, the grains of calcirudites and calcarenites are transported as part of the bedload and thus exhibit most of the sedimentary structures found in terrigenous sandstones of a corresponding grain size. The grains in a cross-stratified oolitic calcarenite (oosparite, or oolitic grainstone) (Figs. 2.36*a* and

FIGURE 2.35 Classification of carbonate rocks. This figure should also be viewed as a translation box that compares terrigeneous terms with those used in carbonate classifications by Folk (1959) and Dunham (1962).

<table>
<tr><td>Dominant Grain Size</td><td colspan="3">SAND</td><td colspan="3">MUD</td></tr>
<tr><td>Size Terms</td><td colspan="3">Calcarenite</td><td colspan="3">Calcilutite</td></tr>
<tr><td>Terrigenous Equivalents</td><td>Sandstone</td><td colspan="2">Muddy Sandstone</td><td colspan="2">Sandy Mudstone</td><td>Mudstone</td></tr>
<tr><td>Modified from Folk (1959)</td><td colspan="2">Sparite</td><td>Micritic Sparite</td><td>Sparry Micrite</td><td colspan="2">Micrite</td></tr>
<tr><td>Modified from Dunham (1962)</td><td>Grain-stone</td><td colspan="2">Packstone</td><td colspan="2">Wackestone</td><td>Calc-mudstone</td></tr>
</table>

0 10 20 30 40 50 60 70 80 90 100

Percent Carbonate Mud

FIGURE 2.36 (*a*) Cross-bedded oosparite. Precambrian limestone, White-Inyo Mountains, California. (Phototograph by J. N. Moore, ca. 1975). (*b*) Sand-sized grains of an oolitic sediment showing surface texture and highly spherical shape. For internal structure refer to Fig. 3.37. (Photograph by J. C. Anderson, #107, U.S. Geological Survey).

FIGURE 2.37 Photomicrograph of an oosparite illustrating spar cement. Notice the concentric bands coating a nucleus of the ooid grains. (Photomicrograph by Dave Alt).

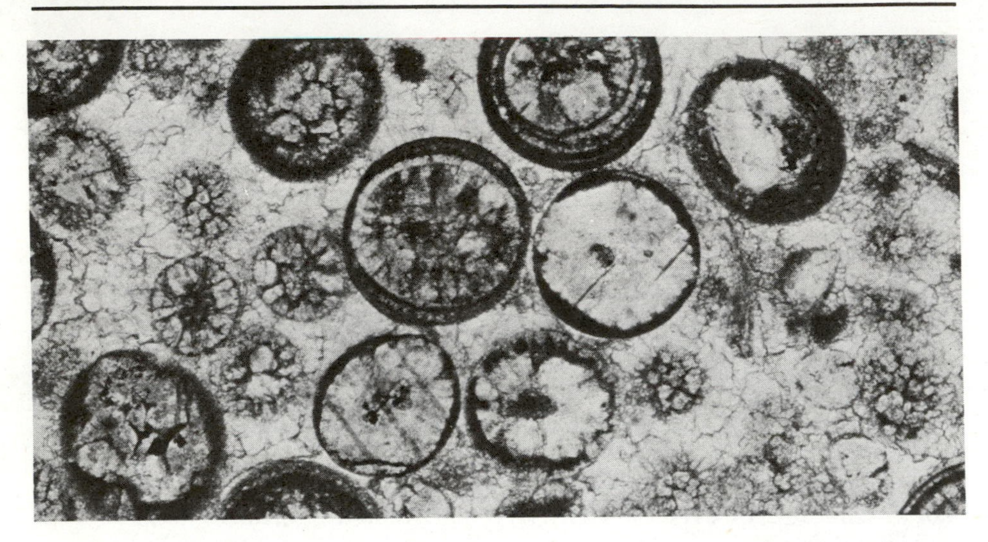

2.36*b*) were deposited by currents just as any sandstone with an equivalent grain size (see Chapter 4). Because of the difference in density of carbonate grains compared with quartz and feldspar, they differ slightly in size from those terrigenous sediments for a given hydrodynamic situation. Such "sparites" (Fig. 2.37) generally result from transportation by moving water that removes all matrix, leaving pores to be filled later with sparry calcite cement during diagenesis. Thus, sparites tend to show sedimentary structures indicative of strong currents. Micrite (calcilutite, calcmudstone), because of the high content of lime mud, indicates deposition by much less energetic processes. Most micrites are composed of pellets and intraclasts (Figs. 2.38 and 2.39), whereas sparites are generally dominated by ooids and bioclasts (Fig. 2.40).

The term **chalk** is used for poorly indurated calcilutite. Chalks are composed of the skeletons of microfossils such as coccoliths or foraminiferans (Fig. 2.41). A general term that encompasses all the varieties of calcilutites is fine-grained limestone, a wonderfully simplistic term that unfortunately transfers very little detailed information. Unlithified carbonate mud is termed ooze. There is a diagenetic transition, then, from ooze to chalk to a better indurated form of limestone.

Because carbonate is easily recrystallized, most limestones are changed by even shallow burial. Often carbonate can be replaced by silica and other minerals during recrystallization. Carbonates thus altered are often named by the size of the new crystals. Care should be taken to not confuse secondary calcite with

FIGURE 2.38 Photomicrograph of a pelmicrite. Cavities filled with clear calcite cement are called "bird's-eye structures" and are caused by infilling of gas bubbles released by algal mats. Mississippian Pennington Formation, northwest Tennessee. (Photomicrograph by W. J. Frazier).

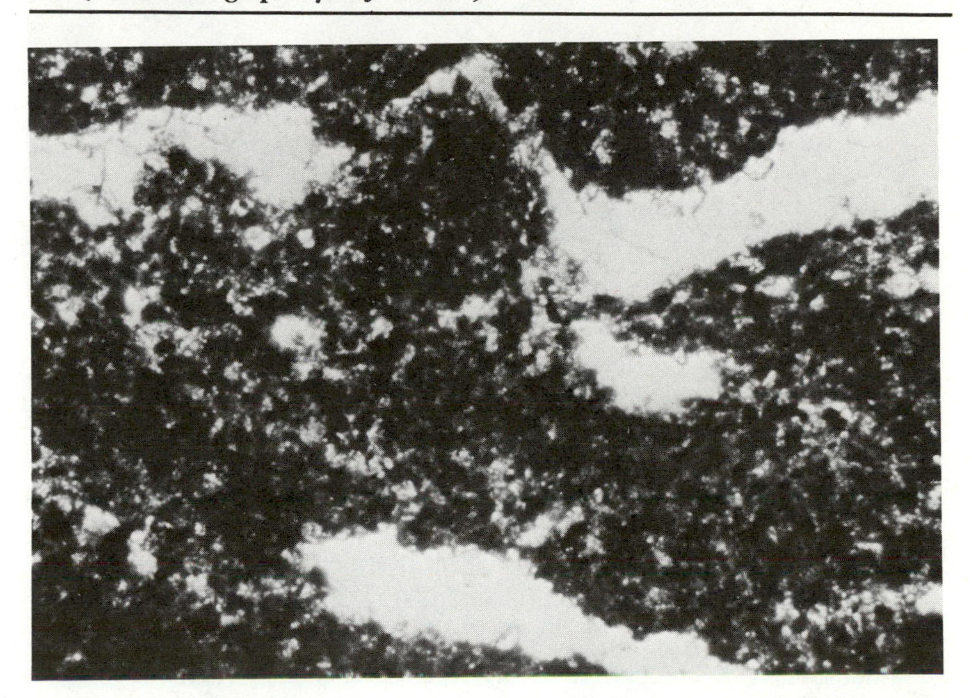

sparry calcite. Often the secondary alteration produces **dolostones,** which are formed during recrystallization of limestones and the addition of magnesium. **Primary dolostones,** those with original **dolomite,** as well as fine-grained recrystallized ones, can be named using the standard carbonate classifications by using a "dolo-" prefix. For example, a rock composed of fine-grained dolomite mud would be termed a dolomicrite.

Boundstones or **biolithites** are limestones that were not transported but rather precipitated in place by the actions of organisms, forming an organic buildup, carbonate mud mound, or **reef.** Other names used in describing either the general boundstone rock type or specific examples include **biolithite, laminites,** and mud mound. Often these form from the buildup of a **bioherm** or organic reef. Bioherms are wave-resistant vertical buildups of the skeletons of any reef-building organism such as corals or oysters, whereas a **biostrome** is a non-wave-resistant layer produced from the *in situ* acretion of skeletal parts and is often tabular in shape. A **biocoenosis** refers to the genesis of this type of rock and is used to describe any group of organisms preserved in their natural community (Fig. 2.42).

FIGURE 2.39 Photomicrograph of a dolomitized intramicrite. Mississippian Pennington Formation, north Tennessee. (Photomicrograph by W. J. Frazier).

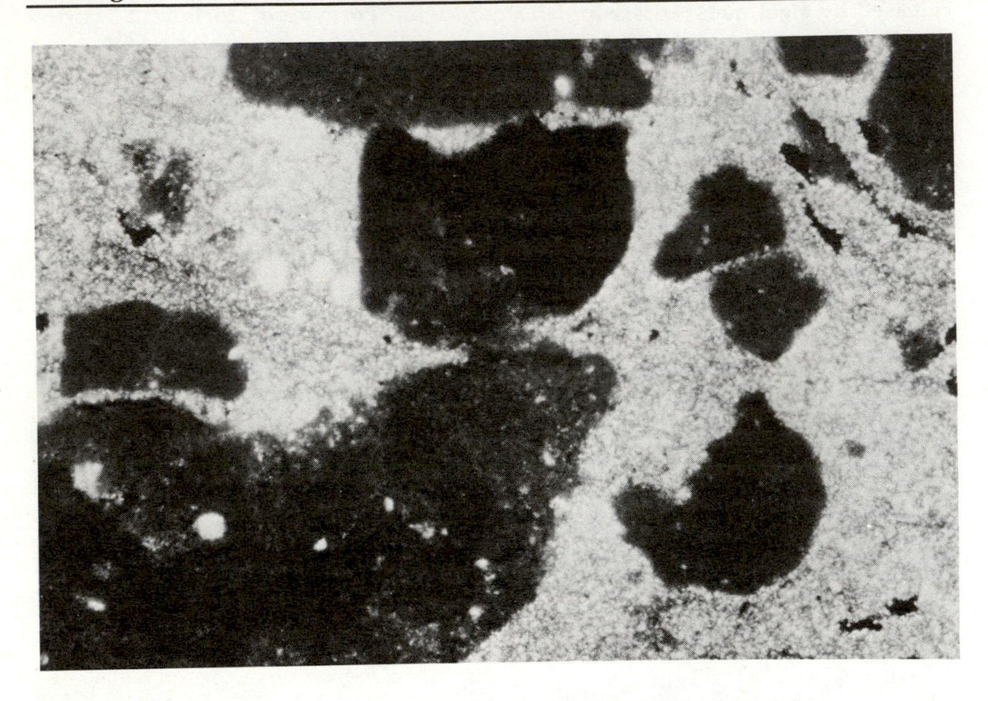

Boundstones can be produced by any organism that helps bind sediment into rock. Two types of boundstone, framework and laminated are common. Organisms responsible for laminated forms have no hard parts, but their metabolic processes either secrete cement or cause the precipitation of carbonate. They also trap sediment and bind it into rock. A good example of the laminated form is an algal or bacterial **stromatolite** (Fig. 2.43). Algal and/or cyanobacteria strands have no hard parts. However, the fine algal threads act as baffles to help trap sediment carried in suspension and the metabolic processes of the organism cause precipitation of calcium carbonate. Stromatolites can build up in vertical mounds or can produce tabular laminations (see Chapter 3). Algae and bacteria are the most common types of organisms that make up this type of boundstone.

The framework type of boundstone is one in which the actual hard parts of the organism make up the bulk of the rock. Some of the more common organisms that produce this type include corals, oysters, bryozoans (Fig. 2.42), rudistids, stromatoporoids, and coralline (red) algae. At first it might seem surprising that coralline algae can occur in sufficient numbers to make a boundstone, however, in many areas of the Caribbean and Pacific, large reefs are made wholly of coralline algae. These algal reefs may have been even more common in the past.

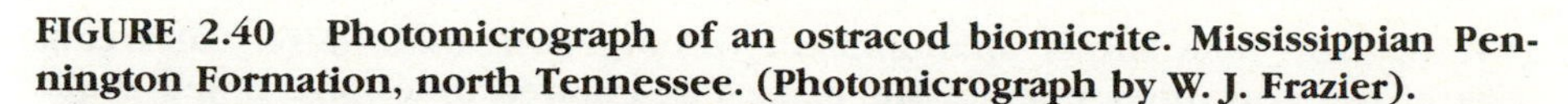

FIGURE 2.40 Photomicrograph of an ostracod biomicrite. Mississippian Pennington Formation, north Tennessee. (Photomicrograph by W. J. Frazier).

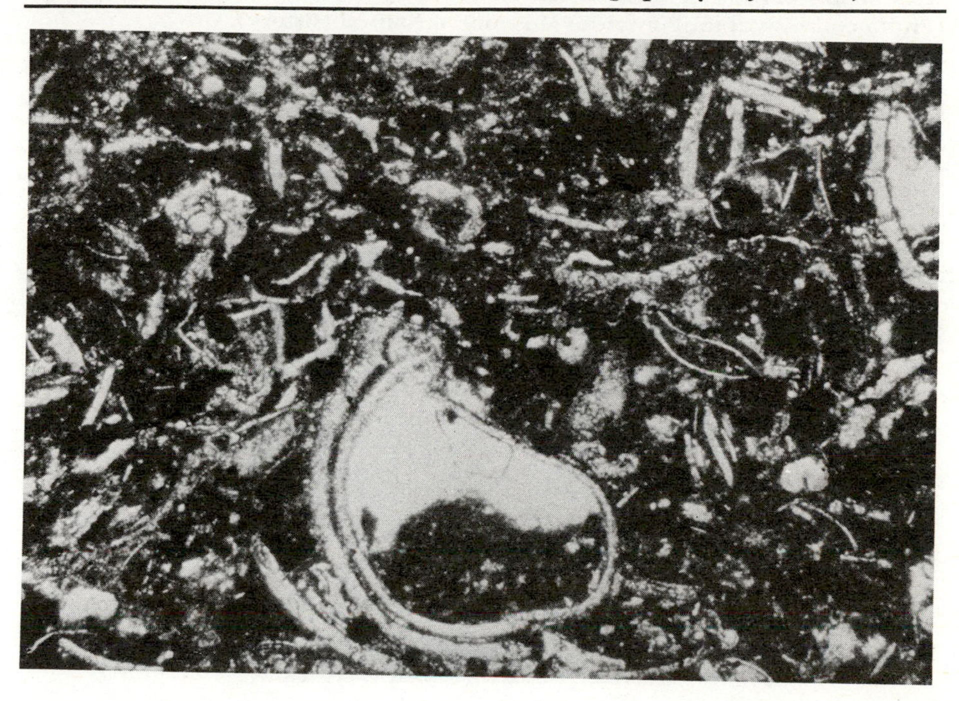

As we have just seen, carbonate rocks share many textures with terrigenous clastic rocks but are classified separately because most of their grains are derived from organic processes. Another group of rocks, the siliceous sedimentary rocks, are also precipitated largely by organisms.

SILICEOUS SEDIMENTARY ROCKS

Siliceous sedimentary rocks are a major group of rocks composed largely of **opaline silica** and **chalcedony** (Fig. 2.4). These rocks can be classified texturally as well as by composition of the grains. However, most are very fine-grained and, therefore, mudrocks in most textural classifications. To understand them fully, one must examine their composition in detail to determine the types of grains present.

Hard skeletal parts from single-celled microorganisms comprise significant quantities of silt- and clay-sized sediment in both deep marine and lacustrine environments. Organisms with siliceous **tests** (shells) are a significant component

FIGURE 2.41 Calcaerous foraminiferan (*Globigerina bulloides*). Calcareous tests of this and other microfossils make fine-grained limestone called chalk. (Photo courtesy of the American Museum of Natural History).

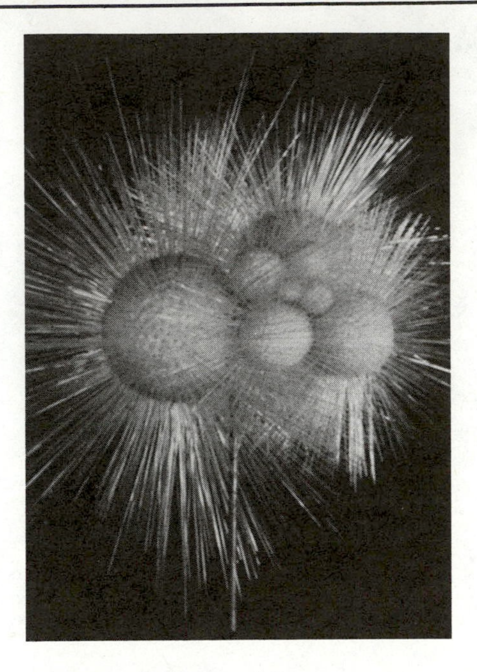

FIGURE 2.42 Bryozoan bioherm from the Ordovician Sequatchie Formation, Ringgold, north Georgia. (Photograph by W. J. Fritz, 1984).

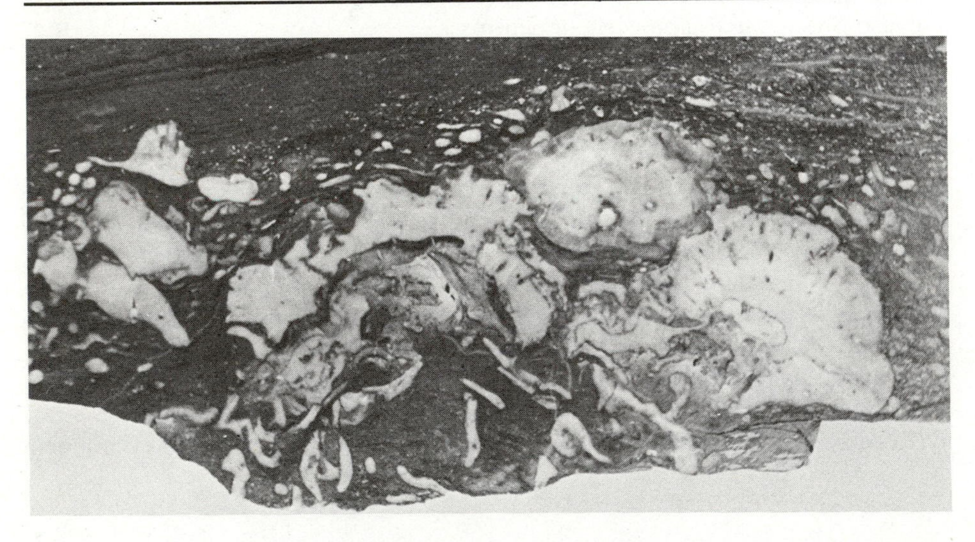

FIGURE 2.43 Stromatolite bioherm from the Proterozoic Siyeh Limestone, Glacier National Park, Montana. (Photograph by Dave Alt).

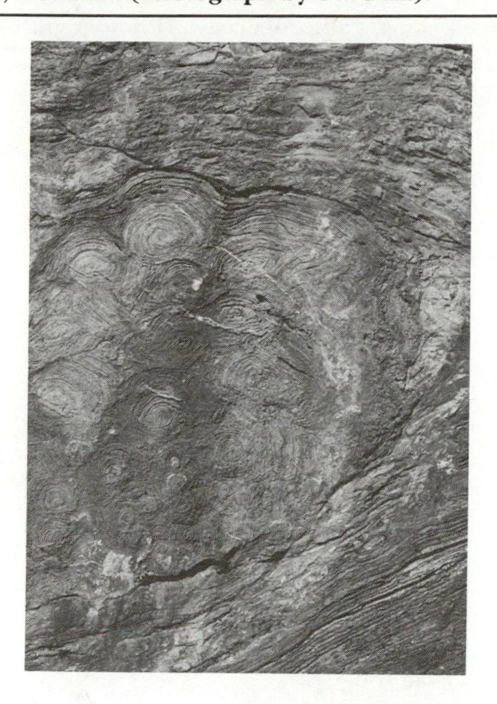

FIGURE 2.44 Radiolarians (coarse mesh), foraminiferans (peaked fine mesh), and sponge spicules (long rods). These microfossils are important sources of biological sediment in marine oozes. Scanning electron micrograph courtesy of Scripps Institute of Oceanography, University of California.

FIGURE 2.45 Bedded chert, California coast. (Photograph by D. W. Hyndman).

of deep marine cherts. The tests of diatoms, radiolarians, and sponge spicules (Fig. 2.44) are common contributors to calcareous as well as siliceous oozes, depending on the taxa of the organism involved.

Sedimentary particles of fine-grained, nonterrigenous silica are produced by several very different processes. Many marine and lacustrine single-celled organisms make tests composed of opaline silica. These mud-sized particles are transported unbroken as part of the suspended load and deposited in **siliceous oozes.** The resulting fine-grained rock is called **chert** (Fig. 2.45); the term "chert" may be modified by the type of organism dominating the sediment. Diatom chert is composed of the tests of single-celled algae called diatoms and radiolarian chert of the siliceous tests of radiolarian protists. An important nonbiogenic source for fine-grained silica comes from volcanic ash, which can also be lithified into chert (Fig. 2.46).

Silica grains can also be mixed with other types of sand or mud, both terrigenous and carbonate, to form various types of sediment (Fig. 2.4), including porcellanite, siliceous mudstone, calcareous chert, or siliceous limestone, depending on the amount and type of the other components present. In common usage, chert is often named by its color. For example, **jasper** is red chert, whereas **flint** is black. Thus, siliceous sedimentary rocks are the result of either biological

FIGURE 2.46 Volcanic ash grains (glass shards) from layers deposited during the eruption of Mount Mazama, Oregon, about 6600 years ago. Grains such as these might also become compacted and recrystallized into chert. (Photograph by R. B. Taylor, #12, U.S. Geological Survey).

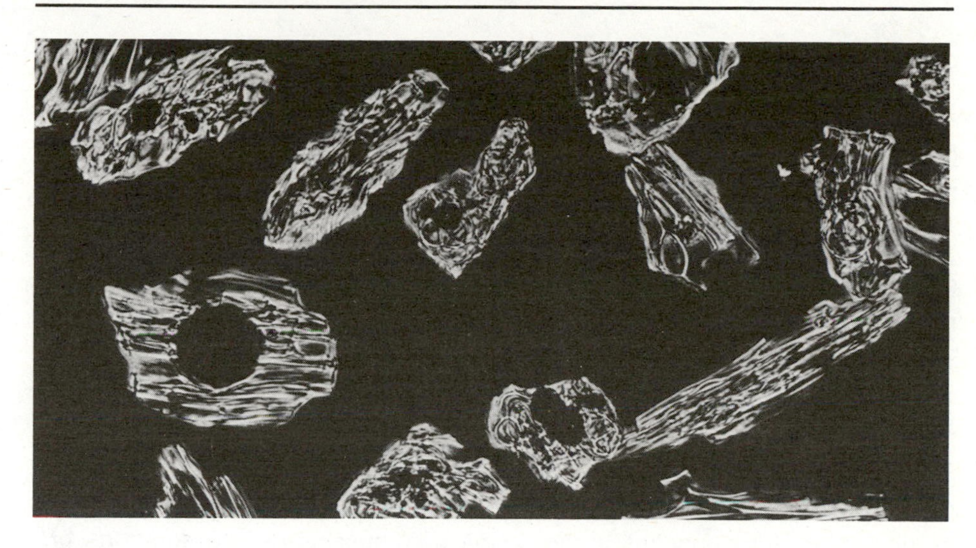

FIGURE 2.47 Travertine forming at Minerva Springs, Mammoth Hot Springs, Yellowstone National Park, Wyoming. (Photograph by W. J. Fritz, 1984).

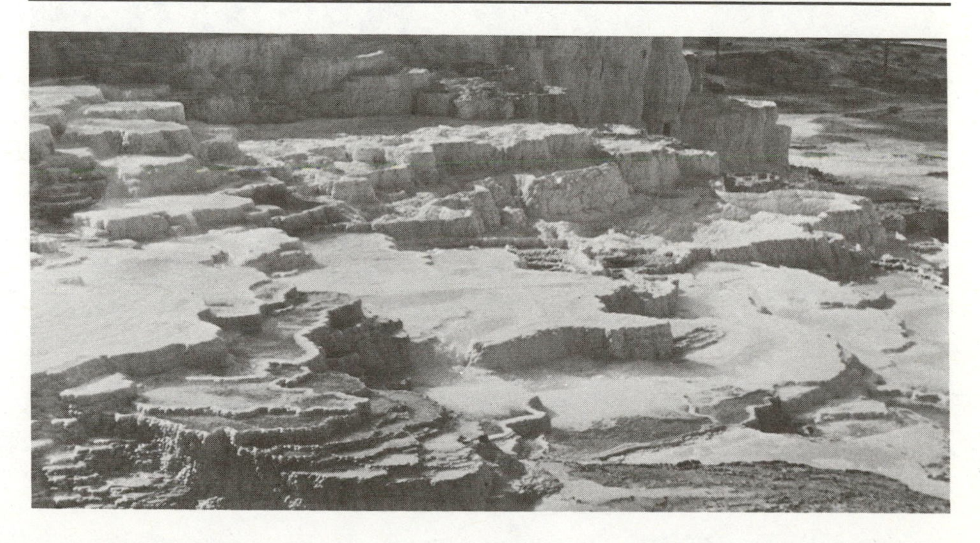

precipitation of silica or volcanic ash accumulation. Another group of similar origin, the chemical precipitates, is composed of minerals other than silica that have been inorganically precipitated from solution.

CHEMICAL PRECIPITATES

Chemical precipitates are rocks that form from minerals precipitated directly from solution by inorganic processes. Another commonly used name for various types of rocks formed by the inorganic precipitation of minerals is **evaporites.** These include **bedded gypsum, anhydrite, halite,** and myriad **borates.** As well, **travertine, tufa,** siliceous sinter, and calcareous sinter (Figs. 2.47–2.49) form wholly or dominantly from inorganic precipitation. As can be seen from this list, chemical precipitates are often named by using the name of the mineral that was precipitated from solution. Other names, such as tufa, travertine, and **sinter,** imply little about the mineralogy but are used in a genetic sense to refer to deposits formed around springs.

Chemical precipitates form as a result of the deposition of salts carried as part of the solution load. Any of the soluble ions in rocks can form chemical precipitates if present in sufficient quantity. The particular compound deposited depends on numerous factors, including the chemical composition of the solutions, pH, Eh, temperature, and climate present in the depositional basin. A common environment for vast chemical precipitate formation is a restricted basin in an arid

FIGURE 2.48 Siliceous sinter forming at Cistern Springs, Norris Geyser Basin, Yellowstone National Park, Wyoming. (Photograph by W. J. Fritz, 1984).

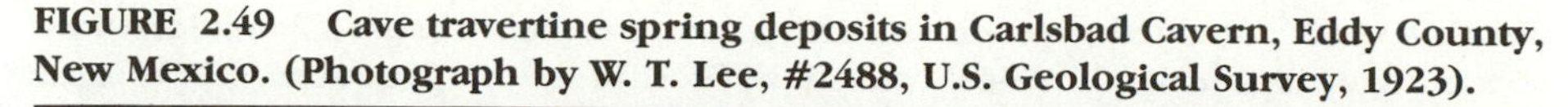

FIGURE 2.49 Cave travertine spring deposits in Carlsbad Cavern, Eddy County, New Mexico. (Photograph by W. T. Lee, #2488, U.S. Geological Survey, 1923).

environment, where evaporation is greater than the inflow of water. However, chemical precipitates can also occur in deep-water environments where brines are concentrated by hot waters associated with volcanic activity or where surface brines sink into deeper water before the mineral grains dissolve.

Common salts deposited as chemical precipitates include halite (NaCl), carbonates (e.g., calcite, $CaCO_3$, and dolomite, $CaMg\,(CO_3)_2$), myriad borates (e.g., Kernite, $Na_2B_4O_6(OH)_2 \cdot 3H_2O$), anhydrite ($CaSO_4$), and gypsum ($CaSO_4 \cdot 2H_2O$). In addition to the inorganic components, many chemical precipitates also have an organic component. For example, hot-spring travertine was once thought to form in thin laminations entirely as the result of inorganic processes. It was believed that as hot, calcium-charged groundwater neared the surface and cooled and the pressure dropped, carbonate gases were released from solution, causing the deposition of limestone. However, recent work by Chafetz and Folk (1984) has demonstrated that many travertines are deposited largely as the result of the metabolic processes of bacteria growing in the hot water. It may thus be impossible to draw a sharp contact between boundstones and these types of chemical precipitates. (We restrict the use of the term **boundstone** to those rocks where the organic influence is unambiguous.)

Most chemical precipitates are laminated to some degree by variations in conditions that cause the minerals to precipitate. Many gypsum and anhydrite units are distinctly banded in thin laminations that can be traced for many miles throughout the depositional basin (Fig. 2.50). In these cases, laminations of evaporite minerals were deposited over an entire restricted basin that was starting

FIGURE 2.50 Bedded gypsum and anhydrite, Permian Castile Formation, west Texas. Note enterolithic folds caused by compression produced by volume changes in the chemical precipitates. (Photograph by W. J. Fritz, 1984).

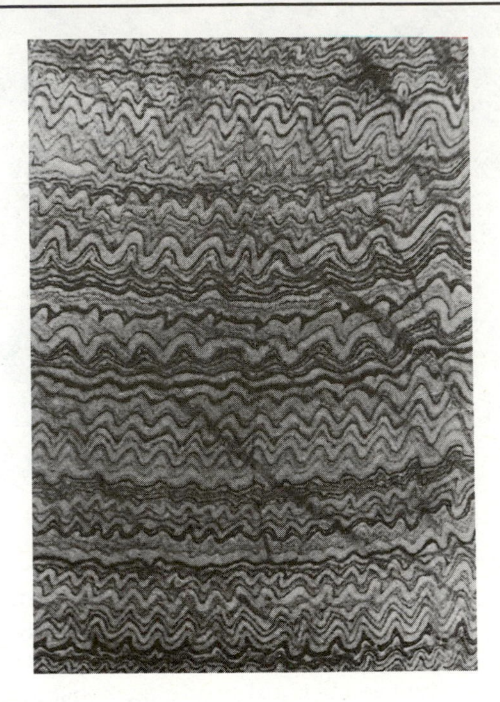

to dry up. In some cases, the alteration of gypsum to anhydrite causes contraction and shrinkage that produces wavy, distorted bedding called enterolithic folds (Chapter 3 and Fig. 2.50). Grain size of chemical precipitates can vary from aphanitic to coarse-grained and this texture can be used, as in limestones, to modify the name of the rock.

PHOSPHATIC SEDIMENTARY ROCKS

Phosphatic rocks are sedimentary rocks of various textures that contain a noticeable amount of organic phosphate in one of several different forms. Texturally, phosphatic sediments can be named by their grain size as sandstones or mudstones or by composition of their grains. Phosphatic sediments are generally rich in the mineral apatite, and some authors (e.g., Murray, 1981) distinguish between phosphatic sediments with greater than 20% apatite and **phosphorites** with over 50% apatite. One type of phosphatic sediment is composed of nodular phosphate and another of phosphate "ooids". These rocks are probably best named by

FIGURE 2.51 (*a*) Outcrop photograph of a phosphatic shale, Cretaceous Pierre Shale, Devils Halfacre, Wallace County, Kansas. (Photograph by J. R. Gill, #60, U.S. Geological Survey, ca. 1967). (*b*) Photomicrograph of a phosphatic sediment of phosphatic pellets in a matrix of smaller pellets, Permian Phosphoria Formation, Idaho. From M. E. Tucker, University of Newcastle-upon-Tyne, England, 1981, *Sedimentary Petrology: An Introduction,* Halstead, New York, Fig. 7.3.

a

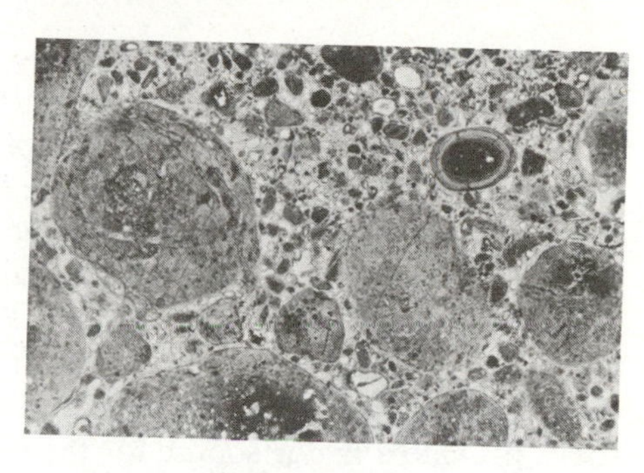

b

FIGURE 2.52 Early Proterozoic banded iron formation from the Lake Superior region with alternating laminae of hematite, magnetite, siderite, and chert. Photograph by W. J. Fritz.

modifying a textural term such as mudstone or sandstone with the term "phosphatic." Examples include phosphatic mudstone, phosphatic oolitic sandstone, and phosphatic shale (Figs. 2.51*a* and 2.51*b*).

IRON-BEARINGS SEDIMENTS

Iron-bearing sediments (Fig. 2.52) are also found, especially in rocks of Early Proterozoic age. These sediments are commonly very thinly bedded and fine-grained with alternating laminations of siderite, limonite, chert, and magnetite and are called banded iron formations. Many appear to have formed in a manner analogous to that of present-day carbonate rocks except that iron rather than carbonate was the soluble ion. Important differences exist between banded iron formation sediments of Archean and Early Proterozoic and ironstones of Phanerozoic ages. The Phanerozoic ironstones commonly are made of iron "ooids" or of various grains replaced by iron (Fig. 2.53). These iron-bearing sedimentary rocks provide a major source for the world supply of iron.

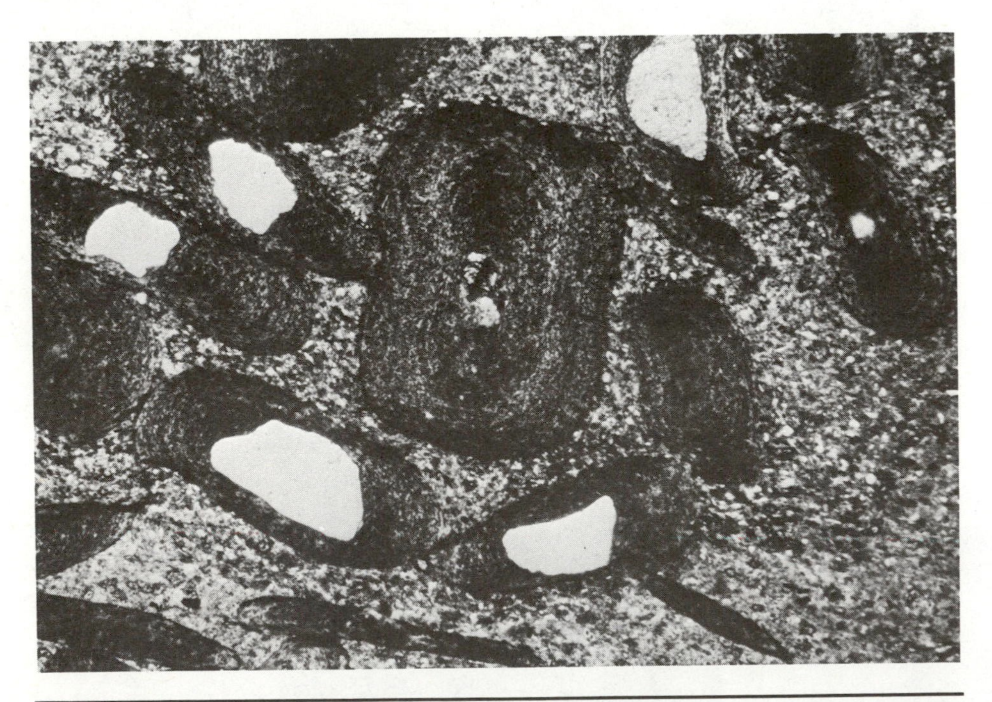

FIGURE 2.53 Iron silicate-rich ooids showing elongated forms. Note that many have cores of quartz and rock fragments in a matrix of quartz silt and finer material. Devonian Mahantango Formation near Amity Hall, Pennsylvania. (Photomicrograph by Bruce M. Simonson).

COAL

Coal is not technically a rock because it has no nonorganic mineral with a definite crystalline structure. However, coal is abundant enough and makes up a significant portion of many sandstone and shale sedimentary sequences so that any stratigrapher should have a working knowledge of a general classification of coal. For this reason it is included here as a rock type. Coals form from the accumulation and compaction of multicellular plants and plant debris and are named for the degree of compaction and alteration of the organic molecules. A mixture of carbonaceous organic sediments exists between terrigenous clay, **peat** (nonconsolidated plant debris), and **sapropel** (fine-grained sediments of algae and bacteria) (Fig. 2.54). The coal classification applies to the compaction of plant material with up to 50% clay, called the ash content of coal, and 25% sapropel.

As plants and wood are compacted they may be reduced in volume by up to 50%. As this compaction takes place, volatile hydrocarbons are released in the form of methane. Eventually, only nearly pure carbon, called **anthracite** coal, remains. The best commercial coals are those with little ash, low sulfur, and compaction to the **bituminous** rank (Fig. 2.55), which is densely packed but leaves highly combustible volatile organic molecules.

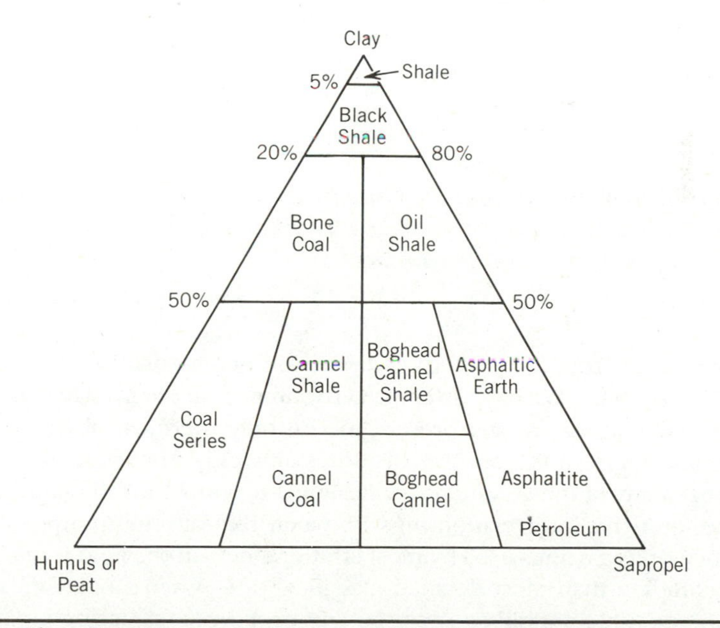

FIGURE 2.54 Classification of organic-rich sediments based on content of humus, sapropel, and clay. Reproduced with permission from J. W. Murray, 1981, *A Guide to Classification in Geology*, Ellis Horwood, London.

FIGURE 2.55 Classification of coal based on the amount of heat, burial, and compaction. Reproduced with permission from J. W. Murray, 1981, *A Guide to Classification in Geology,* Ellis Horwood, London, p. 35. Based on Teichmuller and Teichmuller (1969).

Rank/Compaction	Name	Luster/Fracture	Color
Low	Peat	Dull/Crumbly	Brown
↓	Lignite	Dull/Crumbly	Brown
↓	Subbituminous Coal	Dull/Crumbly	Brown
↓	Bituminous Coal	Bright/Blocky	Black
High	Anthracite Coal	Vitreous/Concoidal	Black

EPILOGUE

It was cold in Missoula that day, and he came at me from nowhere with a Belt Argillite to identify. Quickly I grabbed my trusty rock identifier, and to my dismay, found it wanting.

Anonymous

As you can see from the preceding discussion, sedimentary rock classification is in a confusing state. We have attempted to summarize the myriad classifications proposed in the literature and present to you only the most obvious or widely used systems, but even this at times becomes unwieldy. However, you must keep in mind that many of these systems are designed to show the full range of possible rock types and the interrelationships between the various groups. The actual number of working names used by most stratigraphers in the field is much smaller and less complex than most detailed classifications. When all classifications fail, it is always possible simply to describe any rock by considering grain texture, mineralogy, cement, or bedding and arrive at a very adequate descriptive name.

OUTSIDE READING

General Classification

Blatt *et al.* (1980); Blatt (1982); Compton (1985); Fairbridge and Bourgeois (1978); Folk (1974); Krynine (1948); Murray (1981); Pettijohn (1975); Smosna (1987); Tucker (1981, 1982).

Grain Size, Textures, and Fabrics

Folk (1951); Fritz and Ogren (1984); Harland *et al.* (1966); Krumbein (1934, 1936); Krumbein and Sloss (1963); Mills (1984); Potter and Pettijohn (1975); Powers (1953); Raiswell (1971); Sneed and Folk (1958); Waag and Ogren (1984); Wadell (1932); Wentworth (1922); Wickman (1954); Zingg (1935).

Conglomerates

Koster and Steel (1984); Walker (1975a,b).

Sandstone Classification

Cummins (1962); Dott (1964); Folk (1954, 1974); Hawkins and Whetten (1969); Klein (1963); McBride (1963); Okada (1971); Pettijohn, *et al.* (1972); Scholle and Spearing (1982); Whetten (1966, 1972); Whetten and Hawkins (1970).

Mudrocks

Blatt (1982); Boswell (1961); Grim (1968); Lewan (1978); Picard (1971); Potter *et al.* (1980); Shaw and Weaver (1965); Weaver (1958, 1959).

Carbonates

Bathurst (1971); Chafetz and Folk (1984); Dunham (1962); Embry and Klovan (1971); Esteban (1976); Folk (1959, 1962); Imbrie and Purdy (1962); Lees (1975); Reeves (1970); Scholle *et al.* (1983); Scoffin (1987); Wilson (1975).

Others (Mixed Sediments, Coal, Iron, Volcaniclastic)

Batiza *et al.* (1984); Cloud (1973); Degens (1967); Edwards (1979); Fisher (1961, 1966); Fisher and Schmincke (1984); Folk and Weaver (1952); Fritz (1980, 1982); Fritz and Harrison (1985a,b); Grim and Guven (1978); Harland *et al.* (1966); Honnorez and Kirst (1976); Moore (1968); Mount (1984, 1985); Peltz (1971); Schmid (1981); Simonson (1985); Sparks and Walker (1973); Suthern (1985); Teichmuller and Teichmuller (1968); Tisote and Welte (1978); Wentworth and Williams (1932); Wright and Bowes (1963).

Chapter 3

Descriptive Classification of Bedding and Other Sedimentary Structures

INTRODUCTION

Any sedimentary rock has passed through three main stages on its way to becoming a rock. The first is the actual deposition of the sediment. This can be from physical, biological, or chemical processes or combinations of these three processes. The sedimentary structures formed during this stage are termed primary because they are the first formed. They include features such as bedding, ripple marks, stromatolites, or any feature actually formed during the processes of deposition. Immediately after deposition, the sediment likely undergoes some changes. Processes that act on the sediment after deposition, for example, loading from the weight of overlying sediment, changes in volume resulting from drying or wetting, and many others, modify the original deposit. Structures formed during this stage, just after sedimentation but before the sediment is lithified into a rock, are termed penecontemporaneous, meaning nearly contemporaneous with sedimentation. Such structures include shrinkage cracks, load casts, clastic dikes, and certain types of concretions. Whereas **primary sedimentary structures** record the processes of sedimentation, **penecontemporaneous sedimentary structures** record the modifications immediately after sedimentation within a particular environment. By examining both we can build excellent models of the processes that deposited sediment and later modified it. Finally, once the sediment has lithified into rock, **secondary structures** can form within the rock. These include color banding, some concretions, vein fillings, and weathering rinds.

Krumbein and Sloss (1963) used a somewhat different terminology. Instead

Dominant Process	SYNGENETIC (During Sedimentation) Primary	SYNGENETIC (Immediately Following) Penecontemporaneous	EPIGENETIC (Postlithification) Secondary
Physical/Chemical	BEDDING (Vertical Section) External Shape and Size Tabular — Thick Wedge — Medium Lense — Thin Irregular — Very thin Laminated Internal Relationship Parallel laminated Cross laminated Wavy laminated Convolute laminated Massive Graded Truncated (scour and fill) Imbricated clasts Intraformational conglomerate (transposed bedding) Crystal aggregates Salt teepee structures Isolated crystals and casts	BEDDING Folds Slump folds Convolute bedding Teepee structures Load Features Load casts Flame structures Ball and pillow Water escape structures Dish and pillar structures Clastic dikes Mud/sand volcanoes Compactional faults Convolute bedding Slump planes Other Features Air bubble sand Gas escape structures Crystal growth features Nodules and concretions Void fillings Bird's-eye structures Armored mudballs	EROSIONAL/WEATHERING FEATURES Cut channels Potholes Exfoliation Weathering rinds Paleosols Minikarst surfaces DISSOLUTION Stylolites Nodules Clay seams and lumps Lenses FILINGS Cave fillings Void fillings (crystalline) Void fillings (sediment) Filled ice-wedged cracks CEMENTATION Concretions Color banding Reduction/oxidation spots Halos Hardgrounds

	SURFACE FEATURES Bedforms Ripple marks Dunes, megaripples, and sand waves Plane bed and current lineation Antidunes Hummocks Other Surface Features Swash and rill marks Tool marks Adhesion ripples Salt polydons Scour marks (flutes, etc.)	SURFACE FEATURES Shrinkage cracks Mud/sand volcanoes Clastic dikes Rain/hail imprints Ice crystal imprints Mineral crystal imprints and casts Tool marks Foam impressions Load casts and impressions Salt polygons and teepees Armored mudballs	
Organic	BEDDING Stromatolites Cryptalgal laminae Planar laminae Oncoliths Frameworks (coral, bryozoan, molluscs, etc.) SURFACE FEATURES Mounds Alligned fossils	BEDDING Bioturbation Burrows Resting traces Escape structures Mottling Homogenized bedding SURFACE FEATURES Bioturbation Trails Tracks Body impressions Burrows Mottling	BORINGS Animal Plant

FIGURE 3.1 Table of classification of sedimentary structures.

of the terms primary and penecontemporaneous features, they used **syngenetic sedimentary structures** and the term **epigenetic** for secondary structures. Within these two broad classifications, they further divided sedimentary structures into those formed mainly by physical processes and those formed mainly by organic processes. By combining the previous terminology with Krumbein and Sloss's, we develop a scheme that pigeonholes all sedimentary structures in an existing terminology in relation to when they formed: during sedimentation, immediately following sedimentation, or sometime after lithification. To do this we must also detail the morphology.

In Fig. 3.1 we delineate sedimentary structures of two basic origins: Those formed by physical and chemical processes and those formed by organic processes. Within each of these main subdivisions we have identified the time of formation: during sedimentation, immediately following sedimentation, or after lithification. These criteria define three columns and two broad rows. We have listed most of the structures found in sediments and sedimentary rocks. We will

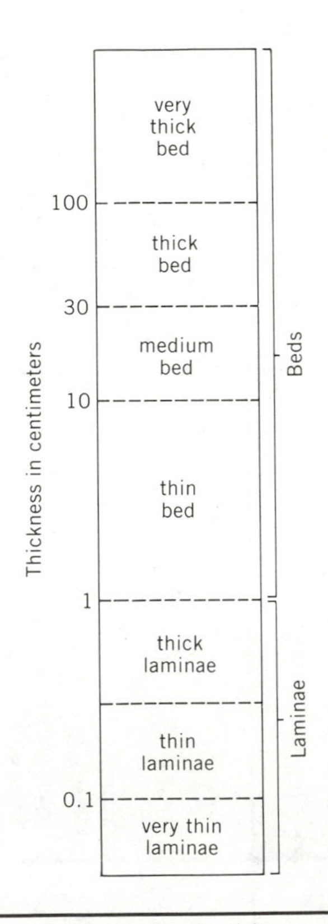

FIGURE 3.2 Thickness terminology for bedding.

FIGURE 3.3 Descriptive terminology for bedding shape.

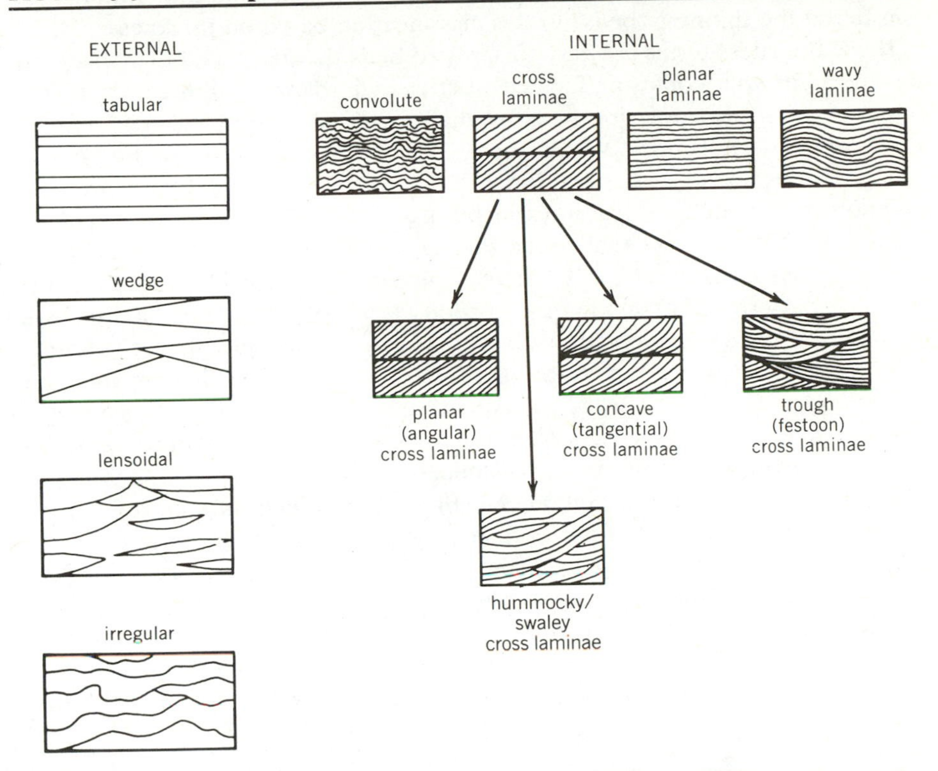

now discuss the morphology of these structures and illustrate them in detail. Keep in mind throughout this discussion that the divisions are for convenience only and that many sedimentary structures cross the boundaries. An example of this is a burrow created by some organism. It could not form unless the sediment was already deposited but it is only preserved when sediment fills it during some later sedimentation event. Because of these subtleties we will use major subdivisions from Krumbein and Sloss (1963).

SYNGENETIC PHYSICAL AND CHEMICAL STRUCTURES

Bedding in Vertical Section

The processes of sedimentation always arrange sediments into packages. Sometimes the packages are quite distinct and other times they are subtle. These packages, termed **beds,** are most often distinguished by changes in texture of the sediment within them. There are a few important criteria to use in describing beds. The first is thickness (Fig. 3.2). Although there is controversy over exact boundaries, all geologists use similar terms: thick, medium, and thin. Laminae are considered the thinnest "beds" in this classification based on thickness.

These thickness terms are easy to apply to beds that have a tabular shape but become somewhat confusing for beds that change thickness along their length. Thus, we must also describe beds by their external shape (Fig. 3.3) and the relationship of the laminae within them. The external shape of the bed can be either **tabular, wedge-shaped, lensoidal,** or **irregular.** All these forms are common in the rock record and can be mixed together in any particular sequence. Each of the external forms may contain laminae of four main types: **planar, cross-laminated** (or bedded), **convolute,** or **wavy.** Cross-laminated beds can be further subdivided into various types of cross laminae (or **cross bedding**). If cross laminae within a bed are planar and abruptly meet the bottom contact, they are termed **angular** or **planar** cross laminae. If they meet the bottom contact tangentially and curve upward, they are called tangential or concave. If the cross laminae curve upward so much that they form troughs, they are termed **trough** or **festoon** cross laminae.

Cross laminae can also form in intersecting wavy laminae, often termed **hummocky** or **swaley** cross-stratification, usually shortened to "HCS" in the literature. Hummocky cross-stratification also lacks clear evidence of directionality as opposed to other types of cross-stratification. By combining the external and internal bedding morphology with the thickness of the bed, the character of the bed can be precisely described. You should also consider that these terms are used to describe cross bedding in two-dimensional exposure and that the same

FIGURE 3.4 Co-sets of cross laminae. Tabular and wedge sets of concave cross laminae. Photograph by J. N. Moore, 1983.

FIGURE 3.5 Special terminology for thinly interbedded mudstone and sandstone.

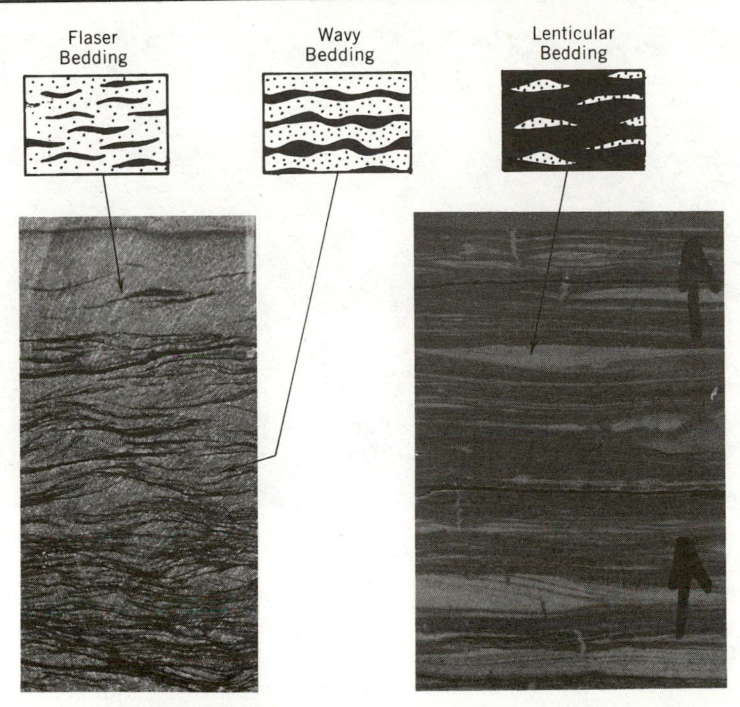

type of bedform could be described in one of several ways depending on the angle of the cut exposing the bedding. For example, trough or festoon cross laminae are most often viewed transverse to the flow of tangential laminae in trough sets.

The term **set** is also useful in describing cross-laminated beds. It is similar to a bed. So one could describe either a "set of cross laminae" or a "bed of cross laminae." If there is more than one set of cross laminae, the term co-set is commonly used (Fig. 3.4).

There are several other factors in classifying bedding that diverge from the scheme just presented that we should discuss at this point. The first classifies thinly bedded intermixtures of mud and sand (both terrigenous and carbonate). Thin beds or laminae of sand forming lenses in mud are termed **lenticular bedding** (Fig. 3.5). If sand forms more continuous layers that pinch and swell in thickness within interbedded mud, then the term **wavy bedded** is used (synonymous with wavy bedding in Fig. 3.3). However, when sand dominates so that small lenses of mud reside in sand, then the term **flaser bedding** is used. These are handy terms to use in specific situations because they cut short the other more descriptive terminology. But because of that, they lose some precision.

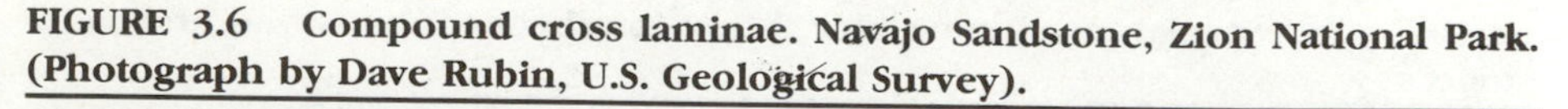
FIGURE 3.6 Compound cross laminae. Navajo Sandstone, Zion National Park. (Photograph by Dave Rubin, U.S. Geological Survey).

Another important variation is the relationship of various types of cross bedding. The classification presented deals only with **simple cross laminae:** each bed contains only one set of cross laminae, in other words, the cross bed does not contain internal erosion surfaces. But there are many sequences where the beds contain co-sets of cross laminae within one bed, that is, the cross beds contain erosional surfaces. Figure 3.6 depicts such cross laminae, which are termed **compound cross laminae.** Such features are very common in eolian sand dune deposits and have been described by Rubin and Hunter (1982) using a purely descriptive classification.

Remember that cross-stratification occurs as the result of deposition on three-dimensional bedforms (discussed in Chapter 4). Most often, deposition occurs on the steep lee slopes and forms **foreset** laminae that dip in the direction of flow. **Backset bedding** (Fig. 3.7) occurs on the upstream, or stoss, slope of some bedforms and others are bounded by bottomset and topset bedding planes.

Within beds there are often textural differences that are useful in deciphering processes. If beds are not laminated, then they may have other features that are used to describe them (Fig. 3.8). Beds that contain no distinct features are termed **massive** or **structureless** (Fig. 3.9). If there is a gradational change in grain size through the bed then the bed is **graded.** Graded beds come in two types, those **normally graded,** which fine upward (Figs. 3.10 and 3.11*a*), and those **inversely graded,** which coarsen upward (Fig. 3.11*b*). These terms are used mostly in coarse-grained sedimentary rocks (conglomerates and sandstones), but graded bedding can even be seen microscopically in many mudstones.

FIGURE 3.7 Backset bedding in gravel deposited by a modern stream near Minas de las Plumosas, Mexico. Stream flow from right to left. (Photograph by E. F. McBride).

Another common and distinct texture is the **intraformational conglomerate** or **flat-pebble conglomerate.** The clasts are generally the same lithology and can be aligned in various patterns (Fig. 3.12), ranging from chaotic to **imbricated.** Imbricated clasts are stacked en echelon, forming an overlapping, shingled pattern. Imbrication can occur in intraformational conglomerates as well as extraformational ones as illustrated in Chapter 2 (Figs. 2.17, 2.18, and 2.19).

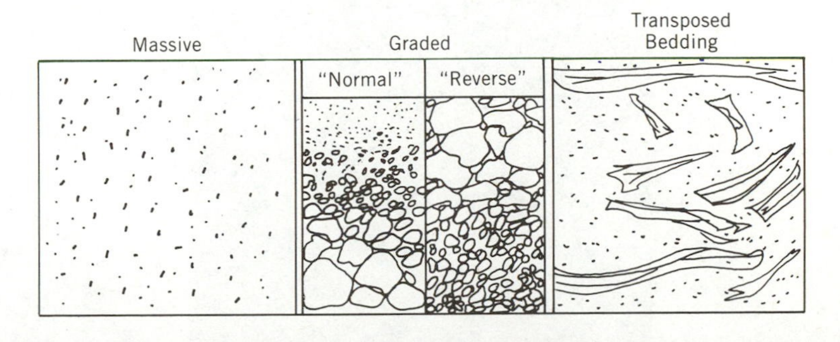

FIGURE 3.8 Terminology used to describe textures of beds. Mostly used for conglomerates and coarse-grained sandstones.

FIGURE 3.9 Massive (structureless) bedding in loess. Huges County, South Dakota. (Photograph by D. R. Crandell, #10, U.S. Geological Survey, 1950).

FIGURE 3.10 Normally graded bedding in conglomerate. Eocene Lookingglass Formation, Oregon Coast Range near Remote. (Photograph by W. J. Fritz, 1981).

FIGURE 3.11 **(*a*) Normally graded bed in Miocene conglomerate. Northeast of Helena, Montana. Platey clasts were derived from slatey argillites of the Precambrian Belt Supergroup. (Photograph by W. J. Fritz). (*b*) Inversely graded conglomerate. Eocene Lamar River Formation, Amethyst Mountain, Yellowstone National Park, Wyoming. (Photograph by W. J. Fritz, 1975).**

Surface Features

Bedforms The complexity and diversity of primary surface features almost defy description, but we will give it a try. The first structures we will describe are **bedforms,** features formed on the bed during sedimentation by currents. The most commonly known are **ripple marks** and **dunes.** When bedforms migrate they produce cross bedding and cross laminae dependent on the shape of the bedform and the amount of accretion. We will discuss the details of such processes in later chapters and concentrate on describing the shape and size of bedforms here. We will begin with ripples and ripplelike forms.

Three important characteristics describe ripple forms: longitudinal profile, plan shape, and size. Bedforms can have a symmetrical (Fig. 3.13) or an asymmetrical profile (Fig. 3.14). In plan view, the crests of the bedforms can be considerably more complex, resulting in several possible shapes (Figs. 3.15–3.29): straight, sinuous, cuspate, lunate, linguoid, and rhomboidal. If more than one form is

FIGURE 3.12 **Intraformational sharpstone conglomerate (sedimentary breccia) in sandstone and shale. Los Angeles County, California. (Photograph by M. N. Bramlette, #45, U.S. Geological Survey, 1931).**

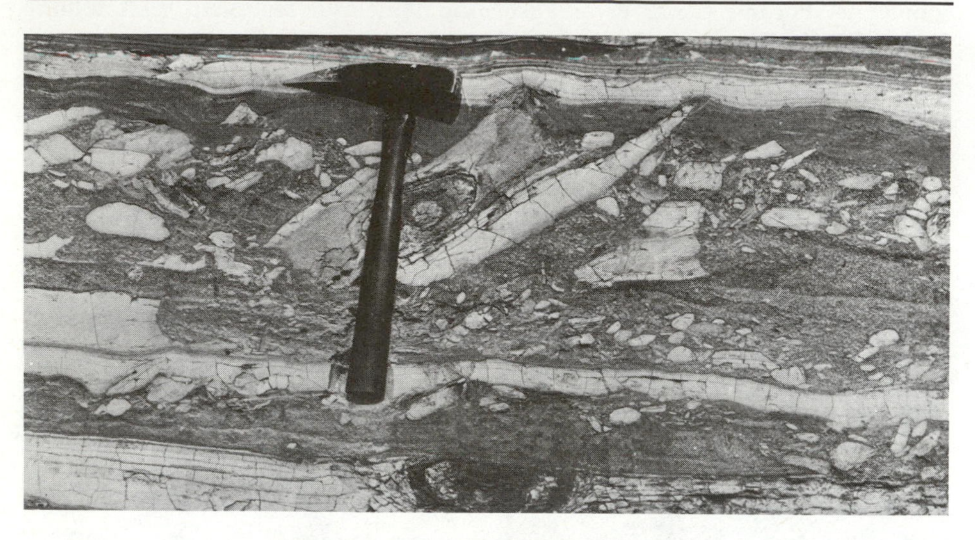

FIGURE 3.13 **Symmetrical ripple marks. Bedding surface of Cambrian sandstone, California. (Photograph by J. N. Moore, 1975).**

FIGURE 3.14 Asymmetrical ripple marks. Sandstone of the Precambrian Belt Supergroup, Montana. (Photograph by J. N. Moore, 1984).

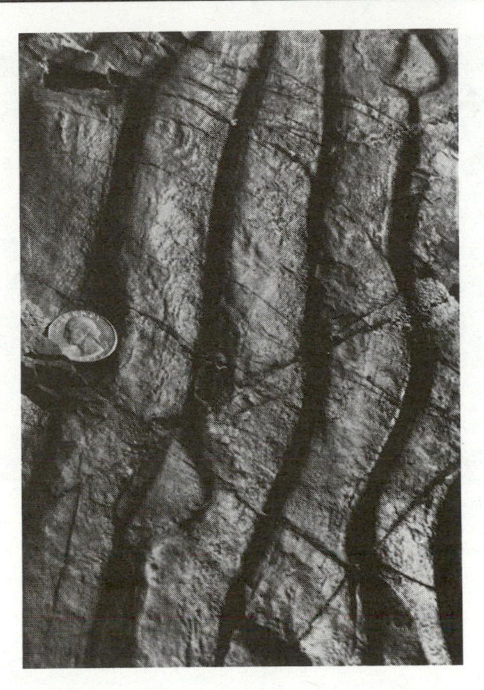

FIGURE 3.15 Straight-crested symmetrical ripple marks. Note the small form in ripple troughs. Coconino Formation. (Photograph by E. D. McKee, #1144, U.S. Geological Survey, 1954).

FIGURE 3.16 Straight-crested to sinuous asymmetrical ripple marks. Precambrian sandstone, Inyo County, California. (Photograph by J. N. Moore, 1974).

FIGURE 3.17 Sinuous symmetrical ripple marks. Agate Beach State Park, Washington. (Photograph by D. W. Hyndman).

FIGURE 3.18 Sinuous asymmetrical ripple marks. Precambrian Belt Supergroup sandstone, Montana. (Photograph by J. N. Moore, 1984).

FIGURE 3.19 Cuspate asymmetrical ripple marks. Gulf of California. (Photograph by E. D. McKee, #1836, U.S. Geological Survey).

FIGURE 3.20 Linguoid ripple marks. Colorado delta, Mexico. (Photograph by E. D. McKee, #1831, U.S. Geological Survey).

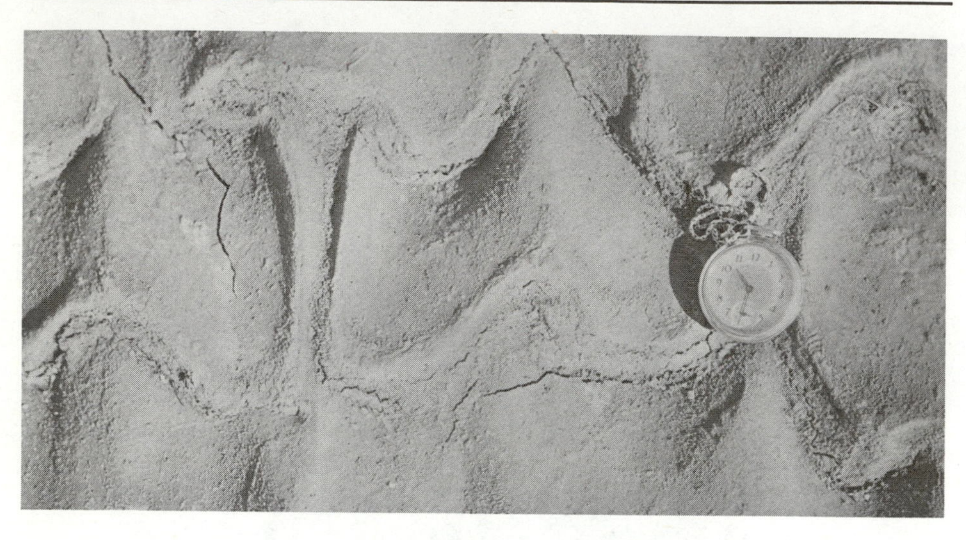

FIGURE 3.21 Rhomboidal ripple marks. Complex shape with linguoidal form. Sapelo Island, Georgia. (Photograph by W. J. Fritz, 1982).

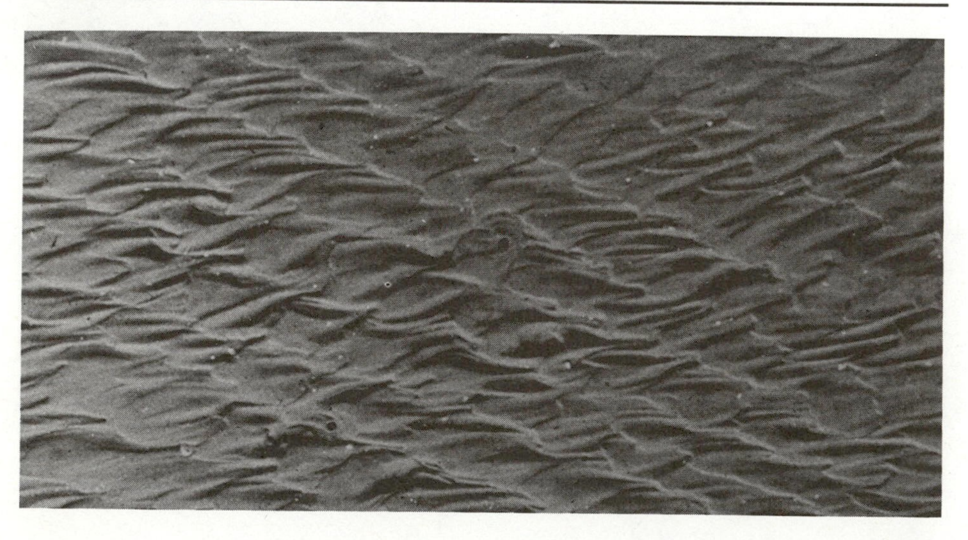

FIGURE 3.22 Rhomboid ripple marks on the lower foreshore, Williamson Island, Georgia. (Photograph by W. J. Fritz).

FIGURE 3.23 Lunate to cuspate large ripple marks. Sandstone facies of the Ordovician Capel Curig Volcanic Formation, Capel Curig, North Wales. (Photograph by W. J. Fritz, 1984).

FIGURE 3.24 Complex bifurcating sinuous ripple marks. Estuarine mudflats, Harlech Point at Tremadoc Bay, North Wales. (Photograph by W. J. Fritz, 1984).

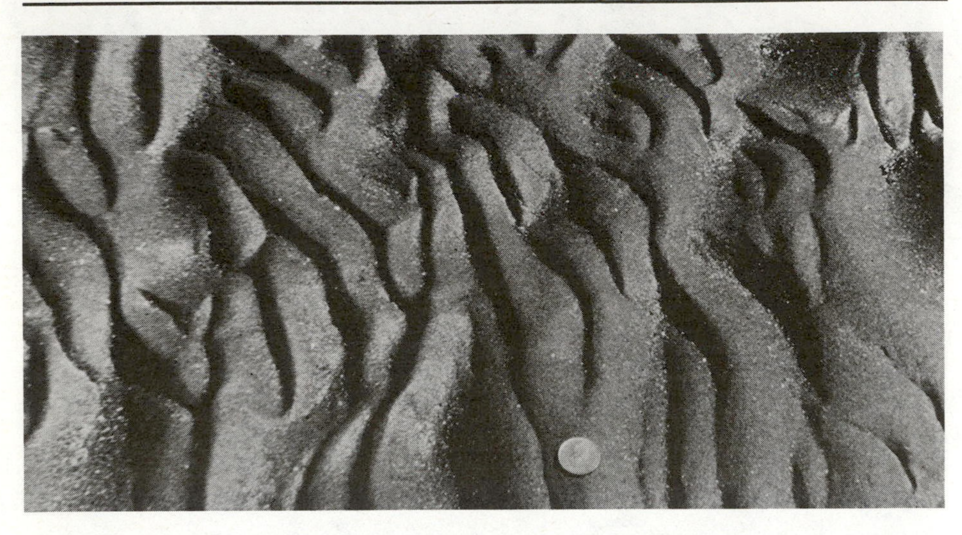

FIGURE 3.25 Lunate ripples, intertidal sand bar in Bogue Sound, North Carolina. (Photograph by W. J. Frazier).

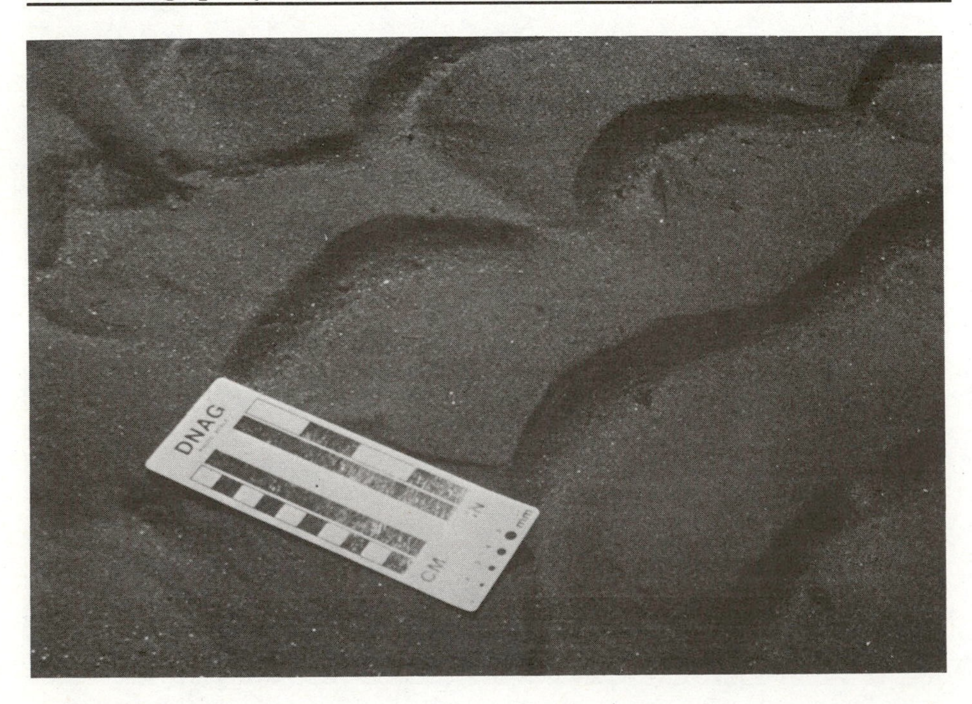

FIGURE 3.26 Linguoid ripples, Jekyll Island, Georgia. (Photograph by W. J. Frazier).

FIGURE 3.27 Interference ripple marks (ladder-back). Moenkopi Formation, Arizona. (Photograph by E. D. McKee, #1143, U.S. Geological Survey, 1954).

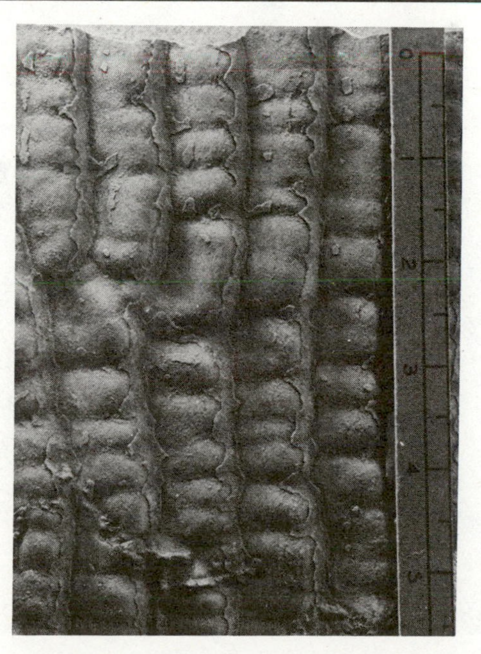

FIGURE 3.28 Ladder-back interference ripples, St. Simons Island, Georgia. (Photograph by W. J. Frazier).

FIGURE 3.29 Irregular interference ripple marks. Precambrian Belt Supergroup sandstone, Montana. (Photograph by J. N. Moore, 1984).

present, the term interference or cross (e.g., **interference ripple marks, cross ripple**) is used. Ripple forms are also classified by size: large scale or small scale. Let us start by defining the terminology used for both asymmetrical and symmetrical forms. Unfortunately, existing terminology is far from consistent and cogent but we have tried to organize it into a coherent conceptual scheme (Fig. 3.30) that relates various terminologies. We first divide ripple forms into either asymmetrical or symmetrical. Under each of these divisions there are small-scale forms and large-scale forms. The shape of the crest is used as a final modifier to delineate forms further. Various other indices or measurements can be used to precisely define these bedforms (Fig. 3.31).

The most common asymmetrical ripple forms are the **ripple marks** (or ripples) and **large ripples.** The distinction between the two is made on size, not shape. Small-scale forms are termed ripples, whereas large-scale forms are termed **megaripples** (or a variety of other terms discussed later). The problem is how big is small. In 1963 Allen collated a large amount of data on the height and length of asymmetrical bedforms, small-scale ripple marks, and large-scale ripple marks. These data suggest a somewhat natural break, albeit subtle, between small forms and large forms. Studies on the formation of ripples suggest that such a break does exist (see Chapter 5), but sizes of the two forms overlap considerably. Costello and Southard (1981) separated ripples from dunes and found that ripples averaged approximately 17 cm in wavelength, whereas dunes averaged between 57 and 84 cm, depending on plan shape. The distributions overlapped though, so that it would be difficult to separate out ripples from dunes by a descriptive measure only. The overlap was even greater for height of the forms (Fig. 3.32), modifying Allen's earlier thoughts on separating ripples from dunes. Given all this confusion about size, it is difficult to pick a distinct break between small-scale ripples and large-scale ripples.

Unfortunately, there are myriad terms used as synonyms for large-scale ripple marks. **Megaripple** is used as synonymous with large-scale ripple marks and was described in studies of ripples in flumes. The term **dune** or **3-D dune** is used for large-scale, lunate and linguoid forms, and very large, straight-crested, asymmetrical forms are termed **sand waves.** So, the terminology is not consistent. Allen used the term "large-scale" to encompass ripple forms up to 1000 m in length and 20 m high. One can use a fairly straightforward terminology by arbitrarily subdividing small-scale ripples from large-scale ripples, but most sedimentologists continue to use the other terms. Because of that usage we propose the following simplification.

The term **small-scale ripple** (or **small ripple**) will be used for ripple forms (either asymmetrical or symmetrical; if "symmetrical" is not used as a modifier the form is considered to be asymmetrical) that have a spacing of less than 60 cm; they will generally have heights less than 5 cm. Forms that have spacings greater than 60 cm and heights greater than approximately 6 cm are termed **large-scale ripples** (or **large ripples**). The crest shape is used as a modifier to complete

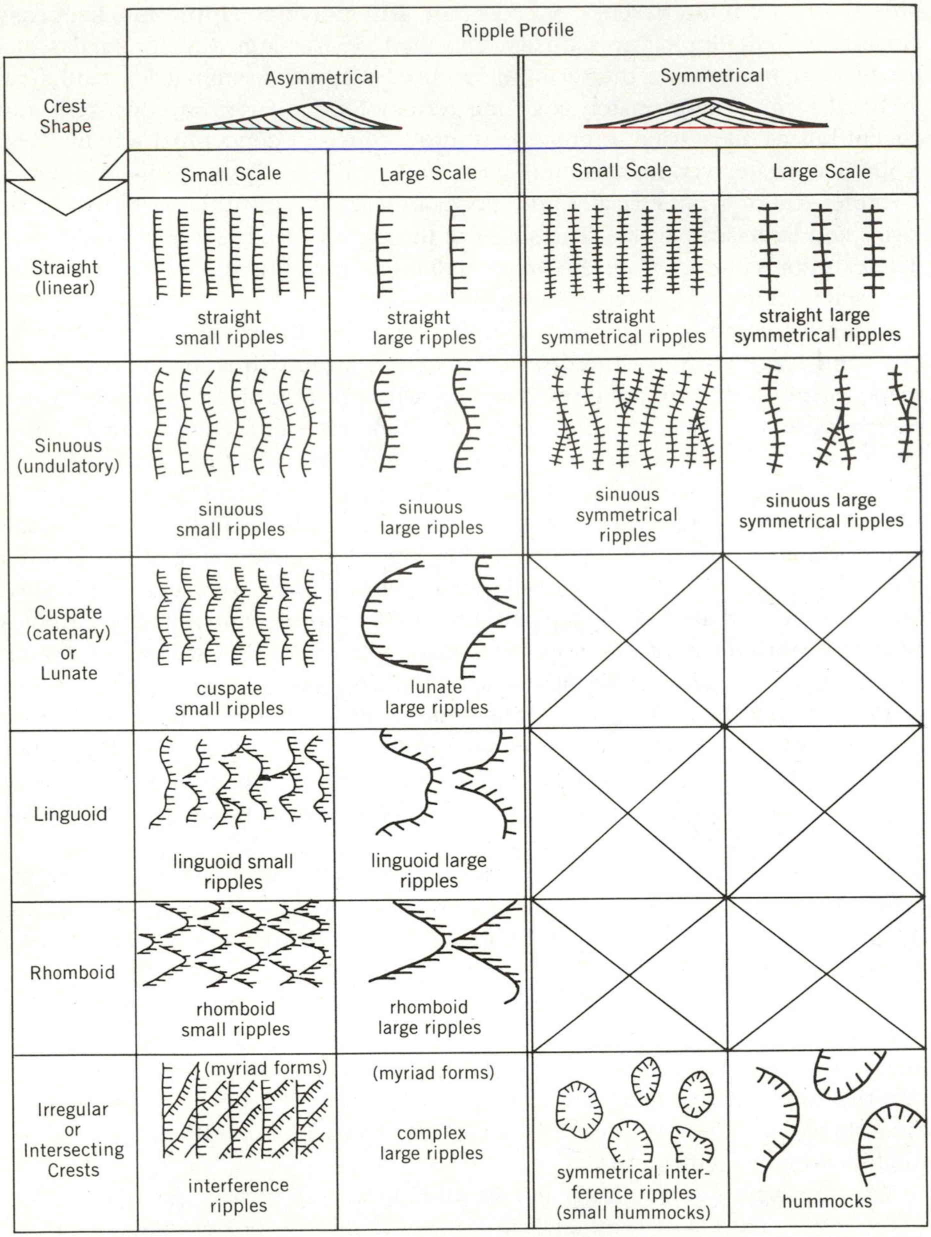

Crest Shape	Ripple Profile: Asymmetrical, Small Scale	Ripple Profile: Asymmetrical, Large Scale	Ripple Profile: Symmetrical, Small Scale	Ripple Profile: Symmetrical, Large Scale
Straight (linear)	straight small ripples	straight large ripples	straight symmetrical ripples	straight large symmetrical ripples
Sinuous (undulatory)	sinuous small ripples	sinuous large ripples	sinuous symmetrical ripples	sinuous large symmetrical ripples
Cuspate (catenary) or Lunate	cuspate small ripples	lunate large ripples		
Linguoid	linguoid small ripples	linguoid large ripples		
Rhomboid	rhomboid small ripples	rhomboid large ripples		
Irregular or Intersecting Crests	(myriad forms) interference ripples	(myriad forms) complex large ripples	symmetrical interference ripples (small hummocks)	hummocks

FIGURE 3.30 Classification of ripple marks by shape.

FIGURE 3.31 Terminology of ripple forms and indices.

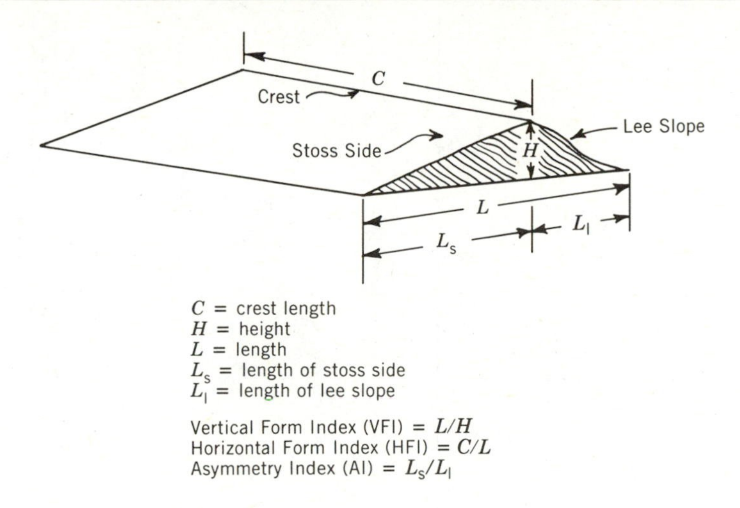

C = crest length
H = height
L = length
L_s = length of stoss side
L_l = length of lee slope

Vertical Form Index (VFI) = L/H
Horizontal Form Index (HFI) = C/L
Asymmetry Index (AI) = L_s/L_l

the name of the form (Fig. 3.33). The term **dune** is restricted to lunate and linguoid asymmetrical large ripples that have heights of several decimeters or greater and spacings of a few meters or more. **Sand wave** is used for very large straight-crested forms with heights in the meters and wavelengths of tens of meters or greater.

The terminology of symmetrical ripples is less confusing. Small-scale **symmetrical ripples** are nearly always straight or sinuous crested (Fig. 3.30). By using the shape of the crest as a modifier, the ripple can be described. The crests sometimes bifurcate or branch (Fig. 3.34). If the crests are irregularly shaped, the term hummocks is used to describe the bedforms (Fig. 3.35). Also note that symmetrical ripples do not show a break in size from small to large forms. Instead there seems to be a continuous gradation from small to large symmetrical ripples.

Another feature that resembles a symmetrical ripple is the **antidune** (Chapter 4). These features are straight-crested, low-relief bedforms that can be nearly symmetrical (Fig. 3.36). They are relatively large (spacing of crests in the decimeters or meters), so they should not be confused with the much smaller symmetrical ripples. Sometimes asymmetrical ripples become flattened as a result of shallowing of water in environments such as intertidal zones. These flat-topped forms may appear symmetrical but are really altered asymmetrical ripples.

Other ripplelike forms result from particular processes and therefore are very useful in determining depositional environments. Small **wrinkle marks** or **Runzelmarken** are tiny interference ripples with heights of only a few millimeters and lengths of centimeters or less (Fig. 3.37). They only form in extremely shallow water. Runzelmarken also form in very shallow water as sediment is blown onto a rippled or wrinkled surface and **adhesion ripples** form where dry sand is

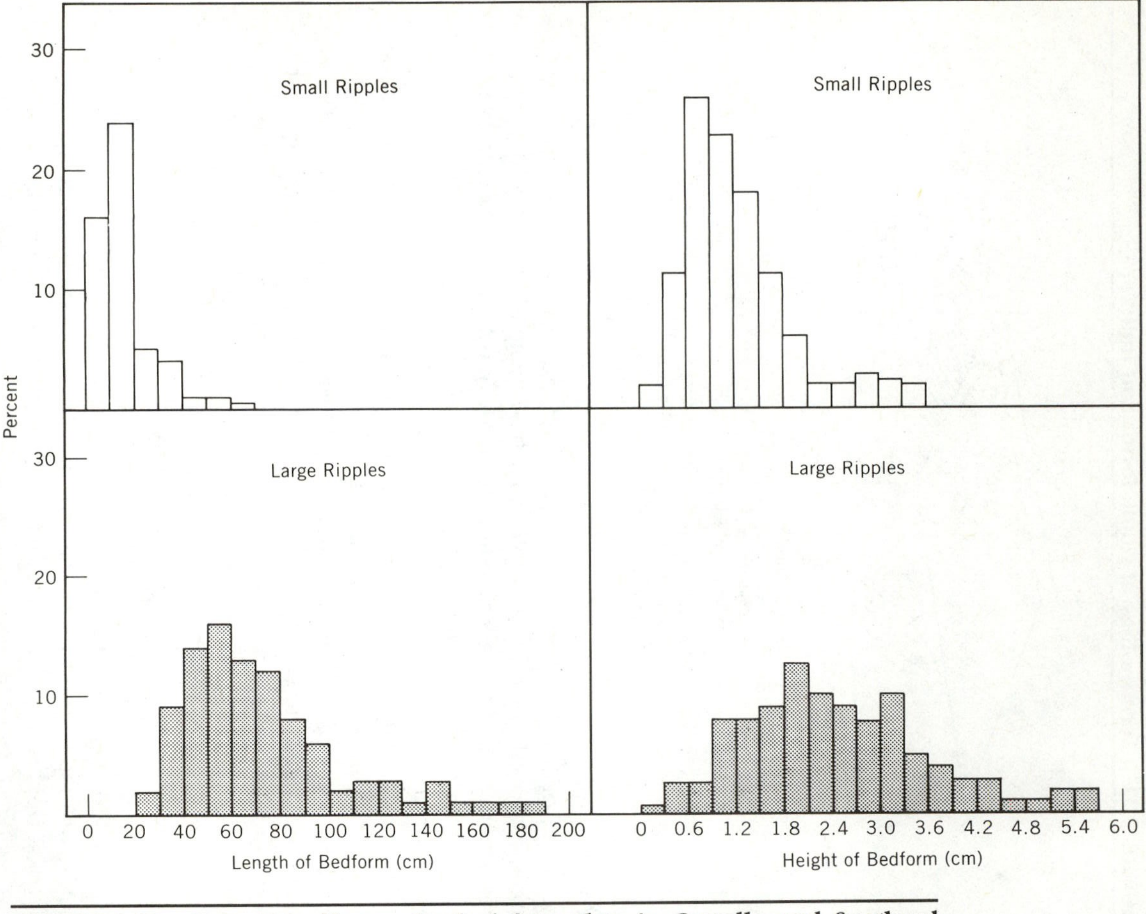

FIGURE 3.32 Size of bedforms. Drafted from data in Costello and Southard (1981).

FIGURE 3.33 Sinuous, symmetrical ripple marks. Moenkopi Formation, Arizona. (Photograph by E. D. McKee, #1137, U.S. Geological Survey, 1949).

FIGURE 3.34 Bifurcating, sinuous, symmetrical ripple marks. Moenkopi Formation, Arizona. (Photograph by E. D. McKee, #1147, U.S. Geological Survey, 1954).

FIGURE 3.35 Hummocky cross-stratification in white sand layer in center of photo. Note rip-up clasts and burrows in sand layer and loaded base of the sand. Cretaceous Eutaw Formation, Columbus, Georgia. (Photograph by W. J. Frazier).

FIGURE 3.36 Antidunes formed by backwash of waves along a beach. (Photograph by Keith Stowe).

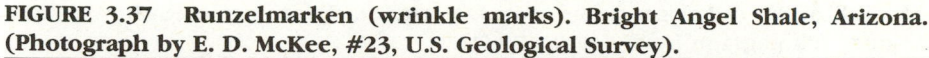

FIGURE 3.37 Runzelmarken (wrinkle marks). Bright Angel Shale, Arizona. (Photograph by E. D. McKee, #23, U.S. Geological Survey).

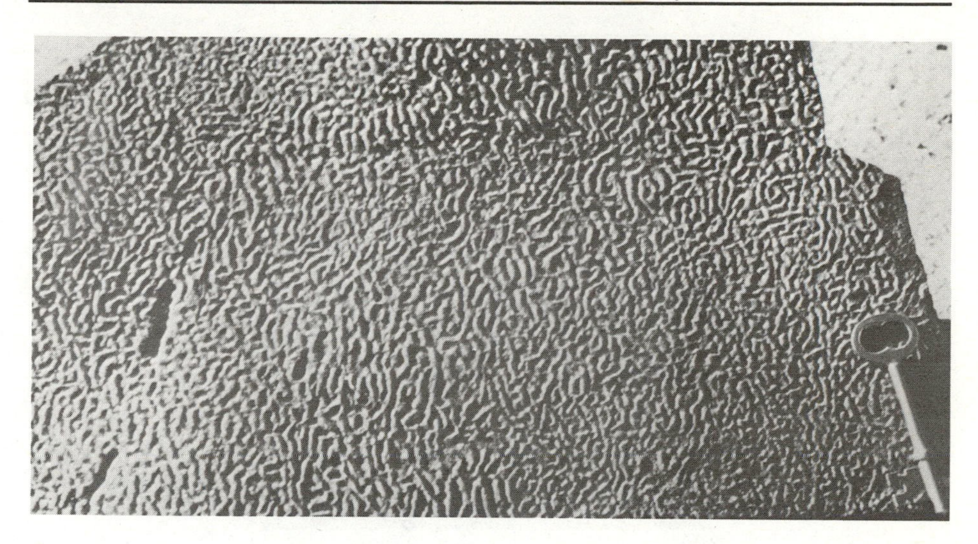

blown over damp or wet sand (Fig. 3.38). The term **ripplet** is commonly used to describe all such tiny, ripplelike forms.

When the surface of the bed is smooth and flat, the term **plane bed** is used. This mainly applies to current-deposited beds and is synonymous with flat bed. Plane beds composed of sand often show a faint current lineation formed of segregated grains of different sizes and shapes (Fig. 3.39).

Erosional Features The features described in the foregoing were formed on the bedding surface by depositional processes. Often, erosion precedes deposition, leaving very distinctive structures. Some such features form when a separate current erodes material, followed later by a current that deposits sediment, whereas others form by continuous processes of erosion and sedimentation, as during sediment flows. We will discuss these processes in later chapters. The general term used for erosional sedimentary structures is **scour marks.** Scour marks come in myriad forms. If limited in extent and "flute"-shaped or triangular, they are termed **flute casts** or flutes (Fig. 3.40). Continuous, linear scours are called **gutter casts** because they resemble gutters (Fig. 3.41). Scour marks or casts can have nearly any form but tend to be elongate. An odd scour sometimes occurs at the base of turbidite deposits (see discussion in Chapter 8) called frondescent sole marks (Fig. 3.42).

Other less dramatic erosional features form from the erosion of sand by very shallow flows. These features, called **rill marks** (Fig. 3.43), are very common on beaches where the wave swash erodes material as it returns down the beach.

FIGURE 3.38 Adhesion ripple marks. Mount St. Helens sediment flows, Washington. (Photograph by W. J. Fritz, 1982).

FIGURE 3.39 Plane bed with current lineation. Cretaceous sandstone, central Montana. (Photograph by J. N. Moore, 1984).

FIGURE 3.40 Flute casts. Precambrian Johnnie Formation, Death Valley, California. (Photograph by J. N. Moore, ca. 1975).

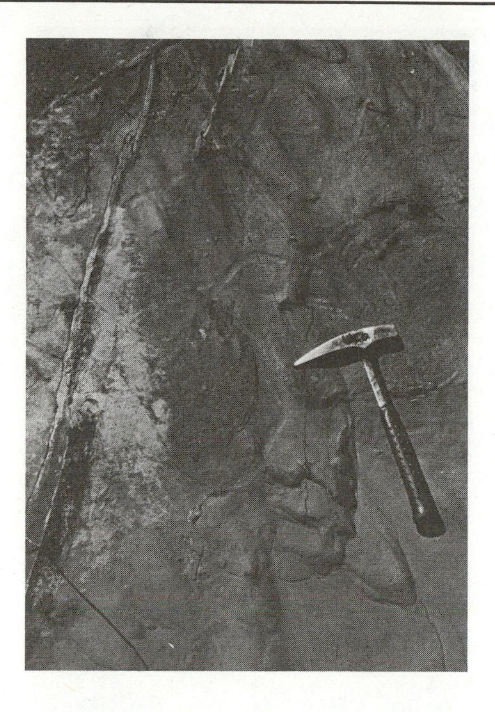

FIGURE 3.41 Gutter casts. Sabah, Borneo, Indonesia. (Photograph by W. B. Hamilton, #568, U.S. Geological Survey, 1979).

FIGURE 3.42 Irregular scour marks. Precambrian Belt Supergroup, Montana. (Photograph by J. N. Moore, 1983).

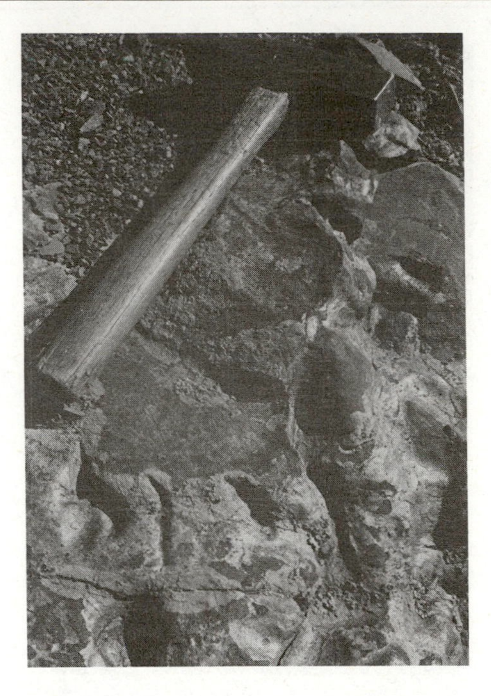

FIGURE 3.43 Rill marks. Modern surface of a beach. (Photograph by Dave Alt).

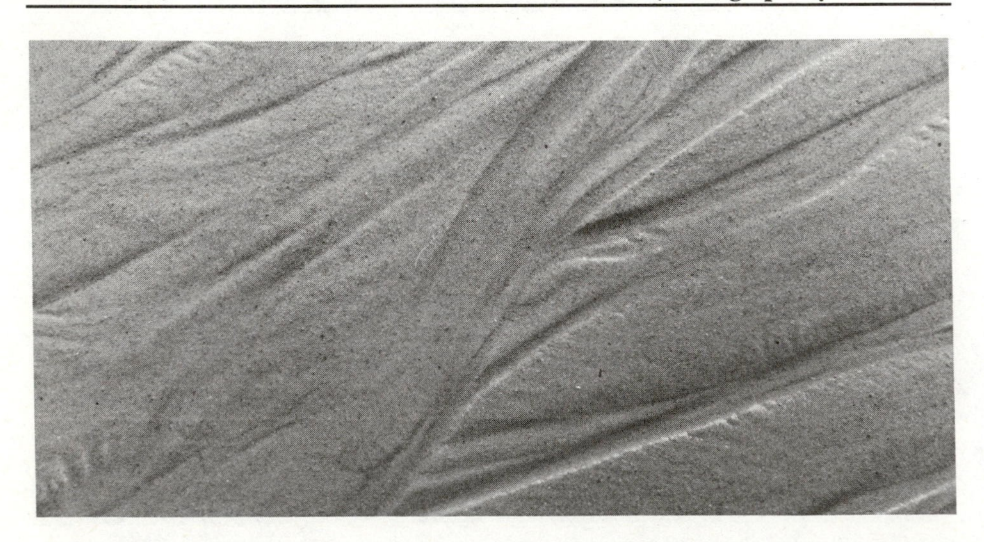

FIGURE 3.44 Swash marks. Modern beach. (Photograph by Dave Alt).

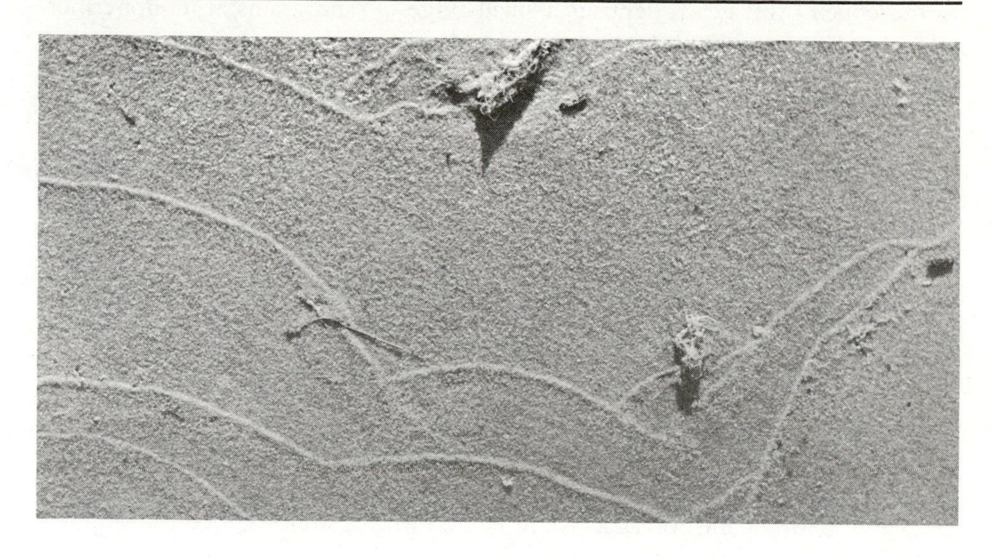

FIGURE 3.45 Ice gouge features in the Beaufort Sea, in 25 m of water. Produced by grounded ice flows. (Photograph by Erk Reimnitz, U.S. Geological Survey).

The depositional counterpart of the rill mark is the **swash mark,** which forms as the edge of the swash deposits a small ridge of sand grains at its uppermost position (Fig. 3.44). Such features are rarely preserved in rocks.

Another feature commonly accompanying erosional structures is **tool marks.** These features result when an object (**tool**) transported by the flow gouges sediment from the bottom. When sediment is deposited on top, they form casts. These features have a strong resemblance to scour marks and are easily confused with them. There are two types: those formed by stationary objects (or nearly so) and those formed by moving objects. Ice blocks commonly form pits, where they wallow around as they melt. These are common on cold-climate tidal flats and lakes. Often when ice grows in lakes it pushes up wedges at the margin that also gouge sediment and form tool marks (Fig. 3.45). Such ice-formed tool marks probably represent a minor component of tool marks in sedimentary sequences.

Tool marks formed by some object moving along the bottom during sedimentation are much more common in sedimentary sequences. Common objects forming them are clasts, sticks, or shells. These marks can be very continuous or quite isolated (Fig. 3.46). A very peculiar sort of tool mark forms in some playa lakes in California and Nevada, where large, moving stones gouge up sediment (Fig. 3.47).

Surface Structures Potpourri Many other sedimentary structures form immediately after sedimentation. One very common and well-known feature is the

FIGURE 3.46 Tool mark casts formed at base of bed by clasts carried during transport. Camp Creek Sequence, Elko County, Nevada. (Photograph by K. B. Ketner U.S. Geological Survey).

FIGURE 3.47 Tool marks formed by stone moving across playa surface. Death Valley, California. (Photograph by Richard Frear, National Park Service).

FIGURE 3.48 Mudcracks. Cracks of different sizes filled with sediment. Playa deposits, Eureka Valley, California. (Photograph by J. N. Moore, 1984).

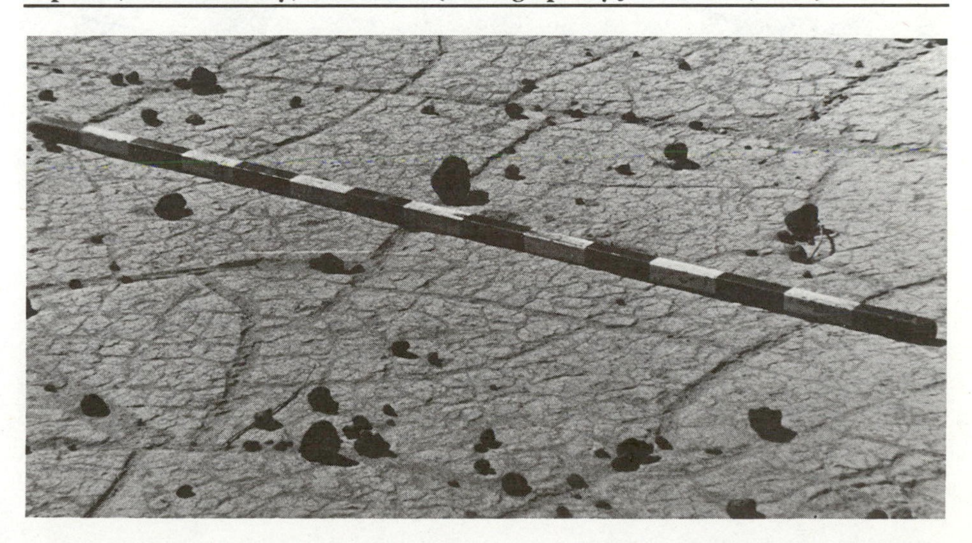

mudcrack or shrinkage crack (Fig. 3.48). These are often preserved as casts composed of sand that filled the crack and was later lithified. Mudcracks can be seen in vertical sections and on surfaces. When attributed to volume changes resulting from desiccation, they are termed desiccation cracks. These often show curled plates of mud and well-developed polygonal shapes (Fig. 3.49). Another type of mudcrack, which is attributed to volume changes and compaction without desiccation, is subaqueous shrinkage cracks or **syneresis cracks.** These have patterns similar to desiccation cracks but often form only incomplete curved portions of polygons (Fig. 3.50). It is often difficult to distinguish between the two types of structures. It is important to realize that mudcracks can form both subaqueously and subaerially. Mudcracks are preserved when filled with other sediment. This filling can be deformed by compaction to form small, squiggly wedges (Fig. 3.51).

Other surface features are common on bedding surfaces. Imprints left by rain are termed raindrop imprints (Fig. 3.52) and even hail imprints have been distinguished from those formed by rain drops. A wind-driven rain can organize the raindrop imprints into small ridges that resemble adhesion ripples.

Structures Formed by Chemical Precipitation

On bedding surfaces, sedimentary structures created during the precipitation of salts form distinctive and complex patterns. Various crystal imprints form in muddy sediments including those from ice and halite. Ice imprints are rare in the rock record but **halite casts** or **hopper casts** (Fig. 3.53) are common.

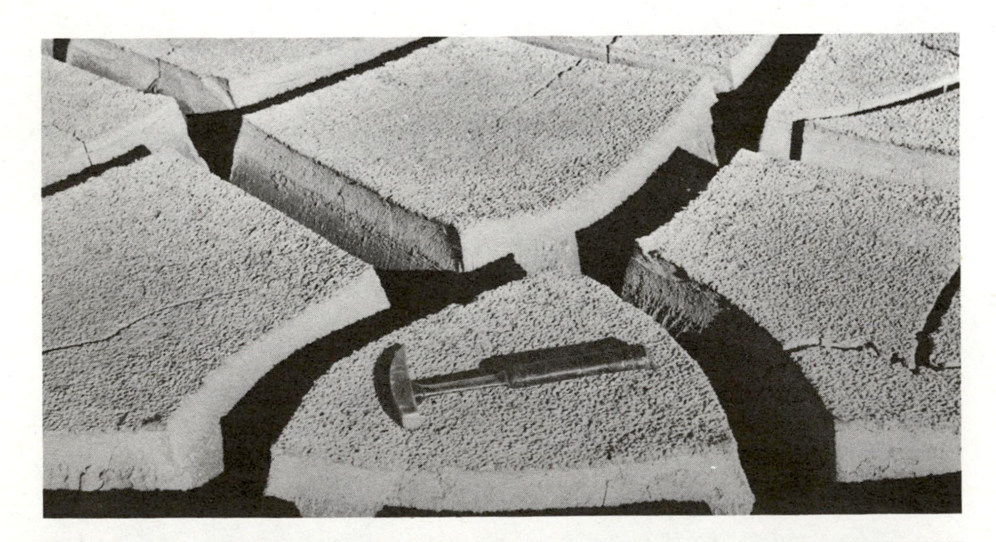

FIGURE 3.49 Desiccation cracks. Plates are curled upward slightly. (Photograph by E. D. McKee, #2196, U.S. Geological Survey.

FIGURE 3.50 Subaqueous shrinkage cracks. Precambrian Deep Spring Formation, Inyo County, California. (Photograph by J. N. Moore, 1974).

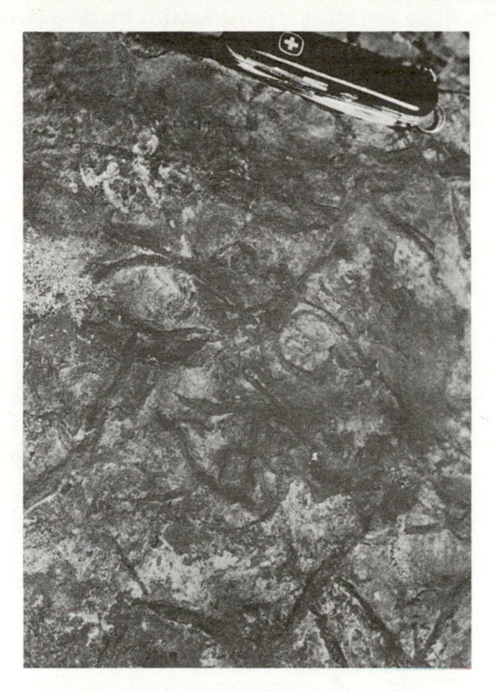

FIGURE 3.51 Deformed cracks filled with siliceous dolomite. Precambrian Siyeh Limestone, Montana. (Photograph by Dave Alt).

FIGURE 3.52 Rain drop imprints. (Photograph by G. K. Gilbert, #948 U.S. Geological Survey).

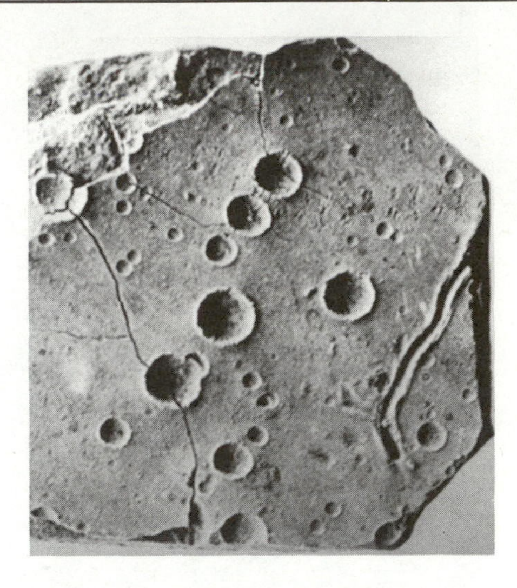

FIGURE 3.53 Hopper casts of halite crystals in mudstone. (Photograph by W. J. Fritz, 1984).

Some evaporite deposits contain extensive primary features related to precipitation and expansion of salts. The most dramatic of these are **salt polygons** (Fig. 3.54). These large, polygonal ridges often grow to several centimeters above the surface and several meters across. Nodular gypsum growing within the sediment in sabkha environments often creates a mottled "chicken-wire" structure.

In cross section, aggregates of crystalline salts form indistinct beds, the morphology determined by the type of mineral and length of growth time.

Soft-Sediment and Water Escape Structures A variety of features form when sediment deforms ductilely or flows like a fluid. These structures can originate from many processes: density contrasts, water escape, volume change, and gas escape. A striking example of soft-sediment features are the **soft-sediment folds.** These are ductile structures ranging from a few millimeters to tens of meters across (Fig. 3.55). Some more coherent folds show a distinct shear direction and have been attributed to slumping or movement of large masses of sediment on a depositional slope. These features, termed slump folds, are found in many different deposits (Fig. 3.56). Soft-sediment folds also form very complex patterns termed **convolute bedding** or **contorted bedding** (Fig. 3.57).

Another foldlike form resulting from volume change is a **teepee structure.** These can be seen best on bedding surfaces, but when cut vertically appear as concave-upward folds joined at a brecciated peak (Fig. 3.58), resembling a Plains Indian teepee. These are folded layers but result from the growth of evaporites in the sediment forming salt polygons and not from strict ductile deformation.

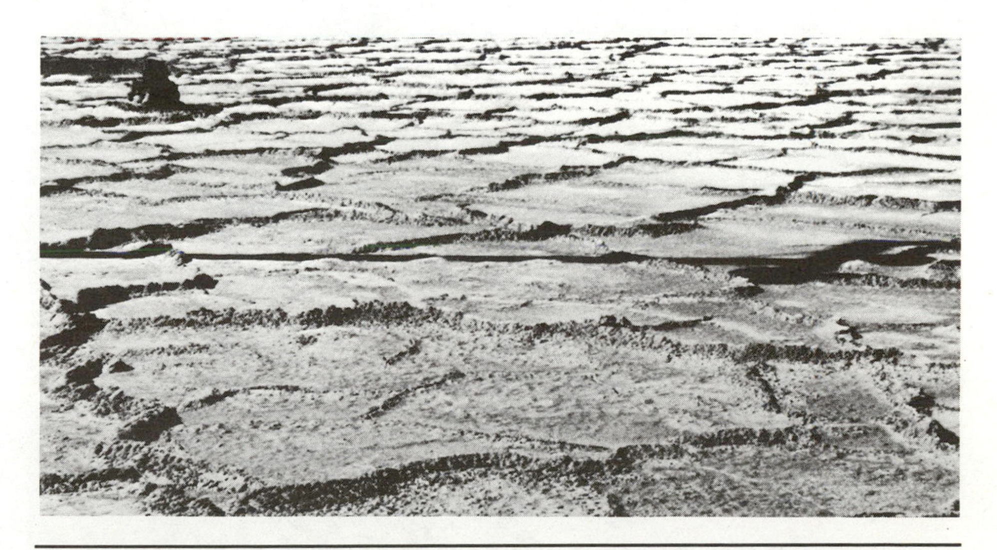

FIGURE 3.54 Salt polygons. Silver Peak playa, Nevada. (Photograph by J. N. Moore, 1975).

FIGURE 3.55 Soft-sediment folds. Lake sediments exposed near the Dead Sea, Israel. (Photograph by E. D. McKee, #2599, U.S. Geological Survey).

FIGURE 3.56 Slump folds. Pliocene lake deposits, Grapevine Grade, California. (Photograph by J. N. Moore, 1972).

FIGURE 3.57 Convolute bedding. (Photograph by Paul E. Potter).

FIGURE 3.58 Teepee structures. Deformation from growth of salt in Death Valley playa. (Photograph by J. R. Stacy in C. B. Hunt, #970, U.S. Geological Survey, 1960).

FIGURE 3.59 Load casts. Moenkopi Formation, Arizona. (Photograph by E. D. McKee, #1127, U.S. Geological Survey, 1951).

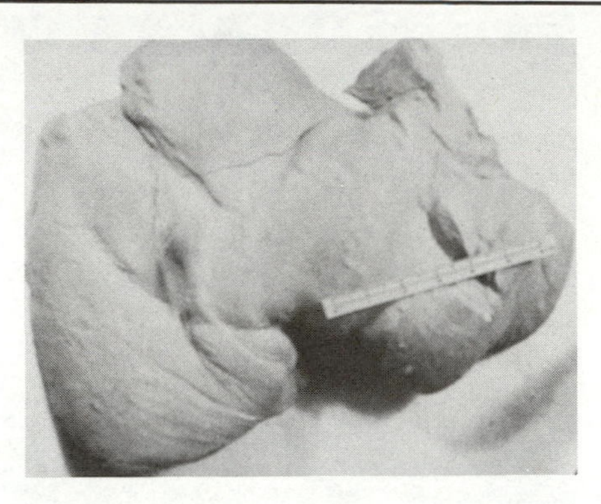

A large number of soft-sediment features fall under the term **load casts** or **load structure.** These form at the base of beds where there is a strong, reverse density gradient between two different sediments. Load casts take on a variety of forms and compositions but are very common in beds of sand overlying mud, either carbonate or terrigenous (Fig. 3.59). If these have a directional nature to them and pinch out upward they are called **flame structures** (Fig. 3.60) because they resemble flames. A particular type of load cast is the **ball and pillow structure.** These features form when a protrusion of sand is isolated in mud because of extreme ductile deformation (Fig. 3.61).

FIGURE 3.60 Flame structures. Tatta-Ferry, west Pakistan. (Photograph by E. D. McKee, #7, U.S. Geological Survey, 1964).

FIGURE 3.61 Ball and pillow structure. Tertiary Astoria Formation, near seaside, Oregon, coast. (Photograph by W. J. Fritz, 1981).

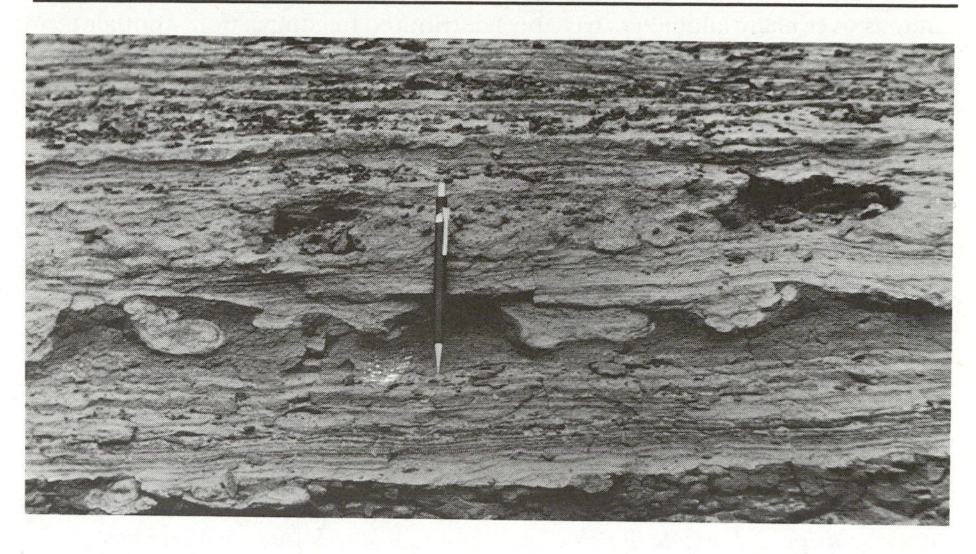

FIGURE 3.62 Small-scale compactional faults in laminated dolomite, California. (Photograph by M. N. Bramlette, #6, U.S. Geological Survey, 1946).

If the sediment behaves somewhat brittlely during compaction then compactional faults (Fig. 3.62) can form. These are often small features, but faults continuous over many kilometers have been attributed to compaction. Another term commonly used for these features, especially the larger ones in which sedimentation proceeds during offset, is growth fault. These may also result from slumping of a large mass of sediment.

Water escape structures are represented by myriad forms resulting from either the actual fluid flow of the sediment itself or the flow of water through the sediment. As sediment compacts and expels porewater, the upward flow can disrupt bedding and leave behind features that crosscut other bedding (Fig. 3.63) or deposits on surfaces of beds called sand volcanoes (Fig. 3.64). One type of water escape structure is termed a **dish structure** because it looks like a vertical section of a dish (Fig. 3.65). Dikes of sediment can also form and cut great distances through sedimentary layers. These **clastic dikes** (Figs. 3.66 and 3.67) have an interesting synonym: neptunian dikes. Some of these clastic "intrusions" can occur parallel to bedding as clastic sills that are difficult to distinguish from a true primary sedimentary bed.

Other Features A whole series of gas bubble features form sedimentary structures. A common one in beach sands is **bubble sand** (Fig. 3.68), which forms when air is trapped between the water table and water at the surface from incoming tides in rapidly deposited sand. Gas released by the decay of organic matter in sediment also produces bubble cavities. These can fill with minerals

FIGURE 3.63 Water escape structures. Pleistocene sedimentary rocks from Willapa Bay, Washington, coast. (Photograph by J. N. Moore, 1980).

FIGURE 3.64 Sand volcanoes. Imperial Valley, from Oct. 15, 1979, earthquake. (Photograph by C. E. Johnson, #61, U.S. Geological Survey, 1979).

FIGURE 3.65 Dish structures. From D. R. Lowe, 1982, *Journal of Sedimentary Petrology,* 52, Fig. 9A, p. 288. Reprinted by permission of SEPM, Tulsa, Okla.

FIGURE 3.66 Compound clastic dike. Pleistocene Touchet beds near Walla Walla, Washington. (Photograph by W. M. Schneck, 1983).

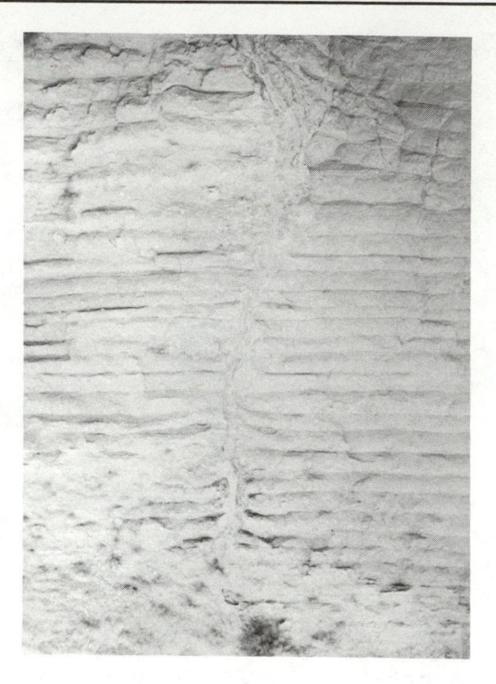

FIGURE 3.67 Photograph and line-drawing interpretation of a complex compound sandstone dike. Cretaceous Eutaw Formation, Columbus, Georgia. (Photograph and drawing by W. J. Frazier).

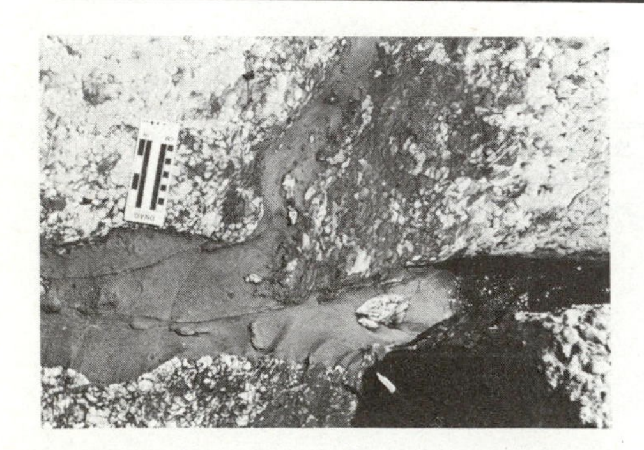

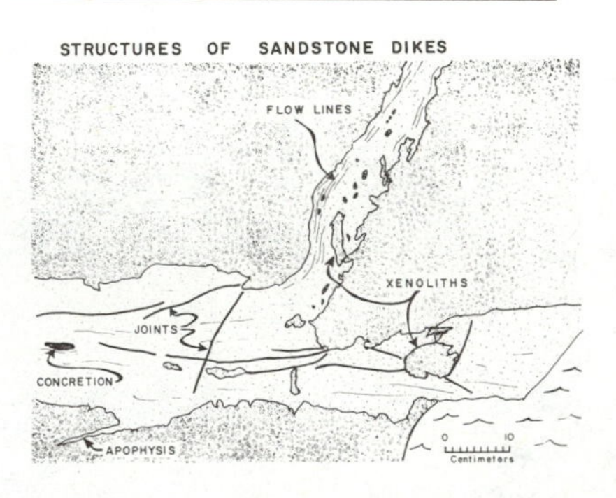

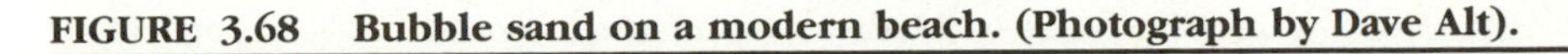

FIGURE 3.68 Bubble sand on a modern beach. (Photograph by Dave Alt).

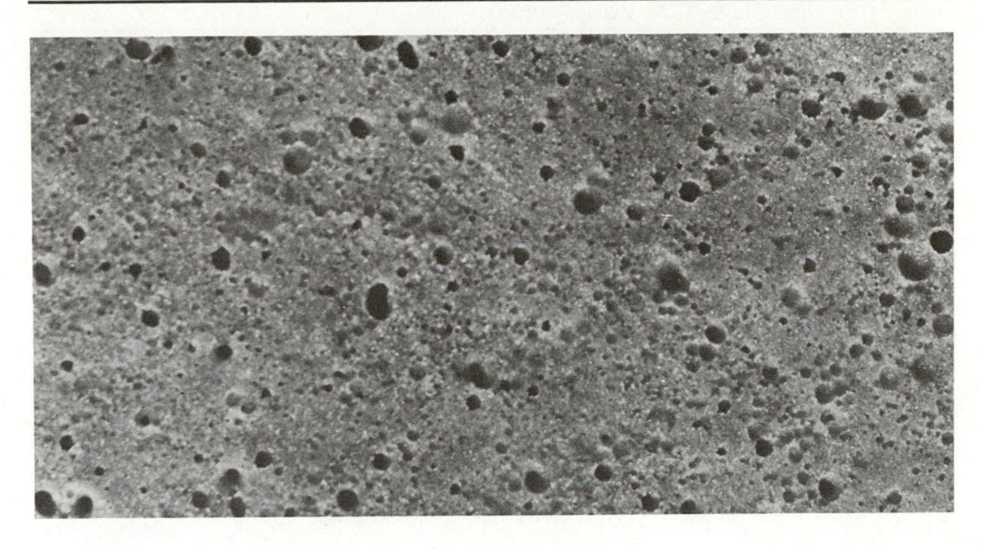

during diagenesis and form fillings termed **birds' eye structure** (Fig. 2.38). Gas escaping upward forms small holes in sediment that can later be filled with sediment. These features are termed **gas escape structures** and may completely homogenize the sediment.

All sorts of minerals can form aggregates within the sediment. **Nodules** (Fig. 3.69) are commonly composed of calcium carbonate, chert, or gypsum. Areas of cement that are more resistant than the enclosing rock are termed **concretions.** Concretions can form very irregular masses or nearly perfect spheres (Figs. 3.70 and 3.71). Some concretions show complex internal features indicative of dissolution and precipitation. The most beautiful of these is the **septarian concretion.** Concretions can also form after lithification as well as before and could thus be classified as either penecontemporaneous of epigenetic depending on when they were cemented. Most often they accompany lithification and represent areas of concentration of cement. Some of these concretions form as the result of shell concentrations that supply a carbonate cement and may be the only site of shell preservation in many sandstones. Also, concretions form where organic decay alters the chemistry of the diagenetic microenvironment immediately adjacent to the decaying organism. Such concentrations often preserve extremely beautiful fossils. In volcanic and volcaniclastic rocks, gas cavities become filled with various minerals such as quartz, opal, chert, and calcite. Because they often form in hot water, rapid cooling at the margins and slow cooling in the centers produces a differential crystal size and a dense outer rind. These concretions are termed "geodes."

FIGURE 3.69 Chert nodules in Madison Limestone, Freemont Canyon, Wyoming. (Photograph by J. N. Moore, 1984).

FIGURE 3.70 Concretions in Miocene sandstone near Echo Lake, California. (Photograph by R. Arnold #228, U.S. Geological Survey, 1905).

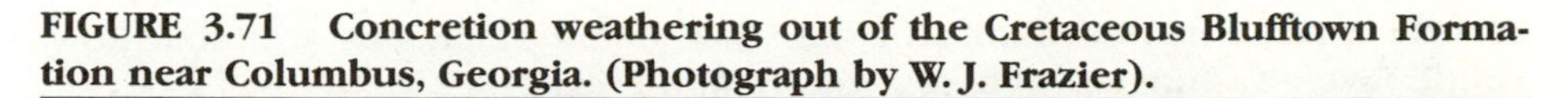

FIGURE 3.71 Concretion weathering out of the Cretaceous Blufftown Formation near Columbus, Georgia. (Photograph by W. J. Frazier).

Another feature that looks superficially like concretions is an **armored mud ball** (Fig. 3.72). These form when chunks of cohesive clay accrete pebbles as they are rolled in streams or on beaches.

SYNGENETIC ORGANIC SEDIMENTARY STRUCTURES

Accretionary Organic Sedimentary Structures

Some sedimentary structures result from biological activity combined with physical and chemical processes of sedimentation. Certain organisms build mounds and frameworks; the most obvious are the large reefal organic frameworks built by coral and calcareous algae. These features can attain huge dimensions but may also form smaller mounds within other deposits. The term **bioherm** is commonly used to designate such features. Bioherms can be formed by many different organisms. Corals and algae are common bioherm-forming organisms today. Similar structures have been constructed in the past by archeocyathids, bryozoans, mollusks, and diverse assemblages of animals and plants (Fig. 3.73). These accumulations occur throughout the rock record from the Cambrian to the Recent.

FIGURE 3.72 Armored mudball from 1982 Mount St. Helens sediment flow, Toutle River, Washington. (Photograph by W. J. Fritz, 1982).

FIGURE 3.73 Stromatolite bioherm. Precambrian limestone, Inyo County, California. (Photograph by J. N. Moore, 1976).

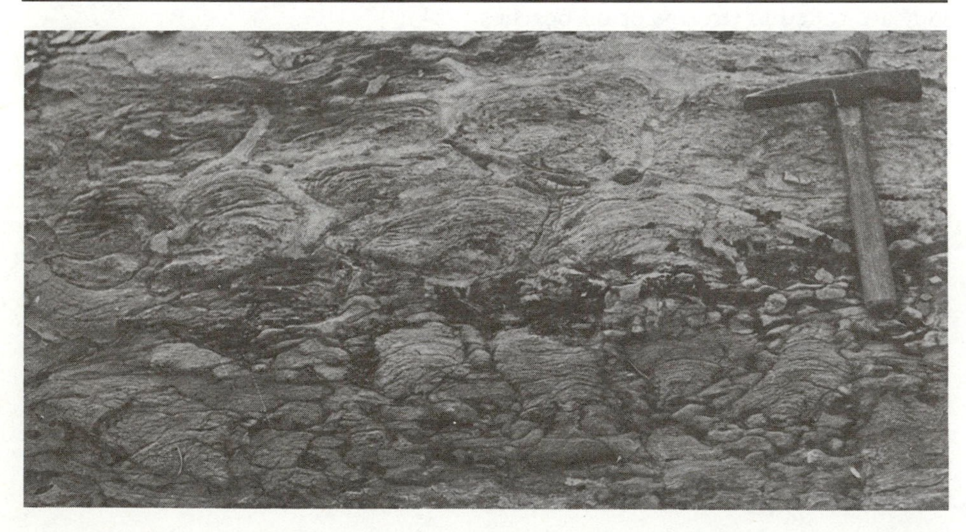

FIGURE 3.74 Stromatolites. Bedding plane (on left) and cross-sectional view (on right) of Alcova Limestone, Wyoming. (Photograph by J. N. Moore, 1984).

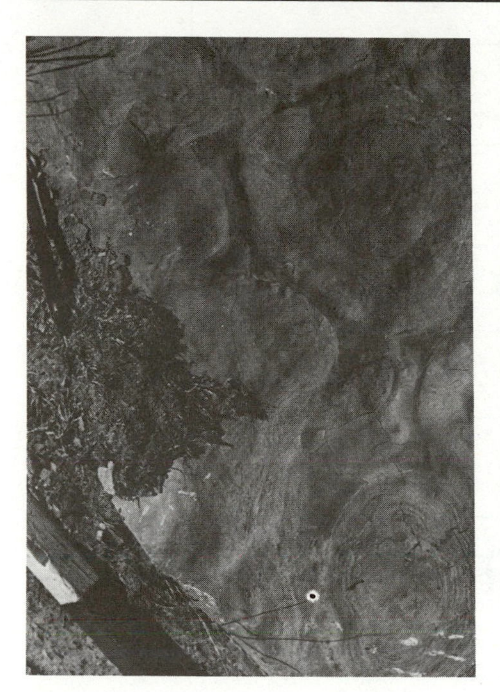

Before the advent of hard-shelled metazoans approximately 570 Ma ago, blue-green algae (cyanobacteria) formed distinct, laminated bioherms termed **stromatolites.** These structures can be found in many younger deposits and are even observed today in places like hypersaline areas of the Great Salt Lake, Utah, and Shark Bay, Australia. Stromatolite mounds and columns can be complex (Fig. 3.74) and require a detailed classification scheme (Fig. 3.75). Stromatolites are very common in Proterozoic rocks throughout the world but are rare in younger rocks. Possibly the evolution of grazing snails that ate the algal mats limited their distribution, for today they flourish in the harsh hypersaline water where snails cannot live.

Some algal accretions form isolated spherical masses rather than mounds. These are often called **algal biscuits** if forming in modern environments or **oncoliths** if found in rocks (Fig. 3.76).

Bioturbation

Sediment is commonly disrupted by the action of organisms living and moving through and on it. The generalized term for this disruption is **bioturbation** and the resulting structures are called **trace fossils.** We can classify bioturbation

Stromatolite Type	Description
	Tufted Mat: thin mat underlain by poorly laminated sediment; tufts form subparallel ridges.
	Convoluted Mat: 3 to 4-cm-high domes separated by flat mat; domes often hollow; mat thin and underlain by moderately laminated sediment.
	Smooth Mat: smooth, continuous felt of algal filaments; mat commonly up to 4 cm thick.
	Low Domed and Discoidal Forms: flat disk shapes separated by scour channels from surrounding flat, continuous mat.
	Sinuous Domes: ellipse-shaped to sinuous structures; laterally linked hemispheres at bottom and separate at top; top not expanded.
	Intergrown Club-Shaped Forms: discrete columns at base with convex upward lamination; expand upward; coalesce at top; elongate forms parallel to current.
	Discrete Columns: subcircular to ellipse-shaped in plan view columns; slightly expanded at the base, restricted above base, then expanded at top; elongated parallel to currents; asymmetric with more growth away from wave splash.
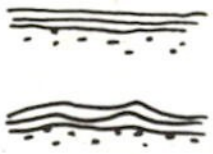	Flat Mats: flat to rippled mats, locally overlying sand.
	Algal Biscuits: oval to ellipsoidal structures 1 to 10 cm by ½ to 6 cm; discrete and attached to substrate by algal filaments; locally elongated parallel to currents.
	Algal Domes: larger than biscuits; commonly 35 cm long; poorly laminated; domes can be isolated but commonly connected; elongated parallel to currents.
	Inverted stacked hemispheres.
	Random orientation of hemispheres.
	Concentric spheres.

FIGURE 3.75 Classification scheme for stromatolites. From B. W. Logan, R. Rezak, and R. N. Ginsburg, 1964, *Journal of Geology*, 72, Figs. 4 and 5, pp. 76 and 78. Reprinted from *The Journal of Geology* by permission of the University of Chicago Press. Copyright © 1064 by the University of Chicago.

FIGURE 3.76 Large "algal" oncoliths from the Deep Spring Formation, Gold Point, Nevada. Note that this photo is of a cross section through a spherical structure. (Photograph by J. N. Moore, 1977).

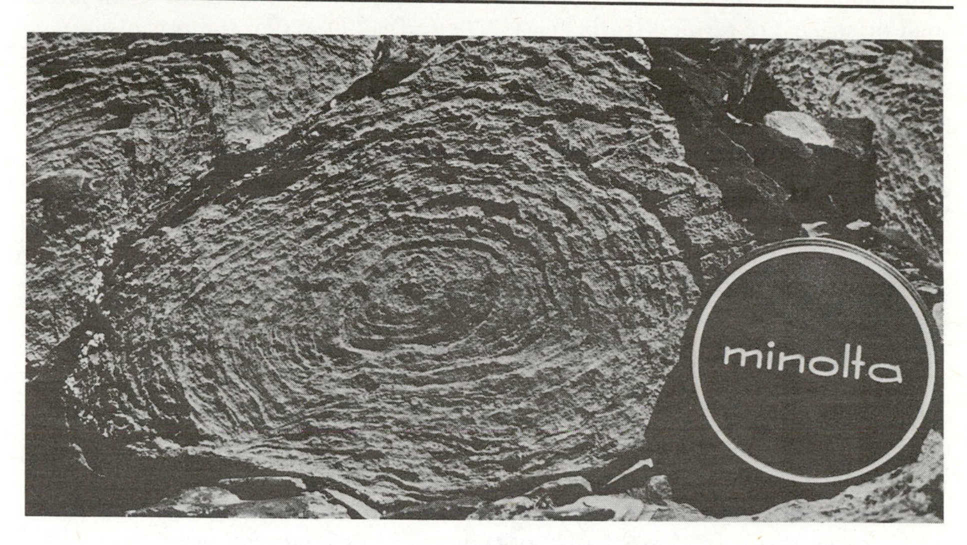

FIGURE 3.77 Vertebrate amphibian tracks from the Pennsylvanian "Pottsville Group," Lookout Mountain, north Georgia. (Photograph by W. J. Fritz, 1982).

structures descriptively once we establish some basic origins. Imprints formed by the feet of an animal are termed **tracks** and **trackways** (Fig. 3.77). The disturbances produced by an animal moving across the surface, for example, a snail or snake, are termed **trails** (Figs. 3.78 and 3.79). When the organism penetrates the sediment and moves through it leaving an open chamber, the resulting disruptions are termed **burrows** (Fig. 3.80). These three main types of bioturbation can be subdivided by examining their detailed morphology. Most of the terms used are representative of some action by the organisms and not purely descriptive. However, some terms can be used without genetic implications to describe the shape of the features. For example, burrows can simply be vertical or horizontal, simple or branching; smooth, striated, or bumpy (ornamented); or straight, sinuous, irregular, spiral, etc. We present such a classification based on the shape and relationship of the features (Fig. 3.81).

Trace fossils are classified differently by different authors. A bipartite system identical to that describing organisms is often used by trace fossil experts. Because trace fossils are also called **ichnofossils,** these names are termed ichnogeneric and ichnospecific designations. Although this terminology is useful for subdividing the myriad forms of trace fossils, it leads to some misunderstanding. First, the names do not represent the organism that formed the trace fossil and, second, trace fossils made by a particular organism during different operations (e.g., traveling, eating, resting) may have entirely different names. The nomenclature

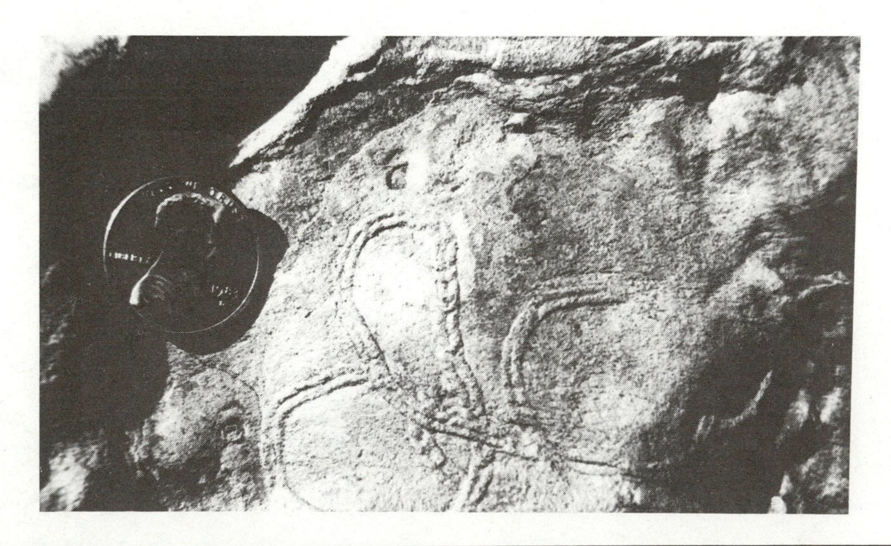

FIGURE 3.78 Arthropod trail in Mississippian Pennington sandstone, McLemore Cove, northwest Georgia. (Photograph by J. N. Moore, 1983).

FIGURE 3.79 Hermit crab (in conch shell) trail on the lower foreshore, St. Simons Island, Georgia. (Photograph by W. J. Frazier).

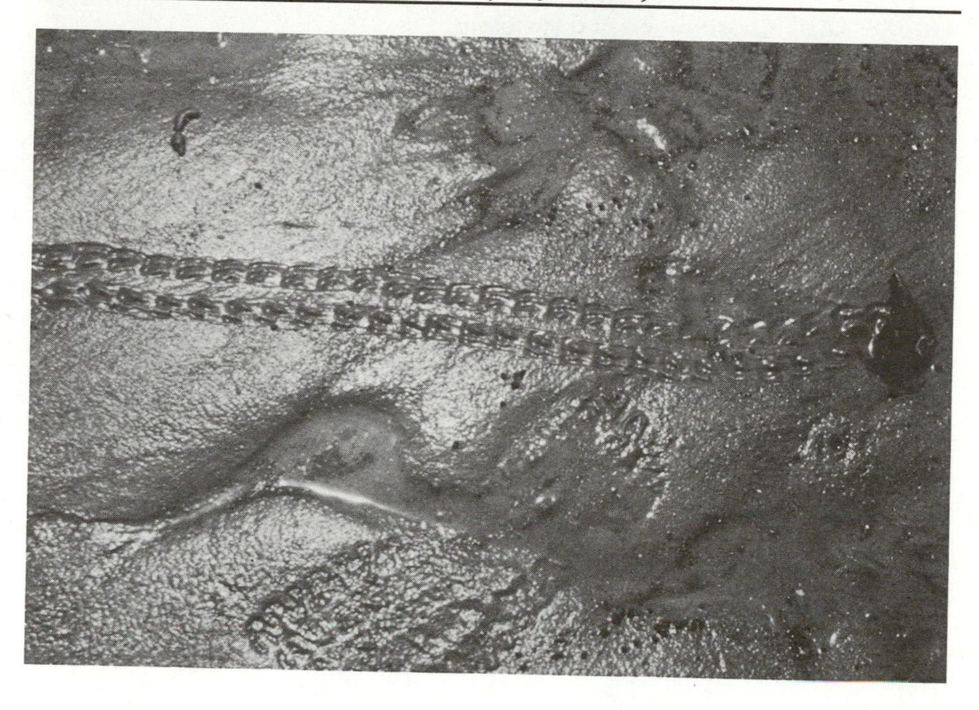

FIGURE 3.80 Worm and miscellaneous burrows in Mississippian Pennington Formation, McLemore Cove, northwest Georgia. (Photograph by J. N. Moore, 1983).

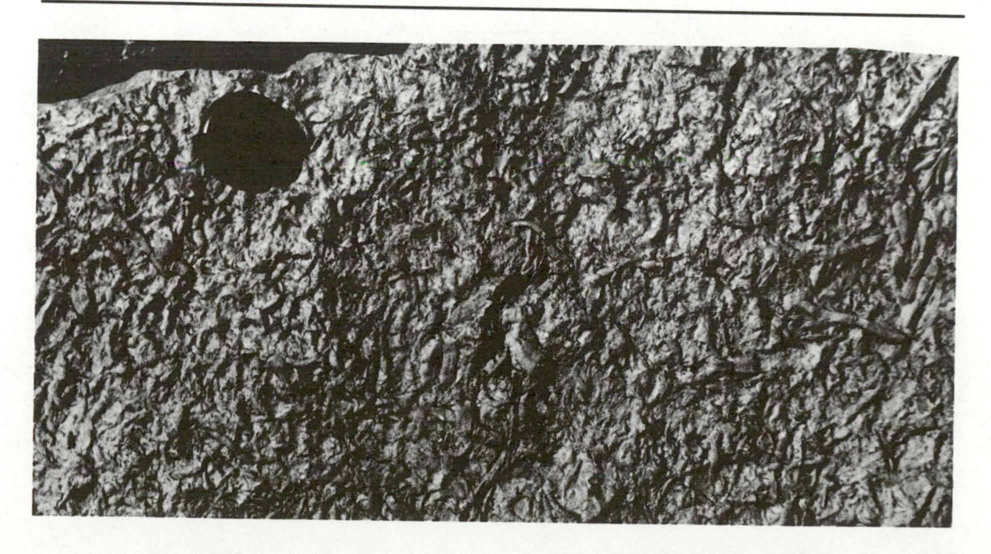

FIGURE 3.81 Classification of trace fossils.

Body Impression	Tracks	Traces		
		Trails	Burrows	
			Simple	Complex
Variety of shapes matching organisms	Digitate e.g., bird dinosaur Circular e.g., elephant, sauropods Tracking	Plan Shape straight curved irregular spiral Cross Section grooved furrowed Internal Detail smooth striated chevron cusped	Orientation Vertical Horizontal Shape straight curved irregular Internal Detail smooth striated back-filled or tabulated ornamented	Branching Spreited

of the traces produced by trilobites is an excellent example of this glitch. The track produced when a trilobite moves lightly over the surface is placed in the ichnogenus *Diplicnites* (Fig. 3.82), whereas the trail formed by the trilobite plowing through the sediment or possibly burrowing beneath it is called *Cruziana.* The resting trace created by a trilobite when it buries itself is termed *Rusophycus*. Because of this complexity in using ichnogenera and ichnospecies, one should be careful not to confuse ichnofossils with actual body fossils of organisms. Thus, the names assigned to trace fossils do not represent unique biological species, but are used to describe the morphology of the trace.

Because all trace fossils are behavioral marks that record specific behaviors, it is only reasonable that different behaviors on the part of a single organism can leave different marks; these different marks get different names. Indeed, it is just this characteristic that makes trace fossils so useful in sedimentology, because

FIGURE 3.82 Various traces of trilobites and associated ichnogeneric names.

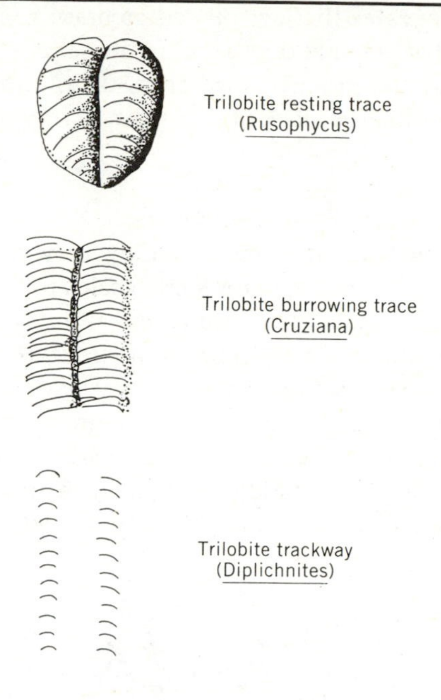

behavior is environmentally controlled and will be produced at different times by different organisms, all of which leave the same marks. Nelson *et al.* (1987) have even found large grooves in shelf sediments formed by the feeding activities of whale and walrus. This functional view can be taken in classifying trace fossils. Some features can be interpreted and given names such as resting traces, body impressions, feeding traces, upward migrating or escape structures, grazing traces, dwelling burrows, crawling traces, and back-filled burrows. Such distinctions require some knowledge of the behavior of organisms that made the traces or must be constructed from the detailed morphology of the trace fossils themselves.

Plants also disrupt the sediment and produce organic structures and bioturbation. These are often called **root structures.** Because roots decay, they are not often fossilized and all that remains is a mottled, bioturbated soil horizon. Sometimes plants and animals so completely bioturbate a sediment that it looks mottled or even structureless.

EPIGENETIC PHYSICAL AND CHEMICAL SEDIMENTARY STRUCTURES

Secondary structures made by sedimentary processes sometimes produce features that can be misinterpreted as having formed during sedimentation. These are common in all rock types but we will consider only those in sedimentary sequences that form after lithification. Thus, we can think of these structures as those produced by processes that act on sedimentary rocks as they weather at the surface and those that occur at depth within the sedimentary package. Erosion and weathering are surface processes, whereas dissolution, filling, and cementation occur at depth within the stack of sedimentary rock or near the weathering surface.

Structures Formed by Erosion and Weathering

Weathering and erosional features are generally easy to tell from syngenetic sedimentary structures. It is common for some shales to fracture into concentric shells (concoidal) that resemble concretions, and at times closely spaced jointing can resemble bedding. The latter is common in structureless beds of sandstone. Weathering rinds form on many rock types and may be confused with concretions or other structures. Spheroidal weathering in fine-grained rocks with a high feldspar content, such as muddy sandstones and siltstones, can often be confused for concretions and even for pillow basalts, an igneous feature.

A very useful but often subtle secondary feature is soils or paleosols. When a

FIGURE 3.83 Dark horizontal bands, red in outcrop, are paleosols developed in volcanic ash deposits of the Tertiary White River Group, Badland National Monument, South Dakota. (Photograph by W. J. Fritz, 1984).

FIGURE 3.84 Channel cut into sedimentary strata, outlined by a volcanic ash/mud drape and filled by younger sedimentary rocks. Miocene gravels, Canyon Ferry Reservoir near Helena, Montana. (Photograph by W. J. Fritz, 1979).

sedimentary surface is weathered during sediment accumulation, a soil profile is developed that commonly can be identified in the rock record (Fig. 3.83). They are extremely useful for paleoenvironmental studies because different soils form very different structures that can be recognized with detailed analyses.

Channels and potholes cut into underlying bedrock are preserved beneath some sedimentary sequences. These erosional features are common along unconformities (Fig. 3.84).

Structures Formed by Dissolution

Structures formed by the dissolution of soluble rock are complex and varied, but they can be simply divided into those that form on the surface and those within a rock sequence. When a surface of soluble rock, for example, limestone, is dissolved by weathering processes, a distinctive surface texture is formed. On a small scale this surface is referred to as **minikarst** because it is a small-scale analog to karst topography (Fig. 3.85). These minikarst features can look very much like ripple marks or scour.

Dissolution within the sediment leads to completely different structures. **Sty-**

FIGURE 3.85 Minikarst features. Dissolution surface at top of limestone bed in lower Tensleep Formation, Wyoming. (Photograph by J. N. Moore, 1984).

lolites are distinctive dissolution structures commonly found in limestone (Fig. 3.86). They are formed by dissolution of the rock and concentration of the residue along irregular interfaces. With enough dissolution, nonsoluble nodules, clay seams, and lenses and lumps of residue and limestone can be produced (Fig. 3.87).

Fillings

Dissolution often produces voids within soluble rocks or organic buildups often form complex shapes that enclose space within the organic framework. Reef limestones commonly contain such voids, which fill with sediment or minerals precipitated from solution. Solution caves filled with sediment form irregular masses of breccia and terrigenous sedimentary material in otherwise massive limestone (Fig. 3.88). On a smaller scale, void fillings or geopedal structures fill small cavities with either terrigenous sediment or mineral precipitates (Fig. 3.89). If banded or layered they can be used to determine the up direction in the sedimentary package, hence the name **geopedal** or, literally, "earth-foot."

A very common type of void filling occurs in glacial areas, where sediment is forced apart by expansion of freezing water and other sediment fills in as the ice melts to form an ice-wedge (Fig. 3.90).

Cementation Features

Structures produced by cementation form a tremendous variety of features and we will detail only a few of the more common ones. Concretions commonly form

FIGURE 3.86 Styolites in Mississippian Madison Limestone, Freemont Canyon, Wyoming. (Photograph by J. N. Moore, 1984).

FIGURE 3.87 Dissolution lumps and clay seams. Cretaceous strata, Mazagan Plateau, northwest Africa slope. (DSDP photograph, 1980).

FIGURE 3.88 **Cave filling, Madison Limestone, Freemont Canyon, Wyoming. (Photograph by J. N. Moore, 1984).**

FIGURE 3.89 **Void fillings. Both crystalline and terrigenous fill in Jurassic limestone, northwest Africa. (DSDP photograph, 1980).**

FIGURE 3.90 Ice-wedge in permafrost soil 5 miles northwest of Fairbanks, Alaska. (Photograph by T. L. Péwé, #3, U.S. Geological Survey).

as a secondary structure that weathers out of the rock in all shapes and sizes (Figs. 3.70 and 3.71). The concentration of various minerals along chemical fronts commonly leads to color banding, which is mostly irregular but can be confused with bedding (Fig. 3.91). A similar type of structure formed in isolated areas is the **reduction spot** or **oxidation spot.** These result from a concurrent change in color associated with a change in oxidation state (Fig. 3.92), leaving a distinctive color difference, generally spherical or nearly so in shape. Features similar to these are halos, where concentric bands form around such spots (Fig. 3.93).

Under certain conditions, sediment is cemented at the surface of deposition and forms **hardgrounds.** These features are very common in marine deposits, where the sedimentation rate is very low and cementation very rapid. They form irregular layers that are better cemented than the surrounding rock and commonly associated with organic buildups. A similar feature is beach rock, a hardground formed in a beach or shallow-water environment by early cementation.

FIGURE 3.91 Bedding and color banding (also called liesegang banding) in coarse sandstone of the Cretaceous Blufftown Formation near Columbus, Georgia. (Photograph by W. J. Frazier).

FIGURE 3.92 Reduction spots. Irregular patches of oxidation alteration of red mudstone in Triassic Chugwater Formation, Wyoming. (Photograph by J. N. Moore, 1984).

FIGURE 3.93 Halo around oxidation spot. Cretaceous sandstone, Shirley Basin, Wyoming. (Photograph by J. N. Moore, 1984).

FIGURE 3.94 Oolitic limestone (oosparite) bored by pholadid bivalve. Numerous ooids along the margins of the boring are sharply truncated, indicating that the limestone was cemented before the trace was made. Boring has been later filled with a pelmicrite. Jurassic Sundance Formation, Como Bluff, Wyoming. Maximum diameter of boring is approximately 9 mm. (Photomicrograph by K. A. Andersson).

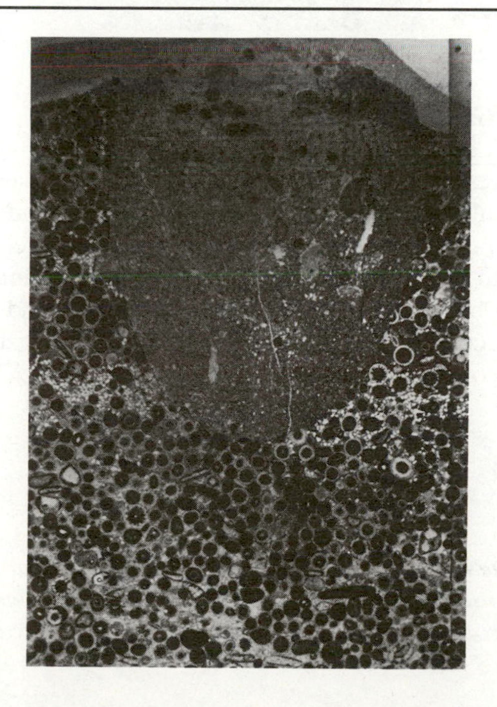

FIGURE 3.95 Well-preserved mytilid bivalve boring with the probable fossilized borer in place. Jurassic Sundance Formation, near Casper, Wyoming. Maximum length of boring is approximately 32 mm. (Photograph by K. A. Andersson).

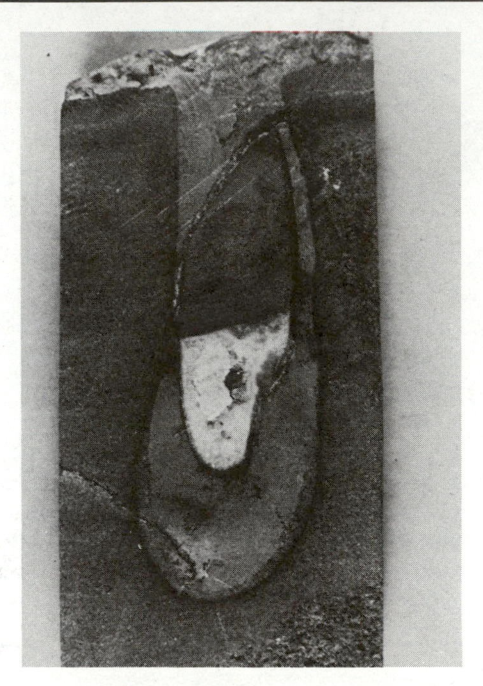

EPIGENETIC ORGANIC STRUCTURES

Secondary structures produced entirely by organic processes are quite limited. Fossils are commonly cemented and reworked from older deposits by currents that form lags, but purely organic secondary features are rare. The most common are **borings** formed by plants and animals actually cutting into rock. Commonly these are found in beach areas, where various mollusks and echinoderms bore into rocks for protection (Figs. 3.94 and 3.95). However, organisms can also bore penecontemporaneously into cohesive mud.

EPILOGUE

Somewhere beneath him, the pre-spice mass had accumulated enough water and organic matter from the little makers, had reached the critical stage of wild growth. A gigantic bubble of carbon dioxide was forming

FIGURE 3.96 Complex structures formed by combination of cementation and weathering. These have also been called aveolar structures and honeycomb weathering features by some authors. (Photograph by C. D. Walcott, #534b, U.S. Geological Survey, 1898).

deep in the sand, heaving upward in an enormous 'blow' with a dust whirlpool at its center.

Frank Herbert,* Dune, *1965

The structures presented in this chapter cover the main range of features commonly seen in sedimentary rocks but does not represent an exhaustive discussion of all forms or variations. Most of the structures discussed have a complex variety of morphologies that often defies simple classification because processes often cross classification boundaries. It is common for more than one process to proceed simultaneously in producing complex forms. For example, loading and dissolution can proceed together in forming distinctive structures that are very complex and interesting (Fig. 3.96). The moral, as always, is to use any classification scheme with caution and describe the structures completely and then make genetic interpretations. Odd processes act beneath the sands.

In Chapter 4 we will move onto the main processes of sedimentation that form some of the structures that we have illustrated and discussed in this chapter.

OUTSIDE READING

Sedimentary Structures—General
Allen (1963a, 1970, 1982, 1985a); Bell (1940); Bouma (1969); Collinson and Thompson (1982); Costello and Southard (1981); DeCelles *et al.* (1983); Fritz and Harrison (1983); Hall and Fritz (1984); Harms *et al.* (1975, 1982); Hoyt and Henry (1963, 1964); Jopling and Richardson (1966); Krumbein and Sloss (1963); Little (1982); Middleton (1965a, b); Mustoe (1982); Nøttvedt and Kreisa (1987); Pettijohn (1975); Pettijohn and Potter (1964); Power (1961); Reineck and Singh (1980); Selley (1982); Skipper (1971).

Trace Fossils and Organic Sedimentary Structures
Andersson (1979, 1981); Basan (1978); Crimes and Harper (1970, 1977); Ekdale *et al.* (1984); Frey (1973, 1975); Frey and Pemberton (1984, 1986); Logan *et al.* (1964); Miller *et al.* (1984); Nelson et al (1987); Sarjeant (1983); Schneck and Fritz (1985).

Chapter 4

Physical Processes of Sedimentation

INTRODUCTION

Physical processes that effect sediments start with weathering and the first currents that transport and deposit sedimentary grains. These currents modify and establish the major characteristics of the sediment, its texture (grain size, shape, sorting, characteristics of the parent material), and primary sedimentary structure (bedding type and bedform). After deposition, physical changes cause compaction and differential loading within the sediment. These processes are often accompanied by modification by organisms (bioturbation) or modification by chemical processes. Then chemical changes take over and cementation turns the sediment into rock. This continuum of processes results in a singular type of sedimentary rock with fixed characteristics. The goal of the stratigrapher is to decipher these characteristics and recreate the sequence of events that formed the rock. Because the sequence nearly always starts with the deposition of sediment, which is commonly by physical processes, we will start by presenting a brief introduction to fluid flow. We will concentrate on developing the tools needed to interpret rock sequences and leave the details of hydrodynamics to other, more extensive treatments on sedimentology. Sedimentologic principles are presented in textbooks by Friedman and Sanders (1978), Reading (1986), Leeder (1982), and Davis (1983). More detailed approaches are given by Allen (1970, 1985a, b) and in publications by the Society of Economic Paleontologists and Mineralogists, especially those edited or written by Middleton (1965a) and Harms *et al.* (1975, 1982).

FLUID FLOW AND SEDIMENT MOVEMENT

Definitions

To understand the affects of fluid flow on sediment we must first understand the principles that govern fluid flow—hydrodynamics. It is probably best to start with some definitions and characterizations of fluid. Fluids are those materials that have no strength under a shear stress, that is, a force per unit area acting parallel to the surface of the fluid. Simple Newtonian fluids, essentially have no strength and conform to the shape of whatever contains them (stream channel, bucket, ocean basin, etc.). These characteristics cause fluids to flow from one place to another under the influence of gravity, resulting ultimately from differences in elevation (hydraulic head). Although fluids do not have strength they do resist flow. The resistance to flow is the **viscosity** of the fluid. Fluids also have a density. From these traits we can characterize the physical attributes of a fluid by its viscosity and density. By combining these with the composition, any fluid

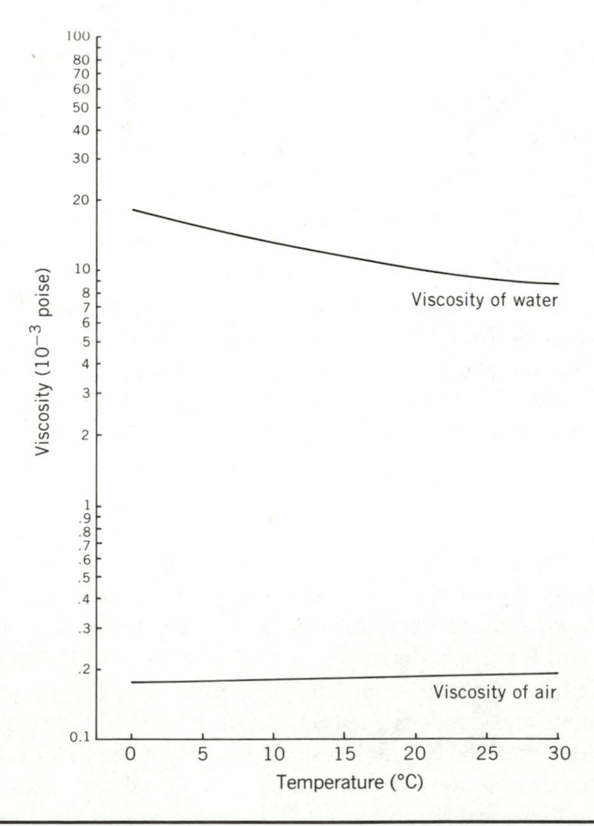

FIGURE 4.1 Graph of viscosity variation of air and water with temperature.

FIGURE 4.2 Graph of density of water versus temperature and salinity.

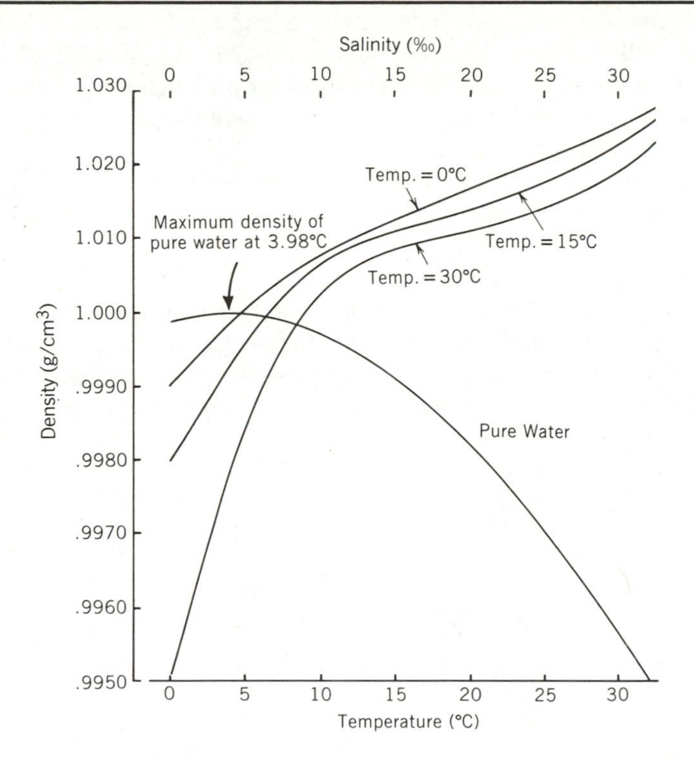

can be categorized and distinguished from any other fluid. The fluids that are important in understanding sediment transport and deposition are those naturally occurring on the earth's surface: water and air.

The viscosity and density of fluids are controlled by several factors. The most obvious is composition. Water is much more viscous and dense than air in all situations where they are both fluids (water freezes at 0°C). At standard temperature and pressure, water is approximately 50 times more viscous than air and up to 2000 times more dense. These differences affect the way sediment is transported by water and air. The viscosity and density of air and water are modified by changes in temperature. The viscosity of air is only slightly affected by temperature change (Fig. 4.1), whereas water shows a much more dramatic effect with changes in temperature. As temperature decreases from 100 to 0°C, the viscosity of water almost quadruples (Fig. 4.1). Temperature also affects the density of water. Colder water has a greater density than warmer water (Fig. 4.2) until just above the freezing point (ca. 4°C). Common compositional factors that affect density of water are salinity and sediment load. As salinity increases, water becomes denser (Fig. 4.2). This process leads to density stratification in large inland seas located in regions of high evaporation, like the Dead Sea. Increased suspended-sediment load, especially of clay, significantly increases the viscosity of water (actually a sediment–water mixture, see Fig. 4.3) as well as the density. Both these changes contribute to the ability of sediment–water mixtures to move vast amounts of material of very large size. We will discuss the power of these mechanisms in Chapter 8.

Now that we have characterized fluids, let us examine how they flow and the controls on the type of flow exerted by density and viscosity.

Types of Flow

One hundred years ago Osborne Reynolds conducted a set of experiments that determined how fluid flows. His experiments remain as the fundamental work on flow of fluids in pipes and his concepts are basic to understanding flow in natural systems and the response of sediment to flow. By measuring pressure

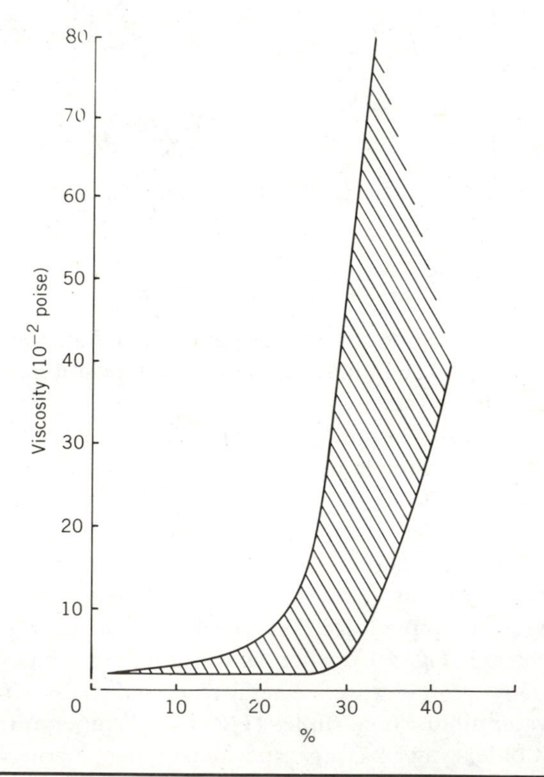

FIGURE 4.3 Graph of viscosity of water with sediment load composed of "native" clay. Shaded area represents variation of typical clays. Redrafted from data in Grim (1968).

drops over lengths of pipe and inserting dye into fluid flowing through transparent pipe, he identified two types of flow: laminar and turbulent. **Laminar flow** is relatively rare in flowing water and essentially absent in moving air, but it does occur in more viscous fluids such as sediment–water mixtures, ice, and lava. Such flows are of relatively low velocity and the fluid moves in "laminae" parallel to the boundaries containing the flow (Fig. 4.4). Reynolds found that dye injected into such flows traveled in straight threads away from the input point and was not greatly deformed by the flow. In laminar flow there are only negligible components of flow in directions other than directly parallel to the average flow direction (downstream).

Turbulent flow, the other type of flow described by Reynolds in 1883, shows more complexity. When flow becomes turbulent the fluid moves in complex and constantly changing directions. These movements form complex eddies that superpose a random nature to the general direction of flow. Velocity of such flows varies so greatly over short time intervals that flow rate is determined by averaging velocity over some time period. Individual masses of the fluid move up, down, and sideways, transferring momentum and mass throughout the fluid (Fig. 4.4). Water moves almost exclusively in this fashion unless flowing very slowly in pipes and channels. In nature, air always flows turbulently. Turbulent flow is more efficient in entraining and transporting sediment because of the complex velocity gradients, so we will discuss it in detail to understand the mechanisms of sediment transport. But first let us look more closely at the results of Reynolds' experiments and how they can be used to decipher some of the sedimentary structures and textures of rock sequences.

The easiest way to understand Reynolds' empiricisms is to examine the formula he used to describe flow:

$$\mathrm{R_e} = UL\rho/\mu$$

where: U = velocity of flow
L = dimension of flow, depth, or pipe diameter
ρ = density of fluid
μ = viscosity of fluid

$\mathrm{R_e}$ in this expression is called the **Reynolds number** and is a dimensionless number that is used to describe the flow. If two different flow systems have the same $\mathrm{R_e}$, then the two flows are the same and have the same degree of turbulence, regardless of the variations in velocity, dimension, density, and viscosity. Let us look at an example that is of utmost importance in sedimentology. Imagine you wanted to model the flow of a large stream or tidal channel, one much too large to build economically to full scale. By constructing a model with a different depth and velocity of flow and using a fluid with different viscosity and density, we could achieve the same type of flow in a much smaller system. By defining the

FIGURE 4.4 Diagram of types of fluid flow.

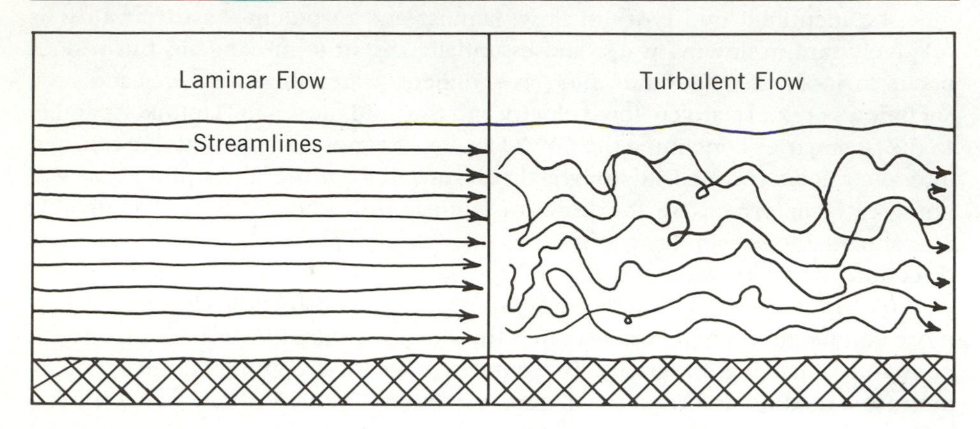

R_e of two systems, we are able to compare flows and know they are the same if R_e is the same.

If we examine the factors of the R_e equation we can use the relationship to explain a large number of flow phenomena. The numerator of the right side of the equation (velocity × dimension × density) represents the inertial forces acting within the system. Think of a stream as a mass of water (represented by the depth and density) moving with a certain velocity. Such a system has a certain inertia that is ultimately created by the acceleration caused by gravity. Also, the system resists flow because of the viscosity of the water. The viscous forces are represented in the lower part of the equation. This means that if the flow system is dominated by viscous forces (the viscosity of the fluid is large with respect to the R_e numerator) then the R_e is small and the flow is laminar. Similarly, when the flow is very slow it leads to a small R_e and the flow is also laminar. If, however, the flow has a high velocity, then the R_e is large and the flow is turbulent, and as you might expect if the fluid is highly viscous then the flow is laminar. The actual value of R_e that defines the change from laminar to turbulent flow is dependent on the system. For fluids flowing in a channel or a pipe, the transition is between R_e values of approximately 500 and 2000. Systems with R_e greater than 2000 are turbulent and ones with R_e less than 500 are laminar; the ones within the range are gradational. For flowing air the viscosity is so low that flow can only be turbulent. Similarly for water, the low viscosity dictates turbulent flow under most conditions. However, if the viscosity is increased by adding sediment, the flow can become laminar. The sediment load must be very high for this to occur, however, and the fluid would more accurately be termed mud than water. There are certain situations where water flows laminarly, but not on the scale of an entire stream. This laminar flow is very important in understanding sediment entrainment so let us consider where and how it occurs before we examine those mechanisms.

In a turbulent flowing system, like a stream, velocity of flow can be represented by a direction and a magnitude (vector) that is the average flow velocity at a certain spot. This results in the familiar diagram depicting the vertical velocity gradient in a stream (Fig. 4.5). As the bottom of the channel is approached, the velocity drops until at the bottom it is essentially zero. If the bottom is very smooth with no protrusions disrupting the flow, the velocity decreases exponentially downward and just before the bottom it decreases linearly (Fig. 4.5, inset). When this sublayer is examined closely we see that the flow is laminar not turbulent like the water above. What is happening in this thin layer that changes the type of flow? The Reynolds number can be used to explain the phenomena of this **laminar sublayer.**

Consider the flow well above the bed, say about a third of the way to the top of the flow. The flow there is fast and turbulent, that is, it has a high R_e. As we approach the bottom, the flow velocity decreases dramatically, in fact it approaches zero. As it does, the numerator in the Reynolds number equation becomes very small. The viscosity of the fluid is not changing, so R_e becomes small and we expect to find laminar flow, which we do. If the velocity of the stream increases, the laminar sublayer is destroyed (eroded) by the turbulent flow above. So, the laminar sublayer is primarily a characteristic of slow flows within smooth channels.

Sediment Entrainment

To understand how sediment is moved from its place of rest and transported by flowing fluid, we must first discuss some important particulars of sediment. Every

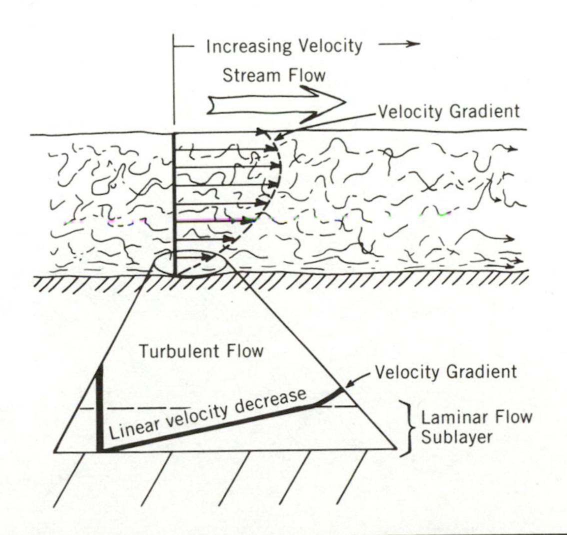

FIGURE 4.5 Diagram of velocity gradient and type of flow for water flowing in a channel.

sediment grain has a size and shape. These two parameters are very important in determining whether a particular grain is entrained and transported in a particular flow. Let us look at size before we discuss the effects of shape.

At first glance it seems easy to determine the size of a grain, especially a large grain like a cobble or boulder. But in reality there is no simple way to describe the size of an irregularly shaped grain. For a spherical grain the size is obviously the diameter of the sphere. But what is the size of a rod- or plate-shaped grain? These considerations are difficult to address even in large cobbles and become nearly impossible to consider for smaller grains such as sand, silt, and clay. What we really need is a way to compare sizes of different sediments, not the shape of each of the millions of grains in the sediment. One such method is to pass the sediment through a set of sieves of a particular size. Any sediment that passes through a certain size but rests on the next smaller sieve falls within a certain range. Unfortunately this does not take into account the shape effects mentioned above. A perfectly spherical grain with the same diameter as the small diameter of a very long rod will be the "same size" (Fig. 4.6); thus, the quantity measured by this technique is the length of the intermediate axis. Also, it seems a bit artificial to screen sediment, determine a certain grain size distribution, and then use that

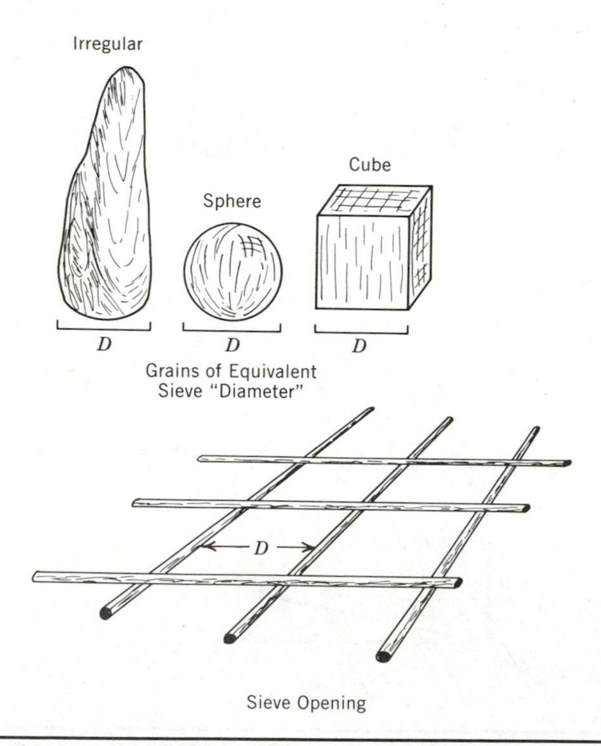

FIGURE 4.6 Diagram of equivalent diameter of different shaped grains.

to interpret how the sediment acts in water. So another approach has been used, settling-tube analysis.

In settling-tube analysis, sediment is introduced into the top of a tube filled with water. The temperature is regulated to keep the density and viscosity of the water constant. The rate at which the sediment accumulates in the bottom of the tube is monitored by one of several ways. In one system, a pan is suspended at the bottom of the tube by a thin wire or line connected to a very sensitive balance at the top of the settling tube. The balance outputs data to a recorder that is activated when the sediment is released at the top of the tube. The balance then records the rate of accumulation of sediment on the pan by measuring the weight change with time. Such devices are very sensitive and have been used by several sedimentologists to determine accurate settling curves for sediments and quartz spheres (Gibbs *et al.,* 1971). But what are these devices measuring?

Settling tubes can be used to determine the **hydrodynamically equivalent grain size** of a sediment. This is a measure of comparison of an actual sediment to the curve produced by a sediment composed of glass spheres in the same settling tube, or by comparisons to theoretical curves. The theory of the velocity of small grains (less than 0.08 mm) settling through fluid was first presented by Stokes in 1851. He predicted the settling velocity of grains by constructing a formula based on simple relationships between sediment and fluid density, viscosity of the fluid, and the size of the grain (for derivations of this formula see Blatt *et al.,* 1980, or Friedman and Sanders, 1978, or the very detailed account by Bogardi, 1974). This relationship is often called **Stokes' law of settling:**

$$w = \left[\frac{(\rho_s - \rho)g}{18\mu}\right] d^2$$

where: w = settling velocity
ρ_s = density of sediment
ρ = density of fluid
g = gravitational acceleration
μ = viscosity of fluid
d = grain diameter

If we examine this equation carefully, we see that the settling velocity of a grain is dependent on the grain diameter if the other factors are kept constant. In a settling tube we can keep the temperature constant and, therefore, the density and viscosity of the water remain constant. If we use grains of the same composition then the grain density term is a constant and the velocity is dependent entirely on the grain size. For a system with water at 20°C and a grain density of 2.65 g/cc, the settling velocity according to Stokes' law is $89.83d^2$. Theoretical and experimental work by many workers since Stokes' time have delineated the limitations on this relationship, especially experiments by Rubey (1933), Rouse

(1937), Prandtl (1952), and Gibbs *et al.* (1971). If we plot the data generated by this work, the relationship between size and settling velocity becomes very apparent (Fig. 4.7). In 1933, W. W. Rubey modified Stokes' work and predicted an actual settling curve quite different from that derived from Stokes' law. This curve took into account the effects of turbulence around the grain by using the relationships established by Newton and is, therefore, often called the Newtonian settling curve. Instead of a linear function (on a log–log plot), Rubey constructed a curve that contained straight-line segments from Stokes' work and that predicted by using Newton's law (Fig. 4.7). Rubey expanded on this theoretical approach, measured the velocity of spheres settling through water, and substantiated his ideas with a number of data points. A few years later Rouse (1937) conducted further experiments that corroborated these ideas. More recent work by Gibbs found some deviations from this earlier work (Fig. 4.7), so that now we have a firm perception of settling velocity and grain size. But what are the factors that

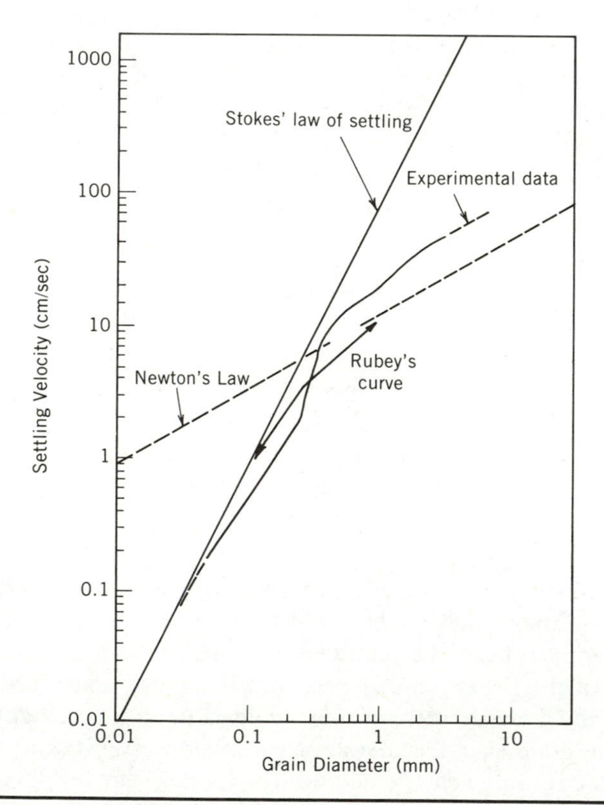

FIGURE 4.7 Plot of settling velocity of grains versus grain diameter. Plots from Newton's law, Stokes' law, Rubey's approximation, and recent experimental data.

control these relationships and why is the curve composed of two segments? We will have to return to the Reynolds number to answer these questions.

Think of a grain falling through water from the viewpoint of the grain, that is, move the coordinate system to the center of the grain. As you fall through the water (you are now the grain), you see only the water passing around you and feel no movement. (This situation is analogous to that described by Einstein in his theory of special relativity.) Now, if you are not very large and the flow around you is slow (small L and small U in the R_e equation), then R_e should be low. The dimension of flow in this system (L) is now the grain diameter instead of the diameter of a pipe or the depth of a channel. Think of this as the width of the flow separated by the grain (Fig. 4.8*a*). On the other hand, if you have a large cross-sectional area presented to the flow and the velocity is high (large L and large U), then R_e is large and the water flows around you turbulently (Fig. 4.8*c*).

Because larger grains settle faster than smaller grains, this is the actual situation of grains falling through water. Smaller grains "have" a low R_e and, assuming

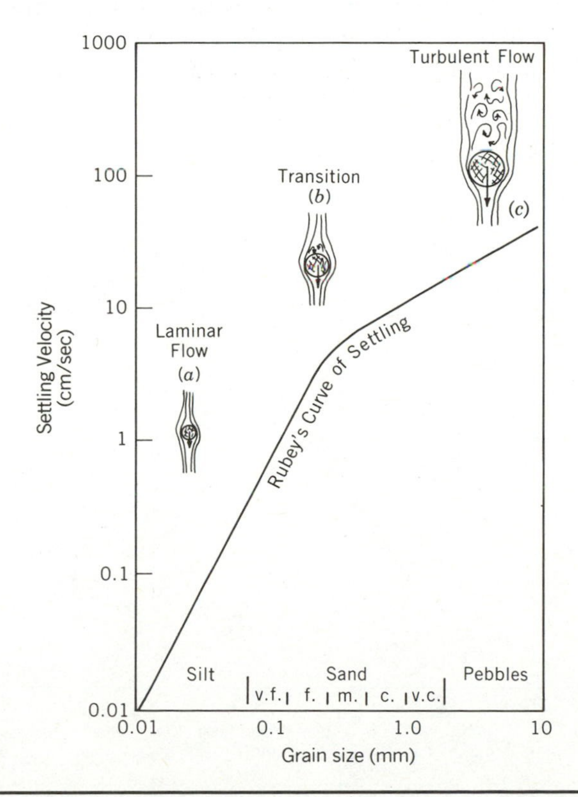

FIGURE 4.8 Settling velocity versus grain size. Diagram idealizing type of flow around grains settling through water.

constant density of grains, the flow around them is laminar, whereas larger grains "have" a higher R_e so the water flows around them turbulently. For the laminar flow there is a larger rate of increase (steeper slope on the settling curve) because there is less drag from the laminar flow, and Stokes' law applies (Fig. 4.8). But as the velocity increases for larger grains, the rate of increase is lower (shallower slope on the settling curve) and Newton's relationships must be considered. This change in settling velocity curves is because some of the grain's momentum is transferred to the fluid, setting it in motion (i.e., turbulence). Loss of momentum causes grains to fall more slowly than predicted by Stokes' law. These two curves meet at the transition from laminar to turbulent flow. This is the bend in the curve (Fig. 4.8*b*) that occurs at approximately the fine–medium sand size. Grains larger than fine–medium sand are large enough so that the flow around them is turbulent as they settle through water. Smaller grains separate the flow less and fall at a lower velocity with laminar flow around them. Observed fall velocity may also depart from Stokes' law behavior in the fine silt to clay sizes because of Brownian motion in the fluid (especially significant in water).

In the preceding discussions we have dealt with spherical grains only. In natural systems there is a distinct lack of ideal particles and, even though some very well worn sands approximate spheres, that is not a typical grain shape. Instead, grains

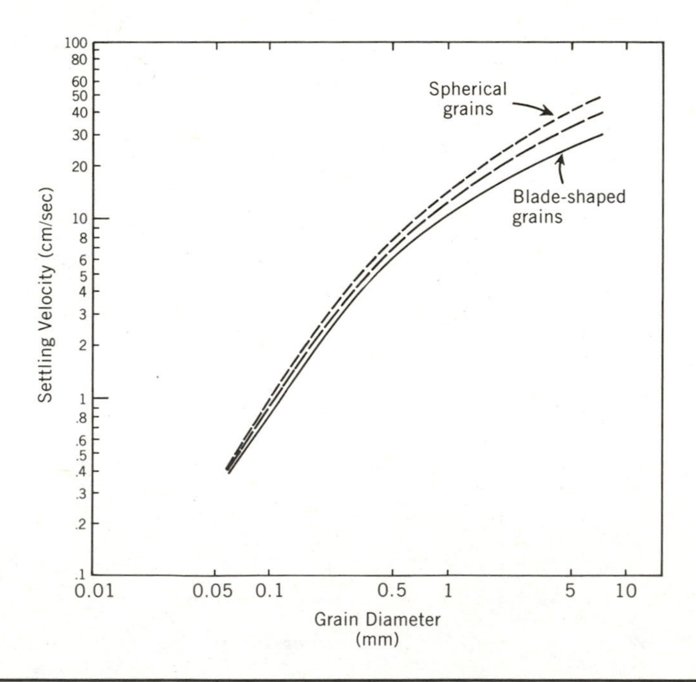

FIGURE 4.9 Graph of settling velocity versus grain diameter for different shaped grains.

FIGURE 4.10 Graph of sieve diameter versus settling diameter for grains of different shapes.

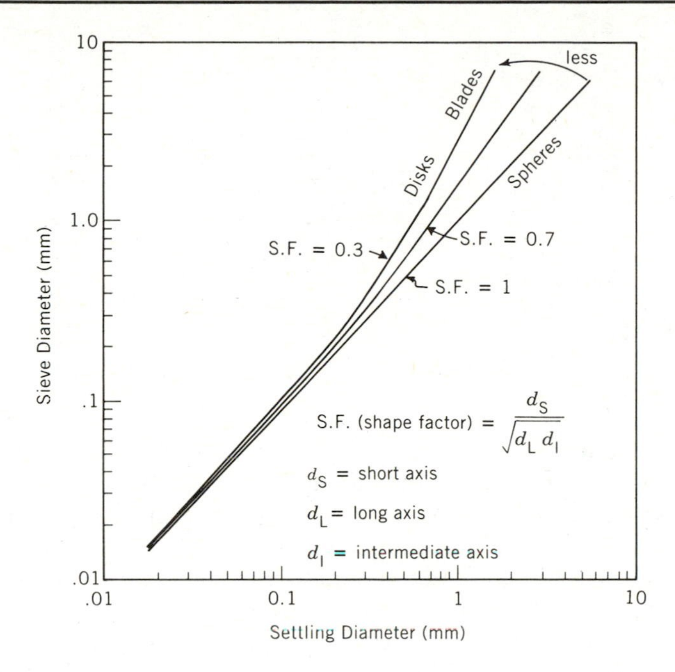

are platey, elongate, or some irregular shape. If we measured the settling velocity of a sediment containing a mixture of grain shapes, the distribution produced would be equivalent to that of spherical grains, which is found in an actual sample. Grain shape does affect settling velocity, but for sand and smaller grains the effects of water temperature (viscosity and density) far outweigh the shape effects (Fig. 4.9), and for very small grains shape can be ignored (Fig. 4.10). In general, the more flattened a sand grain is the slower it settles through water, so in a sense a platey grain has a "smaller hydraulic equivalent grain size" than an equally massed spherical grain (Fig. 4.10). In the case of water flowing over grains on a bed, both the grain shape and size become very important factors in determining whether grains are entrained by the flow or remain immobile on the bed.

To examine these effects let us return to our model of stream flow where a laminar sublayer exists on the stream bed. In this situation the shape and size of the grains forming the bed become vital in determining whether or not the grains respond to the flow. Within the laminar sublayer, the forces on grains of the same mass but of differing shapes are different. Let us look at some examples.

Assume that the laminar sublayer is approximately 1 mm thick so that a spherical sand grain $\frac{1}{2}$ mm thick would lie within the laminar flow but would deflect it (Fig. 4.11*a*). As the flow passes over the grain, the velocity of flow increases

FIGURE 4.11 Diagram depicting the control of grain shape on sediment movement.

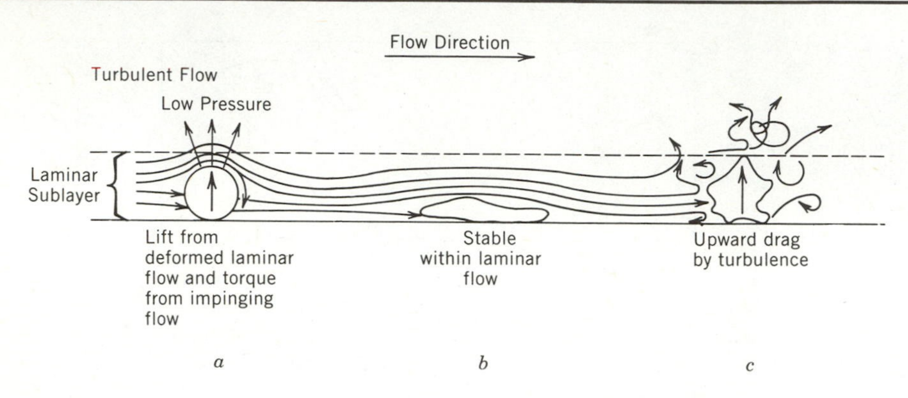

because of the longer travel distance and a low pressure zone is developed, akin to the process that causes an airplane to lift (the Bernoulli principle). Also, there is a force pushing the grain downstream from the water impinging on the upstream side of the grain. If the velocity is great enough to overcome the inertia of the grain, the grain will be lifted off the bottom into the flow; if not, it will remain at rest. A platey grain of the same mass would extend less distance into the flow (Fig. 4.11*b*) and would therefore deflect less of the flow. The same forces from the flowing fluid would act on the platey grain, but, because of the grain's streamlined shape, the lift would be considerably less, as would the push on the upstream end of the grain. It would take a higher velocity to move the platey grain than it would to move the spherical grain, even though they were of equal mass.

Now consider a grain of irregular shape that penetrated nearly into the turbulent flow above the laminar sublayer (Fig. 4.11*c*). Because of the proximity to turbulent flow, any streamlines deflected by the grain merge with the turbulence, and the laminar sublayer near the grain is disrupted. The grain then "feels" the turbulent forces and is more likely to be moved by the flow. The same situation results from grain size alone.

If a bed were composed of grains small enough to lie within the laminar sublayer, the flow over them would remain laminar (Fig. 4.12*a*), whereas a bed of grains large enough to extend into the laminar sublayer would destroy the laminar flow entirely (Fig. 4.12*b*). A bed made of uniformly small grains is smooth hydrodynamically and during low-velocity flow remains protected from the turbulence by the laminar sublayer, whereas under the same flow conditions, a bed composed of large grains or a mixture of grain sizes will disrupt laminar flow so that turbulence will extend onto the bed. More energy is transferred to the bed if it is rough and causes turbulent flow than if the bed is smooth and allows for laminar flow. To examine this effect, let us examine some work done by Filip Hjulstrom in 1939 and then modified and expanded by Sundborg in 1956.

FIGURE 4.12 Diagram depicting type of flow over different grain-sized bed materials.

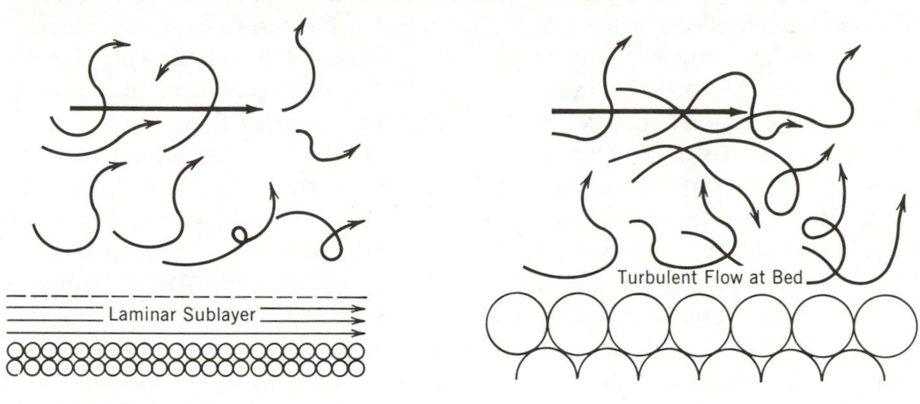

Hjulstrom constructed a **flume** (artificial stream) 1 m deep, filled it with sediment of a fairly uniform grain size, and then allowed water to flow over the bed at a depth of 1 m. He increased the velocity of flow and noted at what velocity sediment was entrained. By changing the sediment in the flume and then repeating the experiment, he constructed a plot of grain size versus velocity or an entrainment curve (Fig. 4.13).

Work by Sundborg and several other authors more recently has expanded the

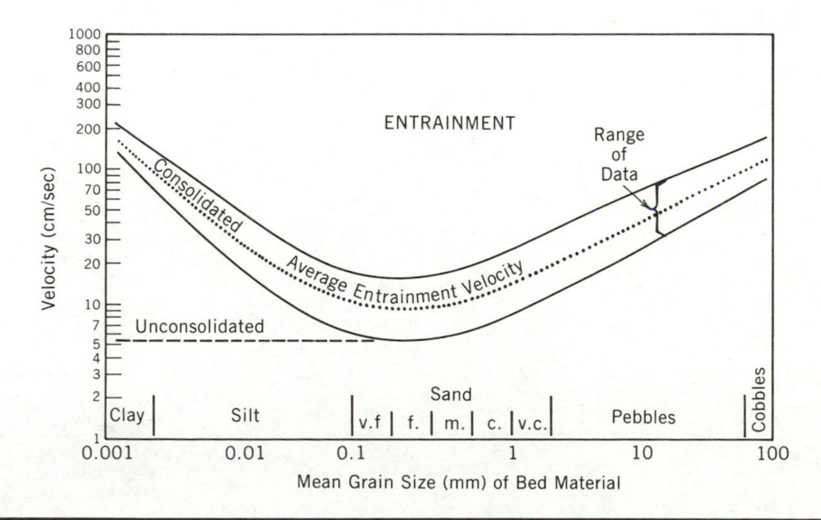

FIGURE 4.13 Plot of velocity of flow versus grain size for entrainment of bed material.

field of these relationships but they all show similar results, that is, the easiest grains to move are not the smallest grains. The minimum velocity of entrainment is for grains of about fine-sand size. Grains larger than that require higher velocity for entrainment, as do smaller grains. This relationship results from the effects of grain size on bed roughness, the weight of the grains, and the relative cohesiveness of grains. Beds composed of sand have very little cohesion and are large enough to disrupt the laminar sublayer, so the turbulent flow over them needs only to overcome the mass of the grain to move it. However, for beds composed of silt and/or clay, the small grains form a smooth bed, which does not disrupt the laminar sublayer, and the bed is very cohesive because of the physical behavior of clay minerals. To move sediment off these beds the velocity must increase to the point where turbulent flow extends down onto the bed. The sediment is then transported, but in natural systems not as individual grains. As the turbulence increases with increased velocity, sediment is eventually eroded from the bed as chunks. The cohesion of the clay makes it more difficult to entrain individual grains. This phenomenon results in the formation of **rip-up clasts** or **flat-pebble conglomerates** in modern environments and the rock record.

When the velocity reaches the **critical velocity** for sediment entrainment of a particular size, the way the sediment moves can vary. For very small grains, for example, silt, once the inertia is overcome by increased velocity the grain is lifted off the bottom and into higher-velocity turbulent flow. Because the grain has very little mass, the instantaneous high velocity within the turbulent flow keeps the grain from returning to the bottom. Instead, it is transported downstream in **suspension** (Fig. 4.14). A larger grain may act differently. If its mass is large

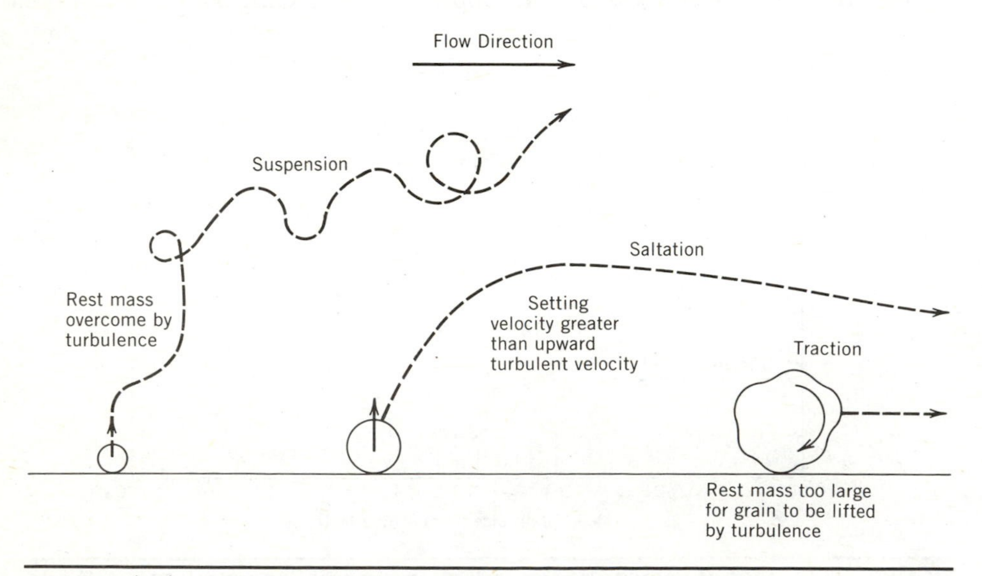

FIGURE 4.14 Diagram of different types of sediment transport.

enough to eventually overcome the instantaneous turbulent velocities that overcome the rest inertia, it will rise at first but then return to the bottom. Because there is a large downstream component of flow, the grain will move downstream as it falls back to the bottom, forming a broad arc. When it contacts the bottom it may bounce or knock other grains into similar motion. This bouncing movement along a bottom is termed **saltation** and represents the main mode of transport of sediment as **bedload** (sediment moved along the bed).

Larger grains may move in an entirely different fashion than either suspension or saltation. Grains with a mass so large that the upward velocity cannot lift the grain can still be pushed along the bottom by the current. Flat grains will slide, whereas more spherical grains will roll (Fig. 4.14). This form of sediment movement is termed **traction.**

The mode of sediment movement depends on several factors but is dependent mainly on velocity and grain size. Grains smaller than coarse silt can never move as bedload. Any flow strong enough to move them will eject them into the turbulent flow above the bed and they will become suspended load. Very coarse grains, cobbles, and boulders can rarely be moved by any mode other than traction in flowing water (we will examine the effects of high viscosity and density of sediment flows later) and so either slide or roll along. The sizes between these two end members act differently depending on the velocity of flow. At certain velocities, sand, and even pebbles, may move as bedload or suspension load, but in most systems the normal flow moves sand and pebbles by saltation. So, in general we can assume that flowing water transports silt and clay as suspension load, sand by saltation, pebbles by saltation or traction, and cobbles and boulders by traction. Now let us combine the principles of settling velocity to build a conceptual model of sediment entrainment, transport, and deposition by flowing water.

If we think of the settling velocity of a grain as representing its resistance to entrainment we can combine the Rubey settling curve with the Hjulstrom entrainment curve (Fig. 4.15). This assumption turns out to be quite reasonable and was suggested by Hjulstrom in 1935 and reproduced and elaborated on by many authors, including Dunbar and Rodgers (1957) and Friedman and Sanders (1978). If you consider the situation of a grain falling through turbulent flowing water then the relationship becomes clear. A grain of a certain size and shape has a particular settling velocity. If the instantaneous velocity upward from the turbulence of flowing water is great enough, it will overpower the downward velocity of the grain due to its mass (settling velocity). As a result, the grain cannot be deposited. As we have seen, there is a distinct relationship between velocity of flow and turbulence (Reynolds number), so let us assume that we can relate settling velocity to entrainment velocity and examine Fig. 4.15 carefully. We have left off the velocity values because we do not want to apply actual numbers in this discussion.

Assume we have a bed composed of medium sand and a very slow or zero

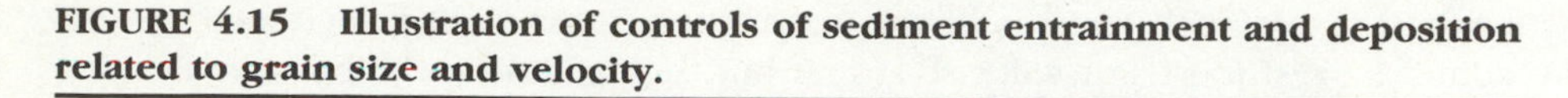
FIGURE 4.15 Illustration of controls of sediment entrainment and deposition related to grain size and velocity.

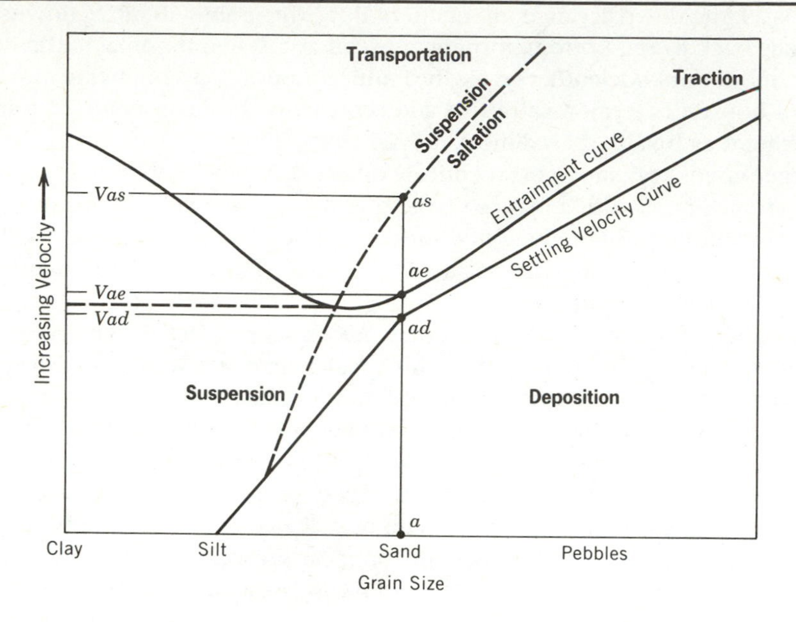

velocity current. (This situation is represented by point *a* in Fig. 4.15.) If the velocity of flow is increased slowly, at the entrainment velocity the grains move (represented by the intersection of a vertical line with the entrainment curve, point *ae* and velocity *Vae* on the velocity axis) by saltation. If the velocity is increased, the grains will eventually move into suspension (if they are small enough and the flow is great enough).

Now, what happens to these grains if the velocity is reduced? As the velocity decreases the turbulence decreases. At the point where the settling velocity of the grain is greater than the turbulent velocities transporting the grain, the grain is deposited. (This occurs at point *ad* in Fig. 4.15.) This velocity is somewhat less than that for entrainment of the grain (velocity *Vad* in Fig. 4.15). Notice the gap between the entrainment velocity and the deposition velocity; this gap represents the "energy" required to overcome the rest inertia of the grains. For very small grain sizes this velocity gap becomes huge, and for clay-sized material the velocity must be nearly zero for the grains to deposit.

These concepts allow us to divide Fig. 4.15 into different regions representing "velocity fields" where grains are either transported or deposited. Within the transport field, grains move by suspension, saltation, or traction. As velocity changes within the field, grains may move from one type of transport to another, but as velocity decreases and approaches the deposition field, the grains must pass through the transport modes labeled at the curve. This suggests that clay to silt can only be deposited from suspension, whereas coarse silt and sand are dominated by saltation and pebbles by traction when they are deposited. Grains coarser than these move only by traction during the last stages of movement before deposition. This relationship becomes very important in understanding the origin of various sedimentary structures created by flowing water, which is the topic of Chapter 5.

EPILOGUE

> *We are led to recognize a property in matter whereby it attracts other matter at a distance, and to refer the planetary motions to a force the same as that whereby an apple falls to the ground, and our own bodies are retained firmly on the seats on which we sit. Before such a theory is finally accepted, it is requisite that we should examine it into its minute consequences.*
>
> ***Sir G. G. Stokes, 1891***

In this section we have applied various principles of physics to understand the "minute consequences" by which particles settle through a fluid and are transported by wind and water. An understanding of such principles of hydrodynamics is crucial to the correct understanding of the processes that transport and deposit sediment and produce sedimentary structures. Using these physical processes of sedimentation we will continue to develop, in Chapters 5–8, an understanding of various processes that deposit sediment. Finally, as in any field study, we must get out of the "seats on which we sit" and examine the results of these processes firsthand in sedimentary structures and relate these to the descriptive forms discussed in Chapter 3.

OUTSIDE READING

Allen (1970, 1985a, b); Blatt *et al.* (1980); Bogardi (1974); Costello and Southard (1981); Davis (1983); Dunbar and Rogers (1957); Friedman and Sanders (1978); Gibbs *et al.* (1971); Grim (1968); Harms *et al.* (1975, 1982); Hjulstrom (1935, 1939); Leeder (1982); Middleton and Hampton (1976); Prandtl (1952); Reading (1986); Rouse (1937); Rubey (1933); Selley (1982); Stokes (1851); Sundborg (1956).

Chapter 5

Bed Response to Flowing Water

FLOW REGIMES

In Chapter 4 we examined the mechanisms of sediment movement as water flows over a bed. But what happens after the critical velocity is reached and sediment is entrained? How does the bed respond? What features form on the bed in response to this flow? These are the questions we will answer in this chapter. To develop those answers we must return to flume studies and detailed observations of the response of beds to unidirectional flow. In general, what we will find is that the bed configuration is dependent on characteristics of the flow and sediment. Let us examine the results of flume studies conducted by Guy *et al.* (1966) from 1956 to 1961 in the U.S. Geological Survey hydrologic laboratory at Colorado State University. Their work expanded on earlier work of many people (see Bogardi, 1974; Friedman and Sanders, 1978; Blatt *et al.*, 1980; Leeder, 1982; Allen, 1985a, b for discussions of the development of the **flow regime model**) and led to the concepts we now apply to deciphering flow based on analysis of sedimentary structures. All these experiments were conducted under extremely uniform conditions and allowed to reach equilibrium. The uniform conditions of these experiments limit their application to modern environments or the rock record, where sedimentary processes are rarely uniform or the bed in equilibrium with the flow. Nevertheless, the flow regime concept was a major breakthrough in sedimentology that greatly expands our ability to decipher the sedimentary record, so we will examine the principles carefully.

Assume we have a flume containing a bed of medium sand. The bed has been smoothed to a flat surface and the flume filled with water to a certain depth, say 20 cm. The velocity of flow is very slow so that the bed is stable (below the entrainment velocity) and no sand is moving. If we slowly increase the velocity, a sequence of distinct types of **bedforms** appears in response to the flow. The sequence we would see is shown in Fig. 5.1.

First, the surface of the bed would be transformed into ripples. They have steep downstream sides and gently sloping backs. The ripples are straight-crested at first, but as the velocity increases they become more complex, changing from straight to sinuous to linguoid. Sediment moves by saltation up the backs and slides down the face so that the ripples slowly migrate downstream. This migration forms cross laminae, which are removed as each succeeding ripple migrates along the bed.

Second, as the velocity is further increased, the complex-shaped ripples are transformed into **large ripples (megaripples** or **dunes** are equivalent terms), which may or may not have small ripples on their backs. As velocity increases the large ripples change from slightly sinuous forms to lunate or linguoid shapes. Large ripples have the same form as ripples but are significantly larger, with wavelengths greater than 60 cm and heights greater than 6 cm. Their migration also produces cross laminae within the large ripple.

Third, the large ripples pass through an abrupt transition at higher velocities.

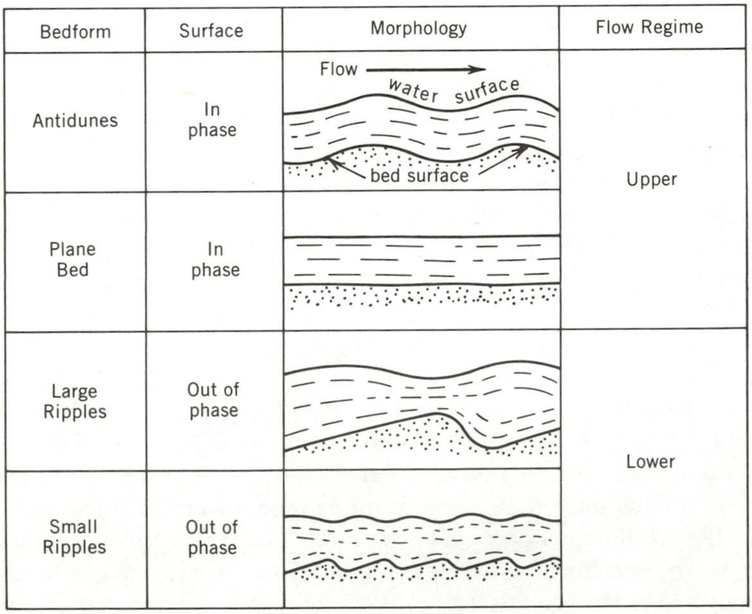

Bedform	Surface	Morphology	Flow Regime
Antidunes	In phase	Flow → water surface, bed surface	Upper
Plane Bed	In phase		
Large Ripples	Out of phase		Lower
Small Ripples	Out of phase		

FIGURE 5.1 Diagram of the flow regime concept.

The large ripples wash away and are replaced by a flat bed. The sediment streams along the bed continuously. The flat bed forms horizontal laminae in the sediment.

Fourth, with continued increase in velocity the flat bed changes into low-relief, undulating forms called **antidunes** (so named for their ability to migrate upstream). Antidunes can actually remain stationary or migrate upstream or downstream. Sediment moves continuously over the antidunes as it did the flat bed, and as they migrate they form indistinct wavy laminae.

The first workers who identified this sequence of forms coined the term flow regime to label two different types of bed configurations and associated flow. The ripple forms (ripples and large ripples) form under the **lower flow regime** and the plane bed and antidunes within the **upper flow regime** (Fig. 5.1). For each of these regimes the flow is different. In lower flow regime the flow is out-of-phase with the bedform. The thickest flow is over the troughs of the ripple forms and the thinnest over the crest (Fig. 5.1). The flow over upper flow regime bedforms parallels the bedform or is in phase (Fig. 5.1). The concept that a particular type of flow molds the bed into a specific bed configuration is the flow regime concept. It is a powerful tool in deciphering ancient deposits, because it allows us to interpret flow from sedimentary structures.

If we were to continue our experiment with flow regimes by changing the grain size of the sediment in the flume and the depth of flow, we would see that the flow regime is controlled by all three factors: depth, velocity, and grain size. Grain size modifies the simple model presented above in several ways (Fig. 5.2). For sand sizes less than approximately 0.2 mm, large ripples do not form in response to an increase in velocity. Instead, small ripples pass directly into plane beds of the upper flow regime (upper plane bed). For grain sizes coarser than approximately 0.8 mm, small ripples do not form, and when sediment first starts to move, it does so in a plane bed (lower plane bed). The lower plane bed passes directly into large ripples.

By considering grain size as a variable we see that both the upper and lower flow regimes are composed of fields of different bedforms. The lower flow regime consists of small ripples, lower plane beds, and large ripples, depending on the grain size and velocity. The upper flow regime contains upper plane beds or antidunes. The same fields are present when the depth is used as a variable and the grain size is held constant (Fig. 5.3). In such a situation we see that the depth can greatly influence the flow regime. At very shallow depths large ripples do not form because there is not enough room for them. If large ripples are present at a certain flow velocity and depth (for example, 60 cm deep and 130 cm/sec in Fig. 5.3) and the depth decreases, then the bed will change into antidunes. The flow regime changed from lower to upper without a velocity change. This change results in a shift from lower to upper flow regimes and shows the dependency of upper flow regime bedforms on depth of flow (Fig. 5.3). Figures 5.2–5.4 exemplify the controls of flow regime by grain size, depth, and velocity. Because all three variables are important in understanding how flowing water

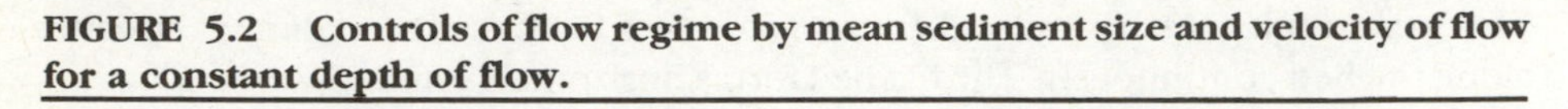

FIGURE 5.2 Controls of flow regime by mean sediment size and velocity of flow for a constant depth of flow.

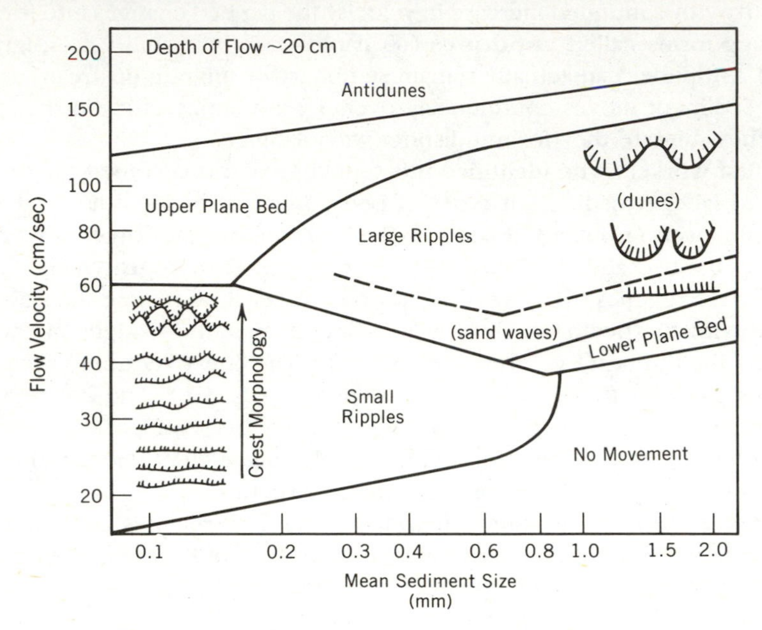

modifies the bed over which it flows, let us look at a diagram constructed by Rubin and McCulloch in 1980 that includes all three variables on a three-dimensional plot.

Rubin's and McCulloch's plot (Fig. 5.4) combines the depth–velocity–grain size data obtained by many workers with their own observations on sand waves (large ripples) in San Francisco Bay. This diagram covers most situations in natural environments where sand is transported. The grain size ranges from slightly less than 0.1 mm to approximately 1.5 mm and the velocity from 0 to a little above 200 cm/sec. By combining their work with the flume information they have increased the depth of flow up to 100 m. These ranges take in most systems transporting sand by stream and tidal currents in modern environments, except for very shallow areas (less than 6 cm) such as beaches. Several interesting associations are apparent from the diagram. Let us first examine the distribution of the upper and lower flow regimes and then the extent of the various fields within each.

The upper flow regime is clearly associated with shallow depths and finer grain sizes. At shallow flows it dominates the diagram for all grain sizes, but is restricted to very high flow velocities at large depths of flow. Within the upper flow regime field, antidunes are restricted to shallow depths entirely. For very shallow depths not depicted on this diagram they are the only possible bedform

(Fig. 5.3). Upper plane beds expand into the finer grain size and deeper flow regions of the diagram. Upper flow regime bedforms require higher velocities to form at greater depths. They are strongly dependent on depth and velocity and somewhat on grain size. Lower flow regime bedforms show much less dependency on depth, but are strongly controlled by velocity of flow. Lower plane beds are restricted to low velocities and large grain sizes (greater than 0.8 mm), but extend through many depths. Ripples mirror this distribution but are associated with smaller grain sizes. Large ripples fill the gap between ripples and the upper flow regime bedforms and cross all depths. They do not extend into the finer grain sizes (less than 0.15 mm).

With the flow regime concept, we have a powerful tool to unravel a sequence of rock containing sedimentary structures deposited under equilibrium condi-

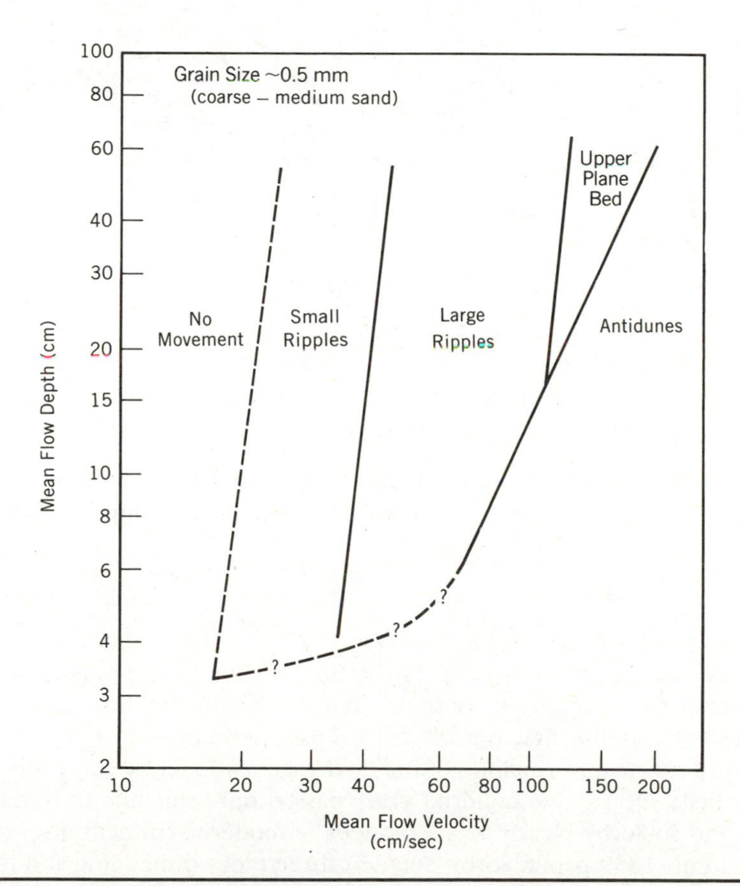

FIGURE 5.3 Controls of flow regime by mean flow depth and mean flow velocity for a constant grain size of bed material.

FIGURE 5.4 Three-dimensional diagram relating flow depth, flow velocity, and grain size to flow regime fields. From D. M. Rubin and D. S. McCulloch, 1980, *Sedimentary Geology,* 26, Fig. 11, p. 224. Reprinted by permission of Elsevier Science Publishers, B.V.

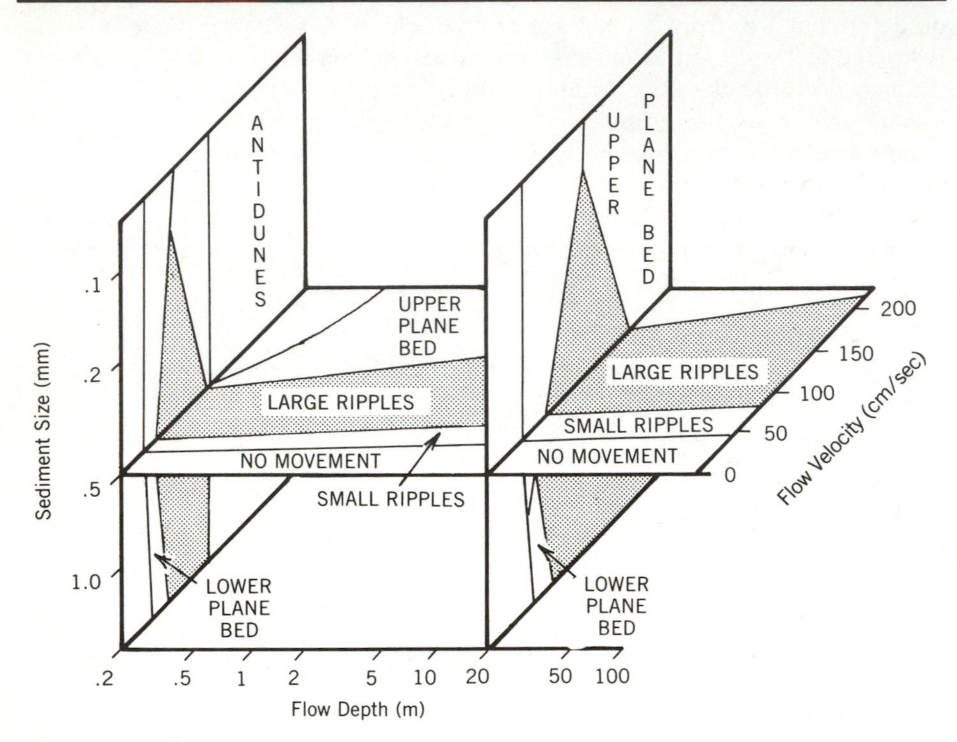

tions. Although such ideal conditions rarely exist in natural systems, we can go a long way toward interpreting the details of flow by applying the flow regime model.

Applications of the Flow Regime Concept

The first step in learning how to apply the flow regime model to rock sequences is to develop the relationship between bedforms and bedding. Henry Clifton Sorby (1859) was the first person to discuss these relationships. In fact, he described all the major principles of the flow regime model by observing streams and their beds. Nearly one hundred years passed until this line of research was resumed and so Sorby clearly was a pioneer of modern sedimentology and stratigraphy. In his 1859 paper, Sorby detailed three types of bedding and the processes that formed them. He recognized that the direction of the current that deposited the sediment forming the bedding was easily determined from cross beds and other structures. Also, he suggested that ripple marks were formed by flowing water and described the origin of cross bedding formed by ripple migration. By carefully looking at Sorby's work and adding some more recent data we can describe all the major bedform types and their resulting bedding. For a historical perspective on these topics, read some of the original articles in the following order: Sorby (1859), Allen (1963), Simons *et al.* (1965), and Harms (1969).

Returning to the flow regime model, let us describe the bedding resulting from each type of bedform, beginning with ripples and large ripples. As a ripple form (ripples and large ripples) migrates, sediment moves up the back of the form and avalanches down the front slipface (Fig. 5.5). Under equilibrium conditions in which no sediment is added to the system, equal amounts are eroded from the back of the ripple as are deposited on the slipface. If a train of such bedforms moves along a surface it would not leave behind a bed of sediment. The only material deposited would be in the ripple form itself and could only be preserved by stopping the flow and covering the ripple form (Fig. 5.5). If sediment were added to the system so that more material was deposited on the slipface than was eroded from the back of the ripple form, then a bed of cross-laminated sediment would be left behind as the form moved downstream (Fig. 5.5). If enough sediment is added, then the ripple forms will not erode at all but instead will climb upward at some angle, the steepness depending on the amount of accumulation of sediment. Such **climbing ripples** (Fig. 5.6) were first described by Sorby, who recognized that they formed a particular type of bedding he termed **ripple drift.** More recently, Allen (1963), McKee (1965), Hunter (1977), Rubin and Hunter (1982), and other authors have carefully described the results of climbing ripple forms and the complexity of the resulting bedding.

As ripples or large ripples migrate, and sediment accumulates (the bed is aggrading), cross-stratification forms. The thickness of the cross-laminated bed depends on the height of the bedform and how much is eroded by succeeding forms as they migrate along the bed. The maximum thickness can only be slightly greater than the height of the form (Fig. 5.5) and is nearly always some small percentage of the form's height (Fig. 5.5). But larger forms produce thicker beds of cross strata so that large-ripple cross bedding is easily distinguished from that formed by ripples. Ripples are mostly less than 3 cm in height and large ripples range from a few centimeters up to several meters high. Another important characteristic used to determine the amount of erosion in a set of cross strata is the angle of the cross laminae themselves. Most cross laminae are steeper at the top of the ripple form and decrease in slope downward until they meet the bottom tangentially (Fig. 5.5). If only a small amount of the ripple were removed by erosion from its upstream counterpart then the steeper upper laminae will be preserved. But if the erosion is deep, then only the low-angle toes of the cross laminae will remain (Fig. 5.7). The size of the bedform and the amount of **aggradation** or **degradation** of the bed control the thickness of beds that are deposited and the morphology of the bedform controls the shape of the cross laminae.

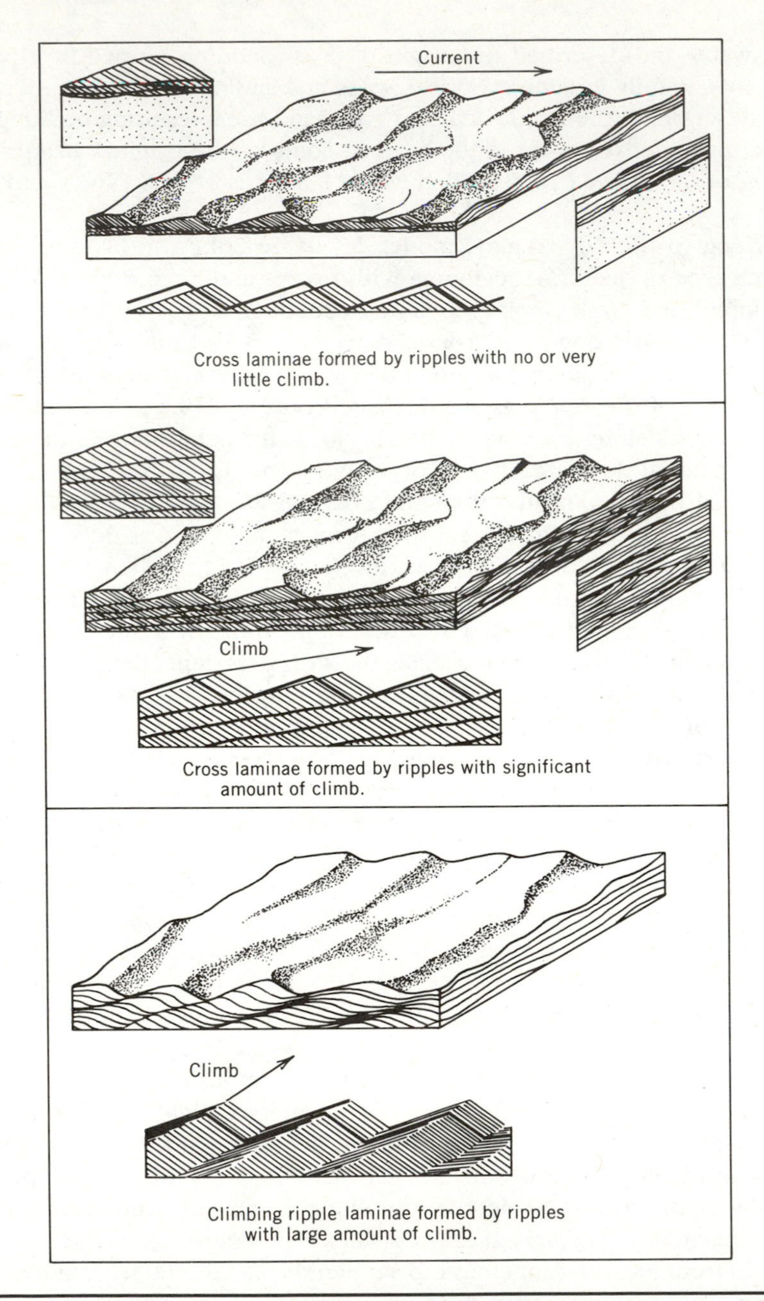

Current
Cross laminae formed by ripples with no or very little climb.
Climb
Cross laminae formed by ripples with significant amount of climb.
Climb
Climbing ripple laminae formed by ripples with large amount of climb.

FIGURE 5.5 Preservation of cross laminae by different amounts of bedform climbing.

FIGURE 5.6 Climbing ripples. (Photograph by E. D. McKee, #3, U.S. Geological Survey).

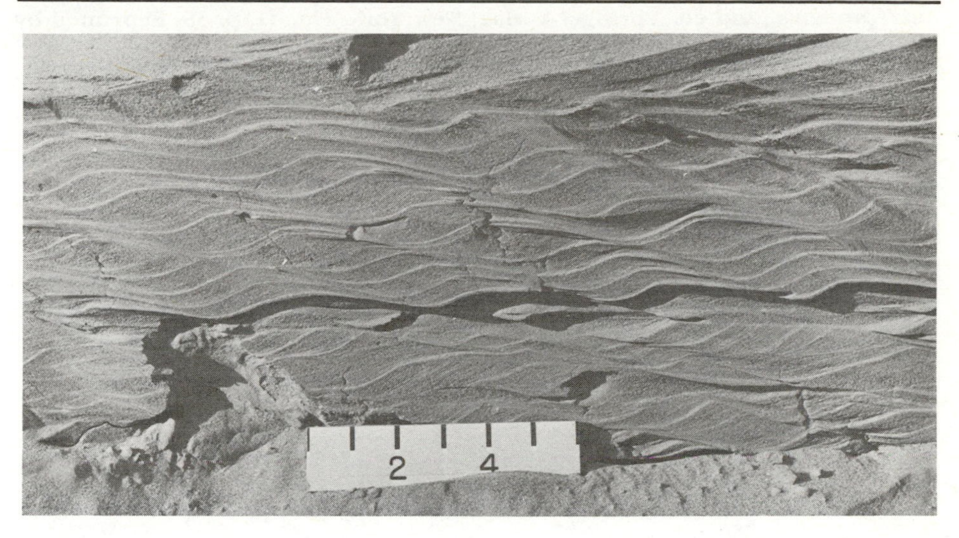

FIGURE 5.7 Deeply erroded cross laminae showing resulting low-angle toes (just above Jacob's staff). (Photograph by J. N. Moore).

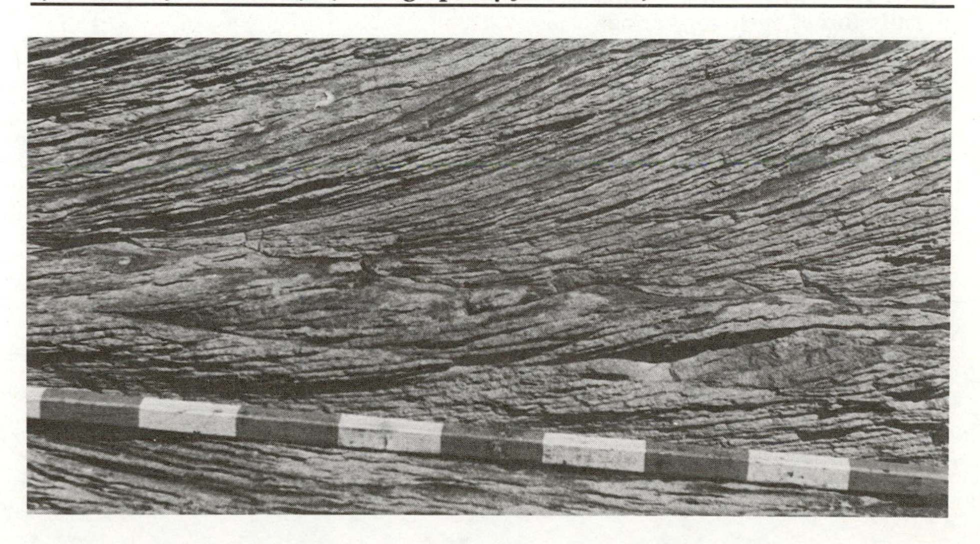

FIGURE 5.8 Type of cross laminae formed by straight-crested ripples or large ripples. From H. E. Reineck and I. B. Singh, 1980, *Depositional Sedimentary Environments,* 2nd ed., Springer-Verlag, New York, Fig. 41, p. 38. Reprinted by permission of Springer-Verlag.

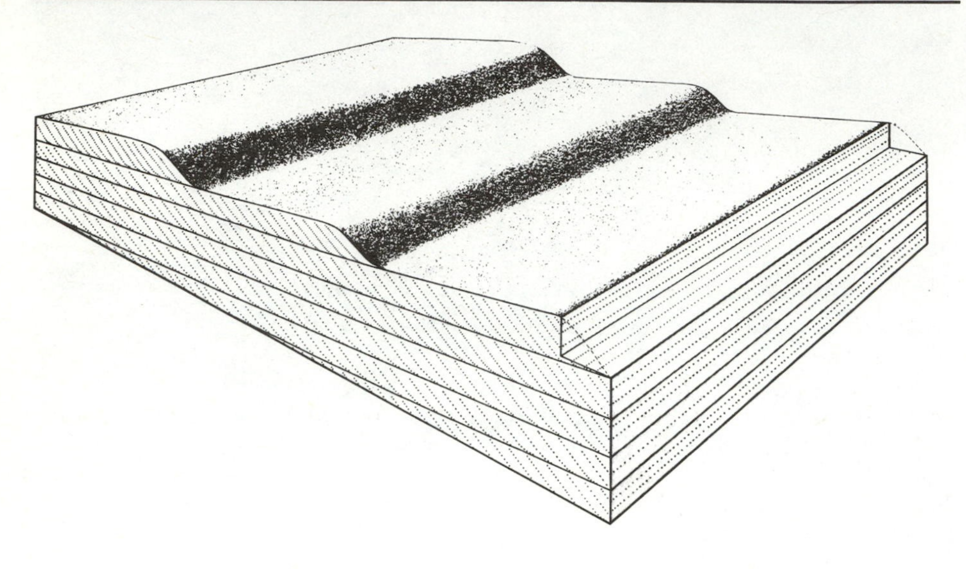

FIGURE 5.9 Type of cross laminae formed by sinuous-crested ripples or large ripples. From H. E. Reineck and I. B. Singh, 1980, *Depositional Sedimentary Environments,* 2nd ed., Springer-Verlag, New York, Fig. 46, p. 40. Reprinted by permission of Springer-Verlag.

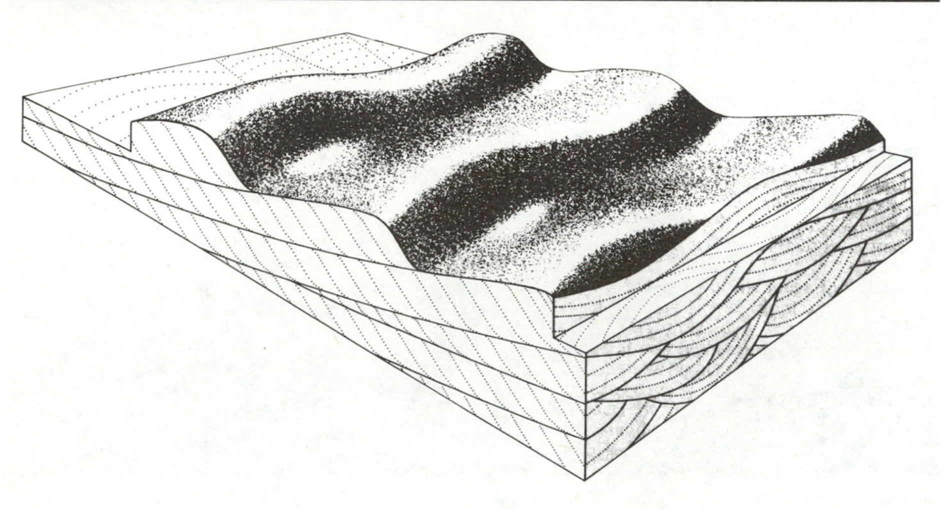

So far we have ignored the three-dimensional shape of the bedform, that is, its morphology. We saw in the flow regime model that, as velocity is increased in both the ripple field and the large ripple field of the lower flow regime, the morphology of the ripple forms becomes more complex. These different types of ripple forms will produce very different types of bedding. The simplest form, the straight-crested ripple or large ripple (**sand wave**), will produce even, nearly tabular beds. Because they deviate slightly from perfectly straight lines, the resulting beds will show some irregularities. If looked at from a direction parallel to flow, then the cross laminae will be quite evident, but on a face viewed perpendicular to flow the steep foresets of the cross laminae will look like parallel laminae or very low angle broad troughs (Fig. 5.8). Both ripples and large ripples produce similar bedding, and both small and large ripples can form very complex cross laminae.

Migrating sinuous ripple forms produce bedding that is more irregular (Fig. 5.9). When examined parallel to the flow direction, the beds are tabular to slightly wedge-shaped but are more troughlike and wavy when cut perpendicular to flow. Very complex ripples, lunate or linguoid forms (Fig. 5.10), produce complex **trough cross laminae.** It is important to note that cross bedding looks different on planes passing through the ripple form at different angles. To determine the bedform from the bedding one must look at a minimum of two directions, preferably one perpendicular to flow and one parallel to flow. From the principles developed in the preceding discussion we can determine the type of bedform by examining the size and shape of bedding. From bedforms we can then determine the flow regime that formed the beds. We can do the same for structures formed in the upper regime flow.

When upper flow regime plane beds form, sediment moves along the bottom continuously. This results in a poorly defined, parallel lamination that is often streaked by small stripes of slightly coarser sediment or aligned elongate grains. Sorby first described this texture as **grained and striped horizontal stratification;** it is now termed **flow lineation.** The resulting deposit from upper regime flow is an indistinct, evenly laminated bed (Fig. 5.11). This bedding is difficult to distinguish from a lower flow regime plane bed (greater than 0.8 mm grain size, remember), but commonly the lower plane bed shows better developed stratification because the flow is slower and therefore grain types and sizes differentiate somewhat as they move along the bed.

Antidunes leave a distinct bedding unlike that formed by lower flow regime ripple forms. The detailed shape of laminae is dependent on whether the antidunes migrate upstream or downstream or remain in place as the bed aggrades. In general, they produce low-relief, wavy beds that mimic the form of the antidune or truncate one another at low angles (Fig. 5.12). Antidunes range in size from a few centimeters in wavelength to several decimeters or larger and, although considered rare in the record, they are really quite common in many shallow-water environments and deposits, including beach, fluvial, and submarine fan

FIGURE 5.10 Type of cross laminae formed by lunate and linguoid ripple forms. From H. E. Reineck and I. B. Singh, 1980, *Depositional Sedimentary Environments,* 2nd ed., Springer-Verlag, New York, Fig. 52, p. 43. Reprinted by permission of Springer-Verlag.

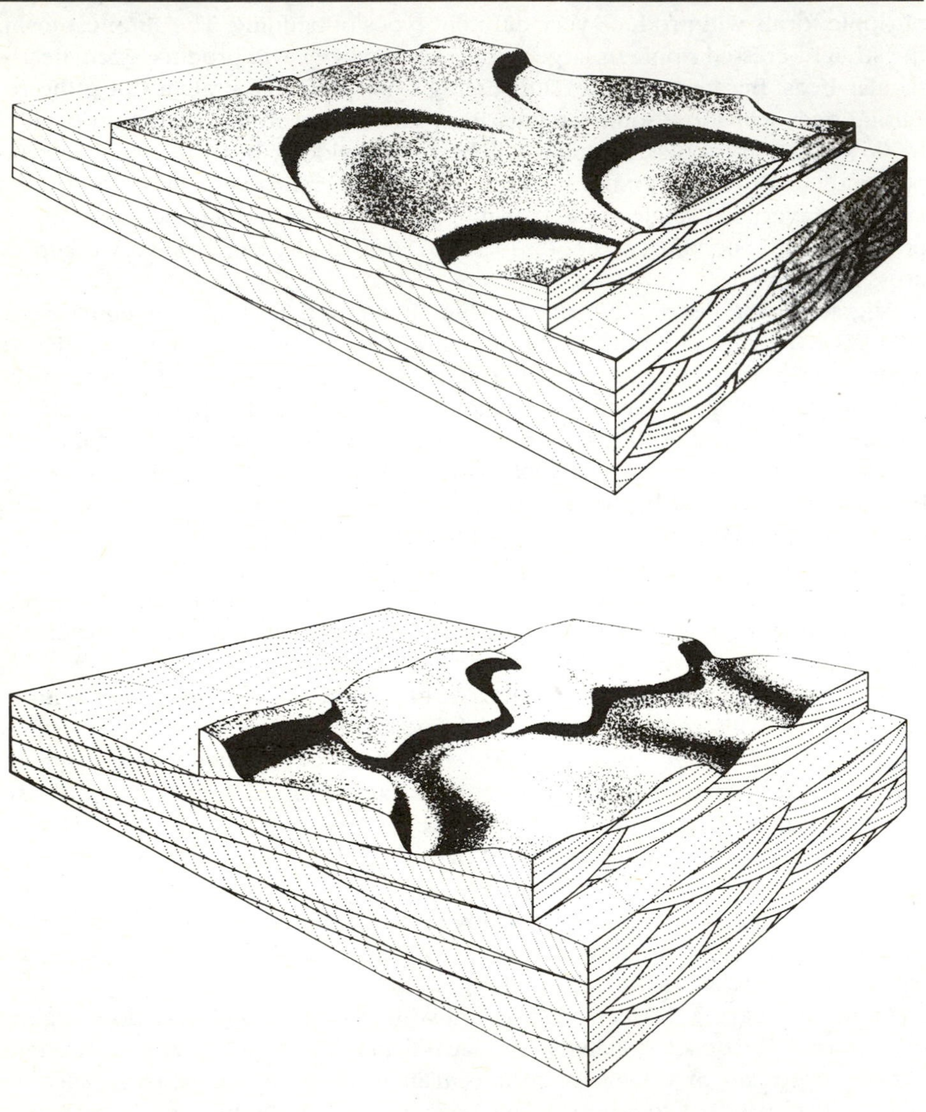

FIGURE 5.11 Type of bedding formed by plane bed. From J. C. Harms, J. B. Southard, and R. B. Walker, 1982, ***Structures and Sequences in Clastic Rocks,*** Society of Economic Paleontologists and Mineralogists Lecture Notes for Short Course No. 9, Fig. 2–19, pp. 2–48. Reprinted by permission of SEPM.

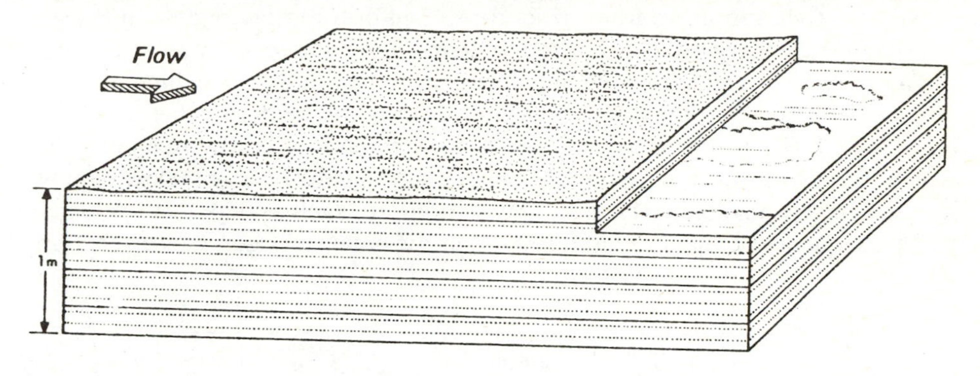

FIGURE 5.12 Bedding formed by antidunes. From H. E. Reineck and I. B. Singh, 1980, ***Depositional Sedimentary Environments,*** 2nd ed., Springer-Verlag, New York, Fig. 59, p. 47. Reprinted by permission of Springer-Verlag.

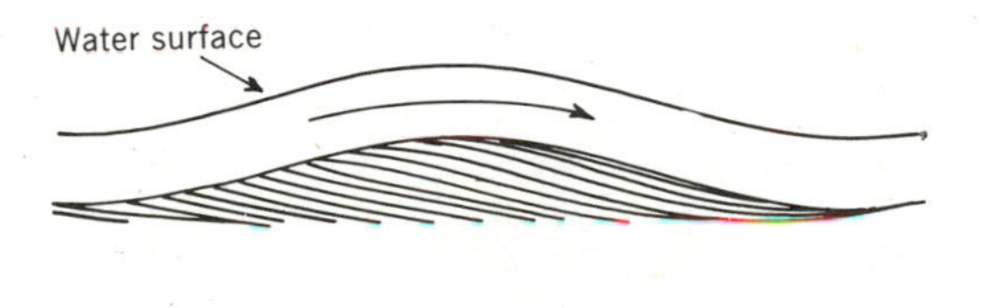

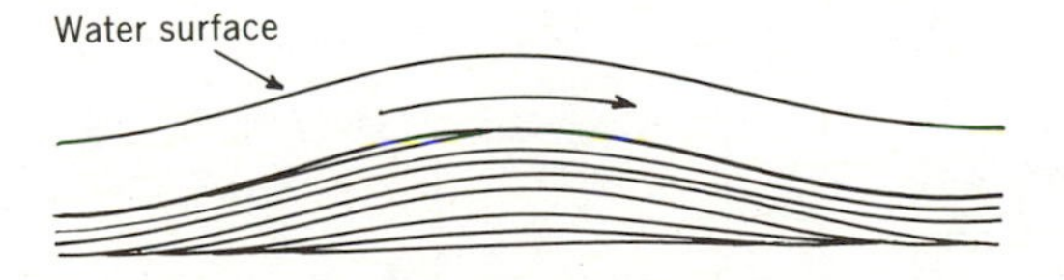

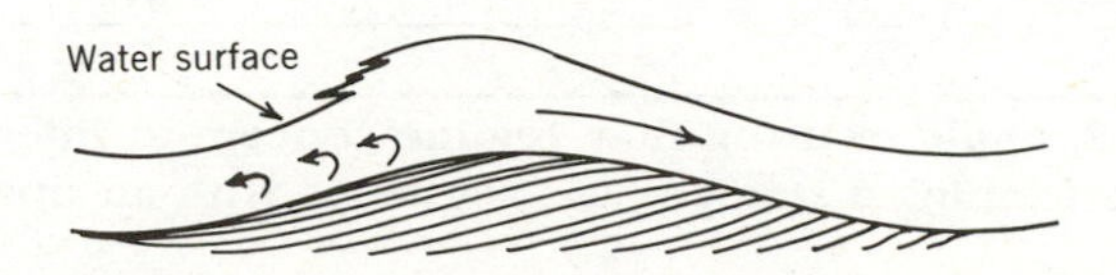

systems. Deposition often occurs on the upstream slope of the antidune to form backset bedding (Fig. 3.7). One reason that upper flow regime bedforms are scarce in rocks is due to reworking by lower flow regime currents following the creation of upper flow regime conditions.

Now that we can move from structure to bedform to flow regime, it is time to apply the flow regime concept to understanding a sequence of strata.

Figure 5.13 is a graphic column representing bedding in a sedimentary sequence. By examining the bedding we can identify the bedforms that produced each "unit." The bedform type allows us to determine the flow regime under which the sediment was deposited. Then, by analyzing the change in flow regime, we can suggest possible changes in flow conditions, such as depth and velocity. For the sequence in Fig. 5.13, notice first that the grain size decreases up section.

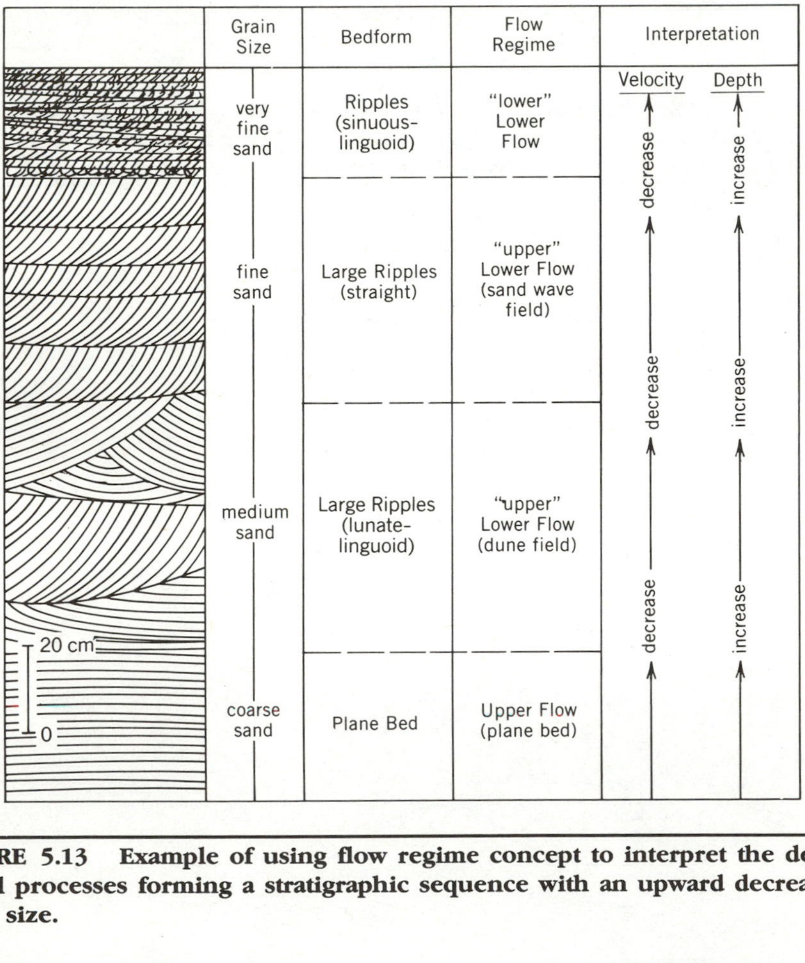

	Grain Size	Bedform	Flow Regime	Interpretation	
				Velocity	Depth
	very fine sand	Ripples (sinuous-linguoid)	"lower" Lower Flow	decrease	increase
	fine sand	Large Ripples (straight)	"upper" Lower Flow (sand wave field)	decrease	increase
	medium sand	Large Ripples (lunate-linguoid)	"upper" Lower Flow (dune field)	decrease	increase
20 cm 0	coarse sand	Plane Bed	Upper Flow (plane bed)		

FIGURE 5.13 Example of using flow regime concept to interpret the depositional processes forming a stratigraphic sequence with an upward decrease in grain size.

This trend suggests that the velocity decreased, because we know there is a direct relationship between velocity and grain size of sediment that is deposited (Fig. 4.15). By using Figs. 5.8 through 5.12, we can determine the bedform that deposited the bedding types in the sequence. From the bedforms we can delineate the flow regime. Then, by presuming there was a velocity decrease up section, we can predict the change in depth that would result in the changes in flow regimes we defined. So, from a set of sedimentary structures we have determined the details of flow and how it changed through time. By establishing such processes we move closer to our goal of deciphering the depositional environment within which the sequence formed.

One must be careful when using the flow regime model, however, and especially careful about the assumptions one makes. For example, the assumption that grain size in sediment is controlled only by velocity can be deceiving. Suppose the source of the sediment is such that only fine- to medium-grained sand is added to the depositional area. Such a situation decreases the precision of interpretations using the flow regime model, but one can still make fairly detailed interpretations. Try it on Fig. 5.14 by covering up the "answers" and making your own interpretations.

In the sequence represented in Fig. 5.14, we see that the grain size remains constant throughout the sequence. If the velocity remained constant during the sequence then we could account for all the changes in flow regime by depth changes. But if we assume grain size was limited by other factors, then the system becomes fairly complex. Let us go through the sequence from the bottom to the top and examine the evidence for the interpretations given.

The lower unit contains plane bed and antidune bedding that must represent alternation of flow within the upper flow regime. By looking at Figs. 5.2 and 5.3 you can see that the alternation could be due to velocity or, just as easily, to depth changes. An example of where this might happen is on a beach where waves swash up on the beach, rapidly changing the depth and velocity of flow. Above these upper flow regime structures, a unit of lunate, large-ripple cross bedding suggests that the depth increased significantly. However, if the lower unit was deposited under deep and very high velocity flow then a decrease in velocity may have produced the same bedforms. A similar argument applies to the ripples in the unit above. If the depth remained great, then the only way ripples could have formed is with a decrease in velocity. Because the ripples become less complex upward within the unit, the flow must have moved toward the lower flow regime within the field. This trend abruptly changed with the deposition of the uppermost unit, which represents upper flow regime plane bed deposition. It seems likely that the depth was shallow here, because to increase velocity alone to get to the plane beds from ripples would require a pass through the large-ripple field. Because large-ripple cross laminae are not present at this boundary it seems unlikely that they existed.

So with these limits it seems that the sequence was generated by combined changes in velocity and depth. The lowermost unit of the sequence was deposited under shallow flow with the depth and velocity changing rapidly. The flow deepened to allow the deposition of the large lunate ripples, which also required a fairly high velocity. The flow depth decreased and possibly the velocity as well to form the ripples. Straight-crested ripples formed on top of the linguoid forms as a result of a decrease of velocity within the ripple field. The uppermost unit formed as the depth decreased and/or the velocity increased. Such complex assemblages are not uncommon in the rock record. To become familiar with these techniques of interpretation, make up some of your own sequences and decipher them using the concepts discussed above, and then try to apply them in the field at outcrops and roadcuts.

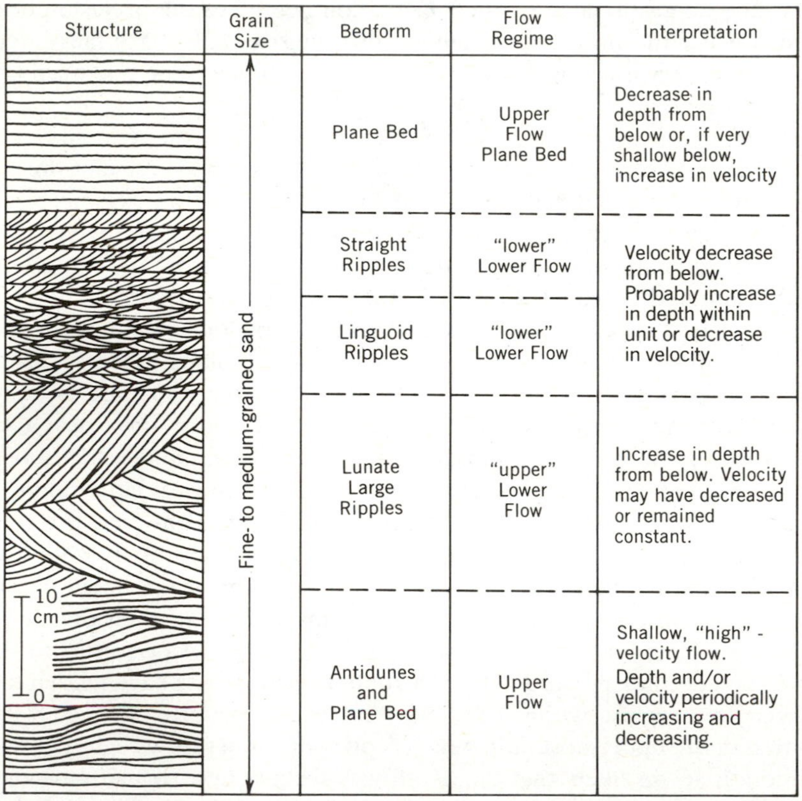

Structure	Grain Size	Bedform	Flow Regime	Interpretation
	Fine- to medium-grained sand	Plane Bed	Upper Flow Plane Bed	Decrease in depth from below or, if very shallow below, increase in velocity
		Straight Ripples	"lower" Lower Flow	Velocity decrease from below. Probably increase in depth within unit or decrease in velocity.
		Linguoid Ripples	"lower" Lower Flow	
		Lunate Large Ripples	"upper" Lower Flow	Increase in depth from below. Velocity may have decreased or remained constant.
10 cm 0		Antidunes and Plane Bed	Upper Flow	Shallow, "high"-velocity flow. Depth and/or velocity periodically increasing and decreasing.

FIGURE 5.14 Example of using flow regime concept to interpret the depositional processes forming a stratigraphic sequence with a constant grain size.

Limitations of the Flow Regime Model

In using the flow regime concept to interpret sedimentary structures in the preceding examples, we had to assume that the sediment was deposited under equilibrium conditions or near to them. But in natural systems equilibrium conditions rarely exist and we must be cautious of applying the flow regime model. One good example of this was presented recently by Crowley (1983) in his work on the Platte River. He defined large bars that migrated downriver during high water and were stable under low water conditions. He constructed a generalized longitudinal section of the complex structure that formed from three different types of bars (Fig. 5.15). The sequence generated by these bars consisted of very different bedding types. The differences resulted from the bar migrating downstream and depositing bedding types characteristic of each individual part of the bar (Fig. 5.16). If we apply the flow regime concept to this sequence, we will come up with some erroneous interpretations.

The lower part of the sequence (Fig. 5.15) is composed of horizontal laminae and ripple laminae. This could represent deposition in the lower flow regime ripple field and lower plane bed field, except that the grain size is too fine for the lower plane bed. So, using flow regime concepts, the lower sequence must have resulted from the alternation of very shallow flows changing from upper flow regime to lower flow regime ripple deposition. The overlying, steep cross laminae, a tabular set, would represent the large-ripple field of the lower flow regime. Because this unit is coarser grained, the velocity must have increased,

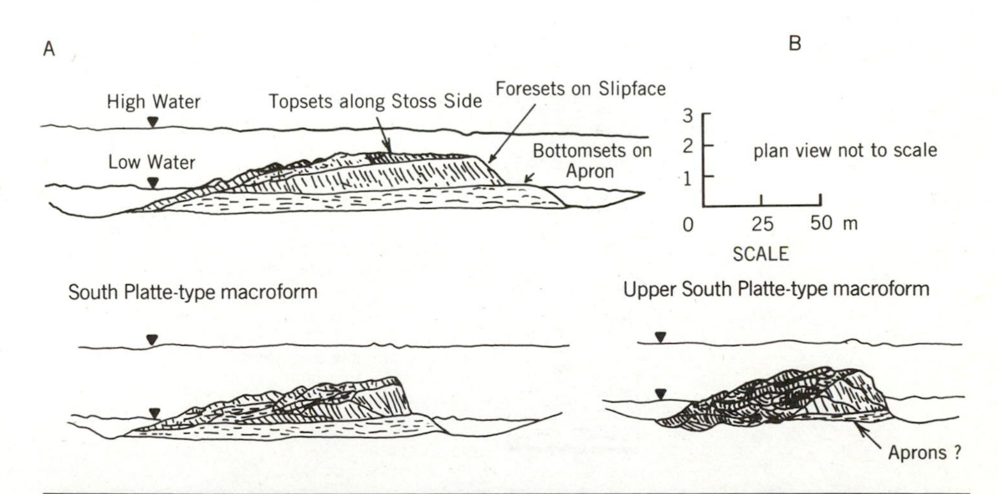

FIGURE 5.15 Longitudinal sections of bars from the Platte River. From K. D. Crowley, 1983, *Geological Society of America Bulletin,* 94, Fig. 10, pp. 126–127. Reprinted by permission of GSA and Kevin D. Crowley.

so that in applying the flow regime concept we would think that flow depth and velocity increased substantially to form the large cross-bed set. Overlying the thick, cross-stratified bed are sets of tabular and trough cross laminae, which suggests migration of complex large ripples, as in the lower flow regime. By using the flow regime model, we develop a complex system of changing velocity and depth of flow to explain the sequence. Unfortunately, the flow in the river was the same during the deposition of all the units and each represents a different part of the bar migrating down the river system. Because we applied the flow regime concept and assumed that each unit represented the migration of equilibrium bedforms, we made errors.

We should have seen our mistake, however, because there is only one set of cross laminae in the middle of the sequence. A single bed could not form from a series of large forms migrating. Such forms would produce more than one bed of cross laminae. Instead, it must represent a large front, almost like a delta, migrating onto the beds below. Sorby, in his 1859 work, described just such a deposit. He saw a deltalike bar migrating into deeper water and producing cross laminae where the velocity decreased rapidly.

The Platte River forms described by Crowley are considerably more complex than those described by Sorby, but the same processes occur. The shallower water on the backs of the bars have a higher velocity and so large ripples migrate downstream along the back of the bar and deposit the smaller sets of complex cross laminae on the top of the bar. As sediment is carried to the front of the bar, where the velocity decreases rapidly (because of the depth change), depo-

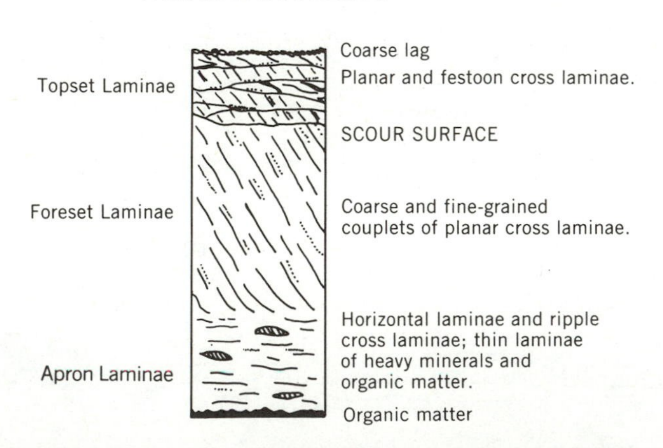

FIGURE 5.16 Generalized stratigraphic sequence generated by the migration of bars in the Platte River. From K. D. Crowley, 1983, *Geological Society of America Bulletin*, 94, Fig. 10, pp. 126–127. Reprinted by permission of GSA and Kevin D. Crowley.

sition occurs and forms a slipface. As a bar moves downstream the slipface deposits one set of steep, planar cross laminae. The lower sequence results from low-velocity currents in front of the bar reworking sediment deposited from suspension as it is carried over the bar front by the strong currents. We can use the principles of fluid flow and some flow regime ideas to decipher this sequence

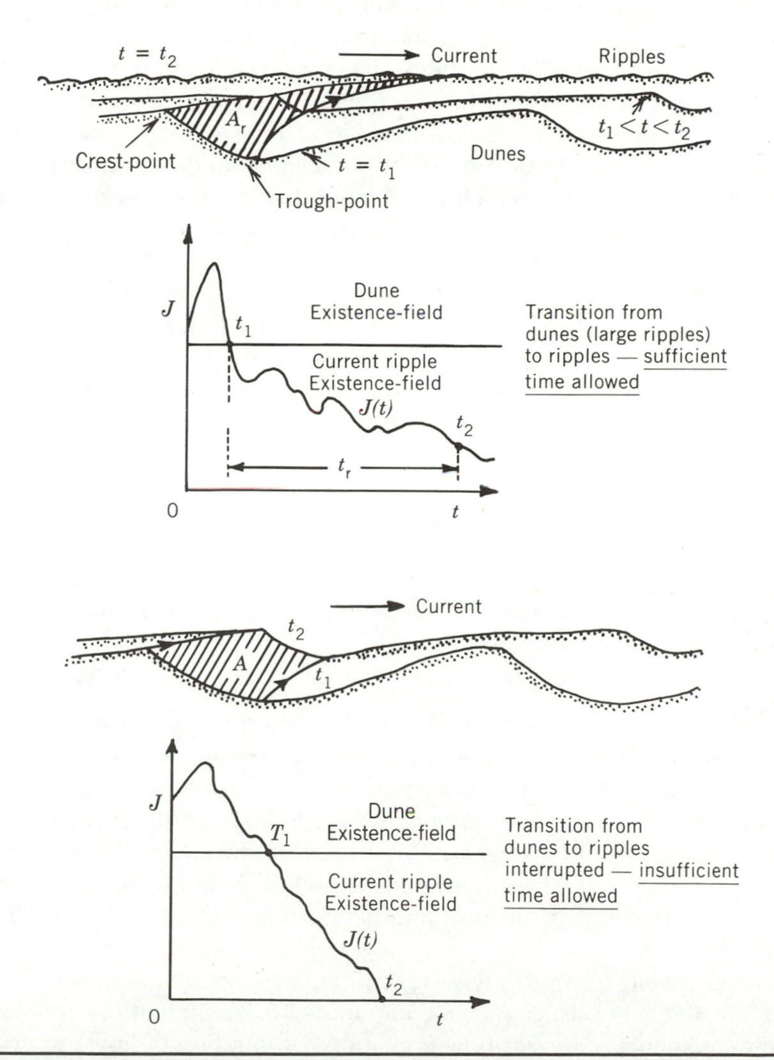

FIGURE 5.17 Diagrams of the time required to transform a bed of one type into another configuration (relaxation time). Reproduced by permission of the Geological Society from J. R. L. Allen and P. F. Friend, 1976, *Journal of the Geological Society of London*, 132, Fig. 1, p. 19.

but we cannot apply the flow regime model directly. The moral of this story is: Be careful of models, they are just that and rarely depict the complexities of natural systems.

Another limitation of the flow regime concept involves the length of time it takes a particular flow to modify its bed. Some excellent work was done on this problem by Allen and Friend (1976) in intertidal dunes (large ripples) off the coast of Norfolk, England. Their work defines the **relaxation time** of large ripples, the time it takes to change them into another type of form. They found that in the tidal areas near Norfolk, the tides changed so that the flow shifts from the large-ripple field at high flows to the ripple field at low flows. But when they examined the bedforms in the harbor they found them to be "permanent features." In other words, they did not respond to the change in flow regime. The reason that the bedforms did not match the flow is that the tidal current did not last long enough to modify the dunes into ripples once the dunes had formed. Because so much sediment must be moved to fill in the troughs of the large ripples (Fig. 5.17), the lower velocities of short duration could not modify the bed. Thus, if we were to examine the sequence produced by the migrating dunes we would see only one flow regime field and miss entirely the lower flow regime ripple field. This is an excellent example of how sediment is deposited mostly by extreme events and not by the everyday processes.

EPILOGUE

However, by carefully attending to minute facts in its structure, it appears almost certain that the actual depth of the water can in many cases be ascertained to within a fathom.

H. C. Sorby, 1855

This chapter has developed the basics of using the flow regime concept as a powerful tool for interpreting bedforms and sedimentary structures. By applying this model, much can be determined about paleoflow conditions given known grain size and bedform type. Although Sorby's hope of determining depth to within a fathom may, in many cases, be overly optimistic, he clearly was one of the first to recognize the relationship between flow conditions (water depth and velocity), grain size, and the resulting bedform and to use this information to interpret rocks.

However, when using the flow regime concept, remember that it is only a model, not the thing itself, and like any model it has limitations. For example, assumptions such as constant flow conditions or duration of flow are not always valid and can lead to erroneous interpretations. Also, because of the mechanical problems associated with large flumes and high velocities, few experimental data exist for grain sizes larger than coarse sand or for grains of varying shape and densities such as bioclastic sediment. To date the flow regime concept applies only to water. It has been suggested that a similar approach can be used to interpret other fluids such as air, hot gases in volcanic base surges, and others. However, much experimental work must be done before rocks deposited by these processes can be interpreted using the flow regime method.

OUTSIDE READING

Allen (1963b, 1982, 1985a, b); Allen and Friend (1976); Blatt *et al.* (1980); Bogardi (1974); Crowley (1983); Friedman and Sanders (1978); Guy *et al.* (1966); Harms (1969); Harms, *et al.* (1982); Hunter (1977); Jopling and Richardson (1966); Leeder (1982); McKee (1965); Middleton (1965a, b, 1977); Reineck and Singh (1980); Rubin and McCulloch (1980); Rubin and Hunter (1982); Selley (1982); Simons *et al.* (1965); Skipper (1971); Sorby (1859); Southard (1971).

Chapter 6

Transport and Deposition by Wind

ENTRAINMENT AND TRANSPORT

The wind blows across every part of the earth, and in areas where the surface soil or sediment lies unprotected, strong winds transport sediment. Sand bounces along the surface and rarely reaches heights greater than a meter, but dust travels to thousands of meters above the earth's surface. Unlike water confined in channels, flows of air are extremely deep and highly variable both vertically and laterally. However, the processes that entrain and transport sediment in moving air are somewhat like those we have discussed for flowing water. The most obvious differences between water and air are the density and viscosity of the two fluids. Air is some 50 times less viscous than water (Fig. 4.1) and nearly 1000 times less dense (1 g/cc versus 0.001 g/cc). The viscosity of air is only weakly controlled by temperature (Fig. 4.1), and the density is controlled by both temperature and pressure (Fig. 6.1). In spite of these variations, in most situations at the earth's surface involving sediment transport by air, density and viscosity never vary more than a few percent and are relatively unimportant in understanding sediment entrainment by wind.

We owe our present understanding of transport and deposition of sediment by wind almost entirely to R. A. Bagnold, who conducted experiments in a wind tunnel and monitored the movement of sand on dunes in the Libyan desert. The details of his work were presented in 1941 and later again in his famous text on the physics of wind-blown sand. We will rely on Bagnold's ideas to examine the

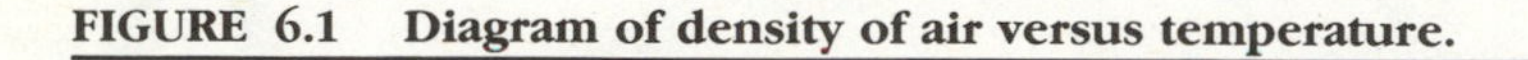

FIGURE 6.1 Diagram of density of air versus temperature.

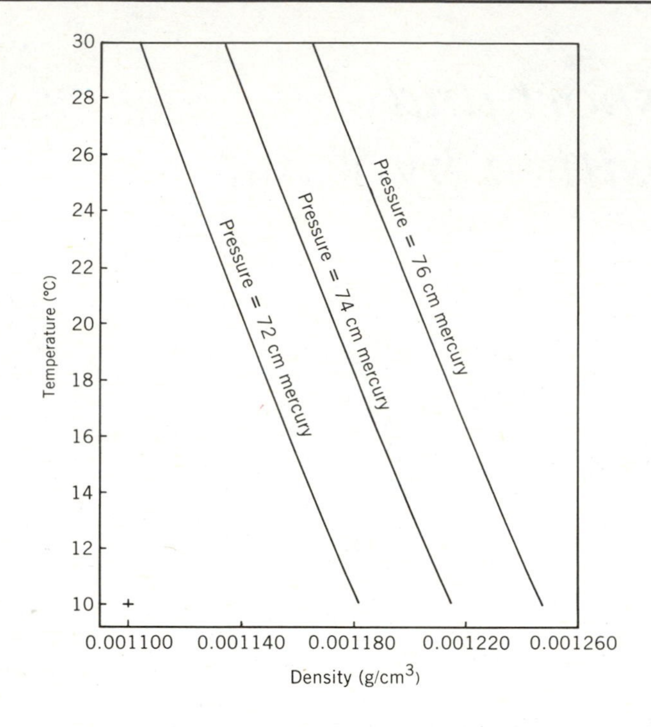

processes of transport and deposition of sediment by wind. Let us begin by examining the differences between deposition of sediment in water and in air.

Returning to Stokes' law of settling (Chapter 4), we see that settling velocity is increased by decreasing the density and the viscosity of the fluid through which a grain settles. For example, a grain of quartz 0.1 mm in diameter would fall through water at approximately 0.9 cm/sec according to Stokes' law. However, the same grain falling through air will fall 70 times faster, at 60 cm/sec. For a distribution of sizes from silt to pebbles, the settling velocities through air show a tremendous range (Fig. 6.2). These differences control sediment movement and deposition by wind in major ways. Using the same arguments we used to understand the entrainment of sediment by water, we would expect that a much higher velocity would be needed to transport sediment of a particular size and shape by wind than by flowing water. This conclusion results from the much lower density and viscosity of air compared to water. Bagnold (1954) showed that these conclusions are correct. The wind velocity required to move sediment of an increasing size does not increase as a linear function but rather as a power function very similar to that for moving water (compare Fig. 6.3 to Fig. 4.12). The easiest sediment to move by wind is very find sand, and both coarser and finer material require a higher velocity for entrainment. Silt requires a velocity

FIGURE 6.2 Plot of settling velocity of grains of different sizes in air.

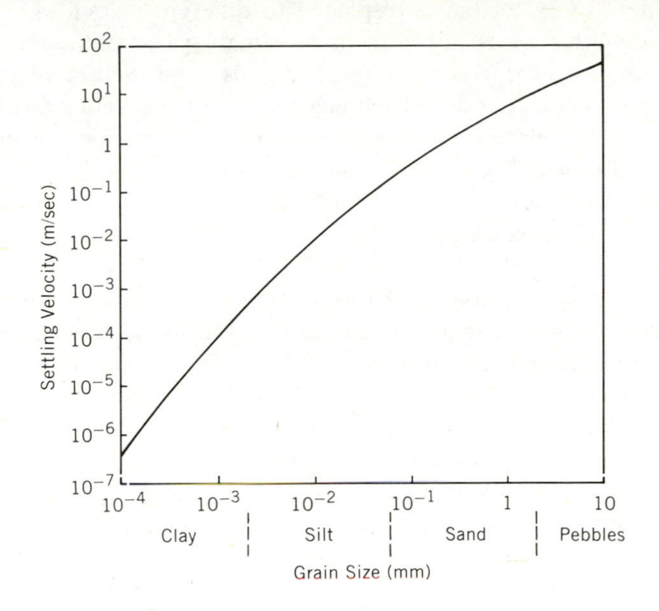

FIGURE 6.3 Plot of wind velocity required to move sediment of different sizes. (After Bagnold, 1954.)

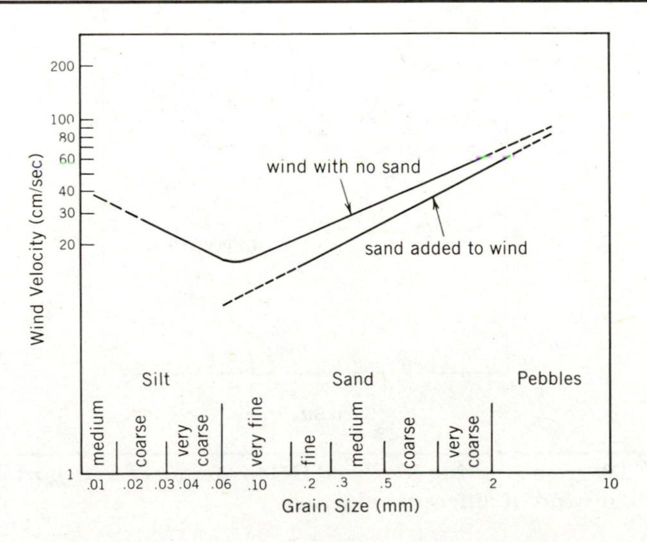

equivalent to that needed to move coarse sand. This curve parallels that of sediment movement by water but is displaced to different velocities. The reasons why higher velocities are required to move silt than very fine sand are similar to those we discussed for flowing water. For beds composed of very fine grains (silt and clay), the surface is aerodynamically smooth and hence protected from the turbulence of the air. For coarser-grained beds, the surface is rough, causing turbulence on the bed. The critical grain size that disrupts the smooth flow is approximately that of very fine sand. Very fine sand moves at wind velocities of only 15 cm/sec, whereas medium silt and coarse sand require velocities of 30 to 40 cm/sec.

In sediment entrainment by wind, the act of grains colliding with other grains on the surface effects sediment movement. Bagnold found that when he added sand to the wind by letting it fall from the ceiling of a wind tunnel, the entrainment velocity was lower than for air free of sand (Fig. 6.3). Once entrained, different-sized grains act very differently because of their differing settling velocities. The finer grain sizes (silt and clay, i.e., **dust**) move directly into suspension. Wind has great amounts of turbulence because of its low viscosity and high velocities (remember the R_e). The upward components of turbulence for wind of even very low velocity are greater than the settling velocity of dust so that dust is kept

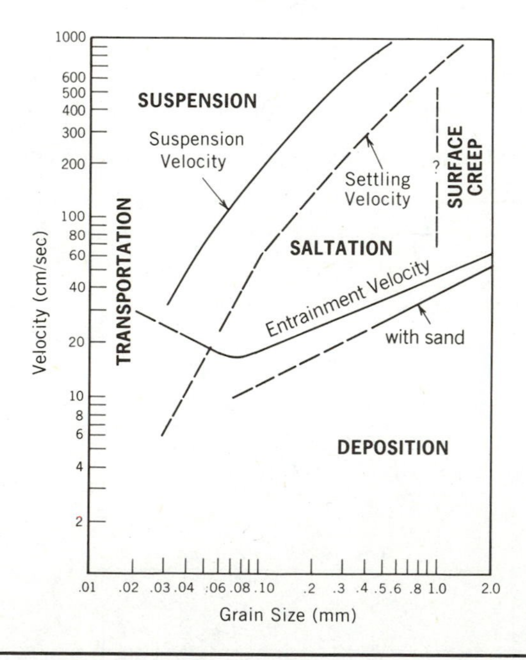

FIGURE 6.4 Diagram showing idealized fields of type of transport for grains of different sizes in wind of different velocities.

in suspension. Bagnold found that the upward eddy velocity was approximately one-fifth of the average wind speed, so that if the wind blows at five times the settling velocity of a particle, the particle will stay in suspension by the wind (Fig. 6.4). Because dust has a settling velocity in air of less than 10–20 cm/sec, a wind of only 50–100 cm/sec (18–36 km/hr) will keep the coarsest dust in suspension. Finer dust can be maintained in the atmosphere by winds of less than a few kilometers per hour. Because of this relationship, turbulence often transports dust to great altitude. Updrafts in storms can reach hundreds of kilometers per hour and transport dust to many thousands of meters above the earth's surface and many hundreds of kilometers from the origin of the dust. Dust originating from the Sahara Desert is routinely collected by ships hundreds of kilometers from shore in the Atlantic Ocean, and it is estimated that from 1 to 4 million tons of dust are removed from the Sahara each year.

Wind does an excellent job of winnowing the sediment, that is, sorting out the fines from sediment and leaving behind the coarser-grained material. As dust is carried away in suspension, sand moves along by saltation, and the coarsest grains, those larger than granules, are left behind as lag deposits (Fig. 6.5). Gravel concentrated in the lag deposits often forms a tight, interlocking layer called desert pavement that protects the underlying sediment from erosion by the wind. When the wind reaches the critical velocity to move sand, grains are entrained

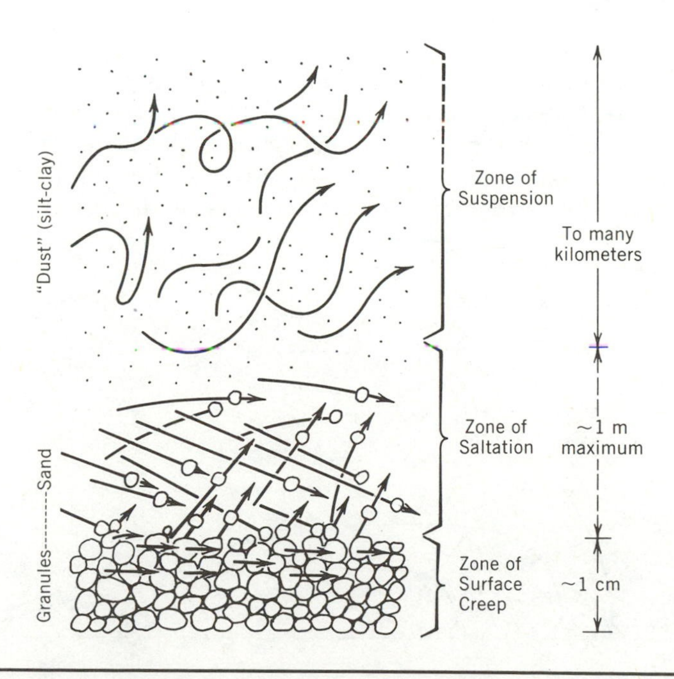

FIGURE 6.5 Mode of transport of sediment by wind.

into the flow by the turbulence and aerodynamic lift over the grains. In his wind tunnel experiments and observations, Bagnold established the paths that these grains take once they are removed from the bed. When movement of grains first occurs, grains rise nearly vertically into the wind and then are carried downwind in a broad arc (Fig. 6.6). If the surface is composed of well-sorted sand, then saltation paths are low, but if an irregular surface is present, grains bounce very high and in extremely random directions. As the grains hit the surface they may dislodge other grains. The combined effect of the wind shear over the bed and the impact of grains forms a saltation blanket of sand traveling rapidly downwind at nearly the speed of the wind near the bed (Fig. 6.5). Because the viscosity and density of air are so low, grains larger than about very coarse grained sand are rarely moved by saltation.

Larger grains can be transported, but their movement is not the direct result of wind shear. During saltation sand grains bombard the surface at very low angles. The force of these myriad impacts is transmitted to the grains lying within the surface that are too large to be entrained by the wind directly. Sand on the bed may be jolted into saltation by the blows but larger grains are just jostled along. Only the upper few millimeters of sediment are affected by this **surface creep** (Fig. 6.5). Wind processes, like those of flowing water, separate different sizes of sediment. Wind is extremely effective at this process so that distinctly different deposits form by suspension and saltation.

Suspension Deposition

If a large supply of silt is available in windy areas, dust will accumulate to form deposits of **löess.** Löess deposits are often associated with glaciers because huge amounts of fines produced by the grinding action of glacial ice (so-called "rock flour") are transported to outwash plains by glacial meltwater. Katabatic winds, caused by dense, cold air flowing down off glaciers and ice caps, blow regularly

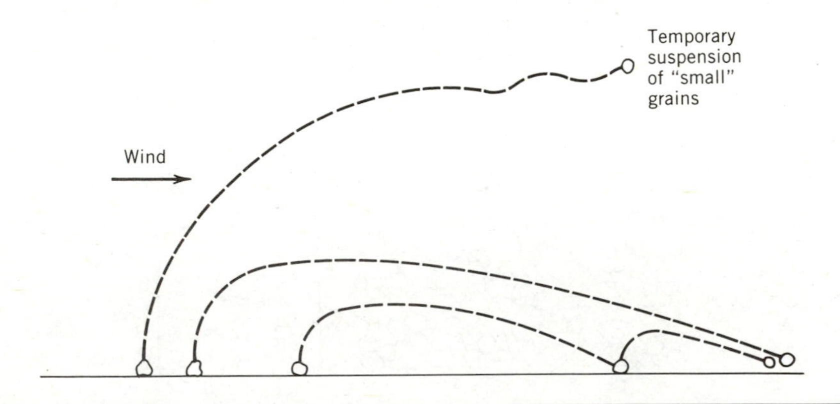

FIGURE 6.6 Paths of sand grains moving by saltation in wind.

FIGURE 6.7 Deposits of löess. (Photograph by R. D. Miller, #84, U.S. Geological Survey).

across the outwash plains and remove the rock flour produced by the glaciers. As the wind dies, the dust settles out of the atmosphere and covers the landscape with a thick blanket of silt. Because loess accumulates from dust settling from suspension, and never moves as bedload, the deposits are very massive (Fig. 6.7). Successive layers deposited from different wind events may result in indistinct layering, but usually the deposits form widespread, massive blankets. Loess deposits are equivalent to suspension sedimentation when flowing water moves silt and clay into standing water. Although dust accompanies wind wherever it moves sediment, loess accumulates only in areas where a huge dust source is available. Wind-blown silt is a common, but very minor, component of many sediments but is rarely detectable unless it forms thick deposits of löess.

Bedload Transport and Deposition

Because the bedload transport of sand by wind is similar to that of flowing water, similar bedforms are created; however, because of the large differences in viscosity and density, the mechanisms are somewhat different. In flowing water we saw that the morphology and size of a bedform depended on the velocity, depth of flow, and the grain size of the bed material. In wind systems, the grain size of

FIGURE 6.8 Wind ripples on modern coastal dune, backshore of Sapelo Island, Georgia. (Photograph by W. J. Fritz, 1983).

the bedload is nearly constant because of the narrow range of sizes that air can move. Nearly all bedload deposits from wind range in grain size from fine to medium sand. Coarser material generally accumulates as lags or moves by surface creep. Because sand moves by the same mechanism at very different wind velocities (i.e., by saltation), velocity has only a slight control on bedform morphology. Also, because wind originates from global or regional pressure gradients, the depth of flow is meaningless in processes of sediment movement at the surface. Thus the factors controlling bedform size and shape are very different in wind deposits. Only for small bedforms (wind ripples) is there an association with wind velocity and grain size.

Wind ripples are very common where sand is transported by saltation, and they form nearly contemporaneously with the first sand movement. They are asymmetrical and have fairly continuous crests (Fig. 6.8). For ripples forming in well-sorted sand, the wavelength closely approximates the average saltation distance; however, the standard deviation of a set of wind ripples shows a range of wavelengths. The standard interpretation of the formation of wind ripples was also developed by Bagnold. He correlated the saltation path with ripple wavelength and suggested that increases in wind velocity caused grains to bounce farther and so form longer-wavelength ripples (Fig. 6.9). However, because wind ripples commonly show a broad range of wavelengths and the saltation distance of grains at a particular wind velocity tends to be spread over a broad range as well, Bagnold's theories do not precisely describe the mechanisms that form wind

FIGURE 6.9 Formation of wind ripples. (After Bagnold, 1954.)

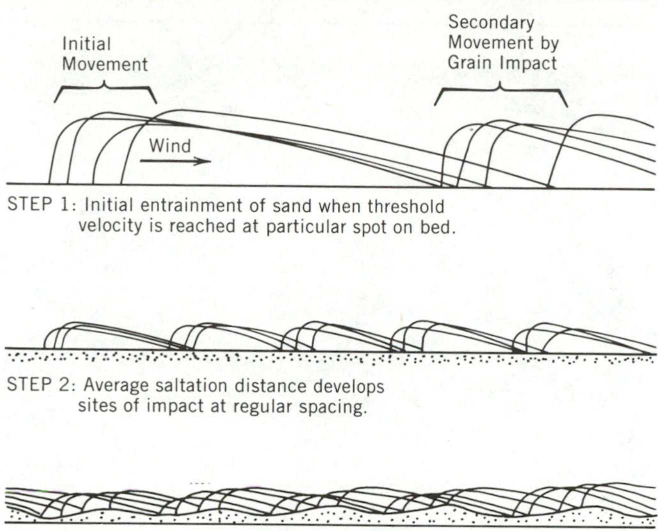

ripples. It is likely that mechanisms similar to those that control the formation of ripples in water also control the formation of wind ripples.

Initial
Movement
Secondary
Movement by
Grain Impact
Wind
STEP 1: Initial entrainment of sand when threshold velocity is reached at particular spot on bed.
STEP 2: Average saltation distance develops sites of impact at regular spacing.
STEP 3: Ripples form at sites of impact as sediment is moved by surface creep and saltation.

More wind tunnel experiments and observations in the field are needed to solve this sedimentologic mystery. In sediment where a wide range in grain size is available, ripple height and length depend on grain size as well as velocity, but in all cases ripple length rarely exceeds 25 cm. Wind ripples tend to have low relief and a much flatter profile than ripples formed by flowing water. Wind ripples will not form in coarse-grained material (very coarse sand and granules), but another bedform, called ridges by Bagnold and granule ripples by Sharp (1963), that resembles sand ripples does form.

Granules ridges are much larger than wind sand ripples but have a similar shape (Figs. 6.10 and 6.11). (Bagnold found granule ridges that were 20 m long and 60 cm high.) They have a concentration of granules on the crest of an asymmetrical form. The granule crests apparently move by surface creep and the sand by saltation; however, there are granule ridges in Wyoming that clearly form by saltating grains during high-velocity winds. Granule ridges have a much more irregular shape than ripples, can be sinuous or cuspate in shape, and approximate the morphology and size of smaller and simpler sand dunes.

Sand transport by wind is not nearly as well quantified as that for flowing water, for which we can use the flow regime model. But we can see similar trends

FIGURE 6.10 **Granule ridges, Eureka Valley, California. (Photograph by J. N. Moore).**

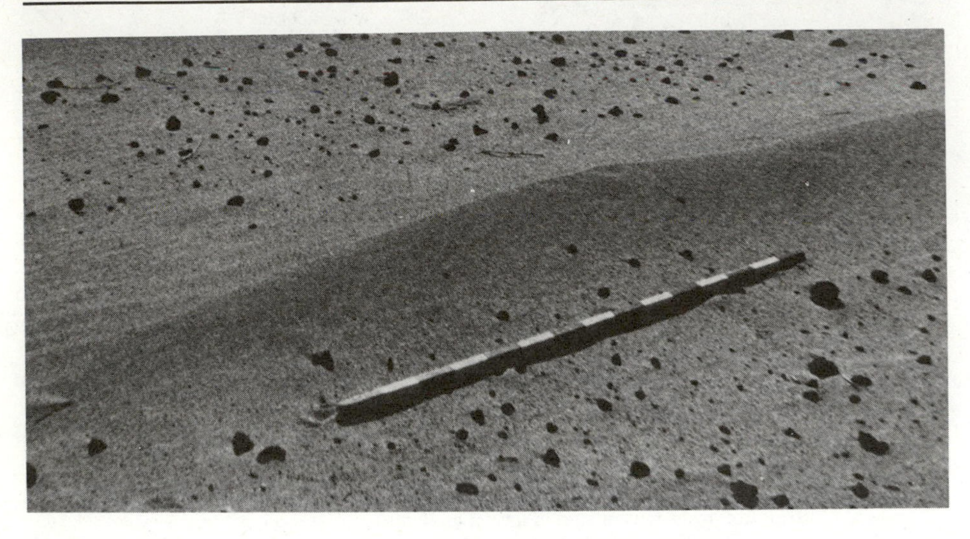

FIGURE 6.11 **Formation of granule ridges. (After Bagnold, 1954.)**

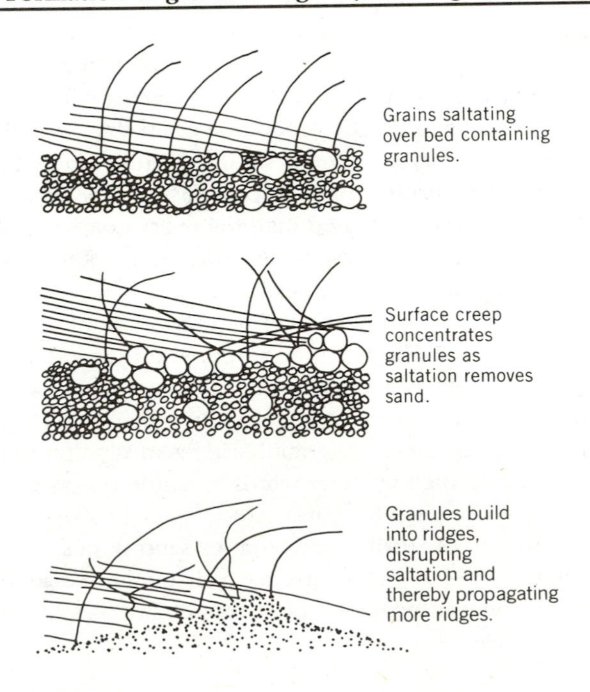
Grains saltating over bed containing granules.
Surface creep concentrates granules as saltation removes sand.
Granules build into ridges, disrupting saltation and thereby propagating more ridges.

in the data gathered by Bagnold (1954), and we can use it as a conceptual flow model for the formation of ripples and other small features. Dunes depend on factors other than grain size and wind velocity for their morphology and, before we examine these processes, let us try to synthesize the rather diverse information on sand transport presented above.

If the bed over which the wind blows is well-sorted, very fine sand, the first bedforms to develop when the threshold velocity is reached are small, very low relief, nearly symmetrical ripples (Fig. 6.12). If the velocity is increased to approximately three times the threshold velocity, the ripples are replaced by a flat bed. If given enough space these flat beds would probably form dunes. Because the experimental work has been conducted only in wind tunnels, this transition has not been observed because of the lack of space and limited sand supply. Within the ripple field, for any constant grain size, the ripple wavelength increases with increasing wind speed. As the grain size is increased and a larger range in grain size is considered, the ripples become relatively taller compared to their length. In general, if the wind speed is increased over a bed of larger grain size, the ripples become larger and taller than those formed in fine sand and under low-velocity winds. If a mixture of sand and granules makes up the bed over which the wind is blowing, then granule ridges form. These are larger than ripples and require somewhat higher wind velocities than ripples alone. The flat bed

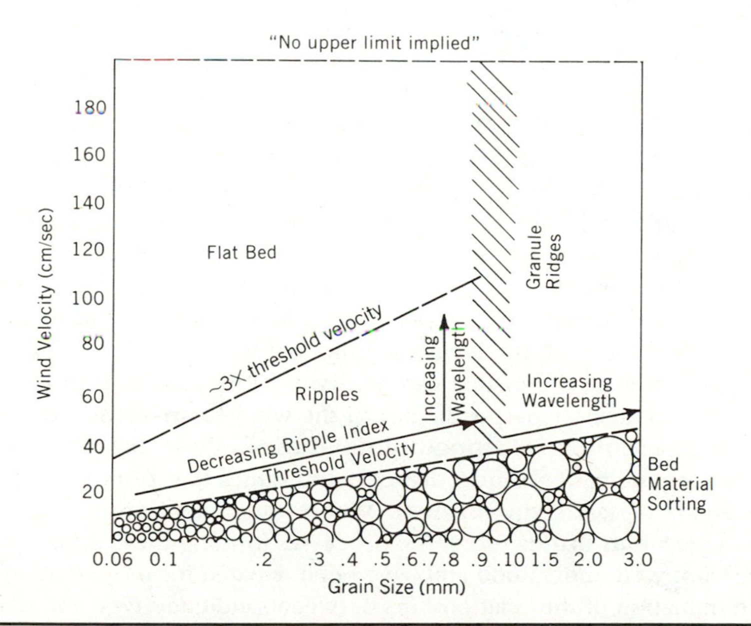

FIGURE 6.12 Idealized diagram of type of bedform under different wind velocities and size and sorting of bed material.

FIGURE 6.13 Idealized diagram depicting the formation of dunes.

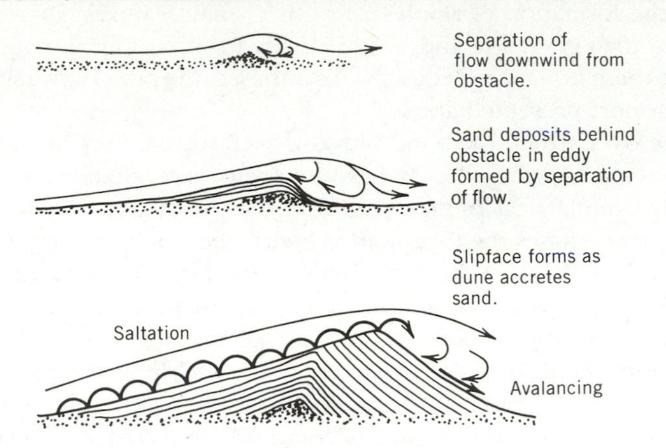
Separation of flow downwind from obstacle.
Sand deposits behind obstacle in eddy formed by separation of flow.
Slipface forms as dune accretes sand.
Saltation
Avalancing

field extends into the coarsest-grained sand but probably does not form if the bed material contains a large fraction of granules. Let us now turn to sand dunes and examine the processes controlling their formation and the resulting internal structures.

When wind-blown sand accumulates in forms much larger than ripples, the deposits are termed sand dunes or eolian dunes. The smallest dunes have crests only 30 cm high, whereas the largest coalesce into complex mountains of sand hundreds of meters high. The supply of sand and the strength, duration, and direction of wind control the morphology and size of sand dunes rather than detailed flow characteristics. Dunes can form wherever an obstruction causes a separation of the wind and a wind shadow can form (Fig. 6.13). Such embryos of sand dunes are often bushes, rocks, or other natural obstructions (Fig. 6.14). As the dune builds in height, sand migrates up the windward side and slides down the lee side to form a slipface. Once the dune is large enough to affect the wind, it will continue to build until it comes into equilibrium with the wind system. Its shape and size depend on the amount of sand available, the direction and strength of the wind, and the length of time that it blows.

These factors form dunes that can be divided into four general types: dunes that form approximately perpendicular to the wind **(transverse dunes),** those that nearly parallel the wind **(longitudinal dunes),** those that are neither transverse nor longitudinal **(oblique dunes),** and dunes that form large, pyramidal accumulations of sand **(star dunes).** Where sand supply is limited, isolated, crecentric **barchan dunes** form. The processes forming these different types of dunes are not well understood and have been debated for more than 100 years. The determination of the relationships between sand-dune type and wind processes is one of the most important unsolved mysteries of eolian sedimentology. For a discussion of some of the present ideas, see texts and papers by Dunbar

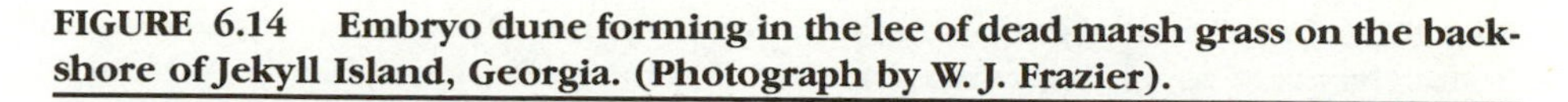

FIGURE 6.14 Embryo dune forming in the lee of dead marsh grass on the backshore of Jekyll Island, Georgia. (Photograph by W. J. Frazier).

and Rogers (1957), Reineck and Singh (1980), Pettijohn *et al.* (1972), Friedman and Sanders (1978), Scholle and Spearing (1982), Galloway and Hobday (1983), and Reading (1986).

One feature that is common to most of these dune types is large compound cross laminae, which form in response to migrating, superimposed dune forms of various sizes. These dunes can accrete vertically, migrate laterally, or erode other dune forms and result in complex scour and fill patterns and large, climbing cross-laminae (Fig. 6.15). These complex cross strata form when smaller dunes migrate along slipfaces of larger dunes. Such cross bedding and its origin have been detailed in papers by McKee (1966, 1979), Hunter (1981), and Rubin and Hunter (1982, 1983).

In many ancient wind-blown sand deposits, the scale of the cross bedding is colossal (Fig. 6.16) as a result of the grand size of the bedforms themselves. Even relatively small transverse dunes often reach heights of tens of meters, so they have the potential of forming cross bed sets that approximate that height. Large longitudinal and pyramidal dunes often reach heights greater than one to two hundred meters and contain a myriad of smaller dunes (Fig. 6.17). The well-sorted sand composing these complex and often immense cross-bedded deposits

FIGURE 6.15 Detail of cross lamination in sandstone deposited in eolian dunes. Pennsylvanian Casper Formation, Laramie, Wyoming. Photograph by J. N. Moore.

FIGURE 6.16 Large eolian cross beds (tree in center of photo is approximately 6 m tall). Zion National Park, Utah. (Photograph by J. N. Moore).

FIGURE 6.17 Complex large and small eolian dune forms on desert terrain in Saudi Arabia. (Arabian American Oil Company).

is a telltale characteristic of wind-blown sand that makes them difficult to confuse with those formed by flowing water. These deposits form some of the most spectacular sedimentary structures in the geologic record.

EPILOGUE

The wind goeth toward the south, and turneth about unto the north; it whirleth about continually, and the wind returneth again according to its circuits.

Ecclesiastes 1:6

Eolian sedimentology is like a dervish, swirling around new ideas and concepts. The processes that form the large, extremely complex cross bedding in ancient sand dunes are finally being deciphered, tools have been developed to distinguish wind-laid deposits from those formed in water, and sedimentary structures can be used to determine the size and shape of ancient dunes and dune fields, but many mysteries remain. How do wind ripples really form? What processes control the formation of granule ridges? What are the controls on the types of dunes that form in different areas? Is there a flow regime concept that can be developed that encompasses all types and shapes of eolian sedimentary structures? These questions, and many others, will occupy present and future sedimentologists, and their answers will lead to a much better understanding of earth history and paleogeography. The basis for this renaissance is recorded in papers by many sedimentologists (see additional reading below). The exposures of ancient sand dune deposits in the Colorado Plateau of the southwestern United States, and similar areas throughout the world, offer a vast natural laboratory to test this new methodology. The mysteries of these ancient sand dunes await the well-equipped sedimentologist, who now includes computer modeling in his or her tool kit. The concepts presented above are no more than a schematic of this vibrant field, but enough, we hope, to start you on your way to deciphering sediments deposited by wind.

OUTSIDE READING

Anderson (1986); Bagnold (1941, 1954); Brookfield and Ahlbrandt (1983); Dunbar and Rodgers (1957); Friedman and Sanders (1978); Fryberger (1986); Galloway and Hobday (1983); Hunter (1981); Kocurek and Dott (1981); McKee (1966, 1979); Pettijohn *et al.* (1972); Reading (1986); Reineck and Singh (1980); Rubin and Hunter (1982, 1983); Scholle and Spearing (1982); Sharp (1963); Smalley (1975).

Chapter 7

Transport and Deposition by Oscillating Currents

INTRODUCTION

Oscillating currents produced by tides and waves are common movers of sediment. Currents produced by tides form periodically opposing, unidirectional flow and create distinct structures because of the long period of oscillation. Waves are a different matter. Oscillating currents produced by waves are significantly unlike tidal flow because the current changes direction and magnitude (oscillates) over periods of seconds instead of the hours required for tidal currents to reserve direction and change velocity. These rapid oscillations generate particular mechanisms of sediment movement and deposition that differ significantly from those occurring during unidirectional flow.

Our understanding of oscillating currents and the deposits formed by them depends heavily on the work of engineers, sedimentologists, and oceanographers. Experimental work on waves and oscillating currents conducted by Inman and Bowen (1963), Carstens *et al.* (1969), Harms (1969), Bliven *et al.* (1977), Komar (1974), Dingler (1974), Dingler and Inman (1977), Lofquist (1978), Miller and Komar (1980a), and many others has led to an excellent understanding of the processes forming wave-generated bedforms. Other authors, including Bagnold (1946), Inman (1957), Clifton *et al.* (1971), Clifton (1969, 1976), J. R. L. Allen (1979), Miller and Komar (1980b), P. A. Allen (1981), and others have developed the constructs of sedimentary processes and bedforms in the field. Many of these ideas and data are presented in excellent summaries by Harms *et al.* (1982), Clifton and Dingler (1984), and Allen (1985a).

Tidal processes and structures have been presented by many authors studying modern and ancient tidal depositional systems, including van Straaten (1954), Evans (1965), Reineck and Wunderlich (1969), Reineck (1972), Knight and Dalrymple (1975), and Klein (1977). These concepts are presented in a compilation edited by Ginsburg (1975) that describes recent and ancient tidal deposits. We will try to collate this diverse information and briefly discuss the processes and bedforms generated by waves and tides.

WAVE-GENERATED CURRENTS

When waves are generated by storms in the open ocean or large lakes, they move away from the storm center and travel unrestricted as deep-water swells. These swells can travel thousands of kilometers without a significant decrease in energy. They have no effect on sediment movement because in deep water they do not interact with the bottom over which they pass. Deepwater swells have a broad, approximately sinusoidal shape and water particles in the waves follow circular orbits with no net forward motion of the water. As they move into shallow water, the waves become asymmetrical, the wave height increases, the velocity decreases, and the crests steepen and become peaked until they finally oversteepen and break. The waves start their transformation from the deep-water forms to breaking waves when they enter water that is approximately one-half their wavelength in deep (Fig. 7.1). If the ratio of depth to wavelength is greater than 4 to 1, the bottom has absolutely no effect on the wave, and the wave is termed a **deep-water wave.** Even if the wave does not begin to break, when the depth to wavelength ratio reaches values of less than 0.05 to 1, the wave is strongly controlled by the bottom and is termed a **shallow-water wave.** Shoaling flattens the circular orbits of the water particles as the wave builds in height. When the wave height reaches approximately 80% of the water depth, the wave breaks, spilling water shoreward. At that point the water movement is translated from orbital motion to unidirectional pulsating flow, with water in the upper part of

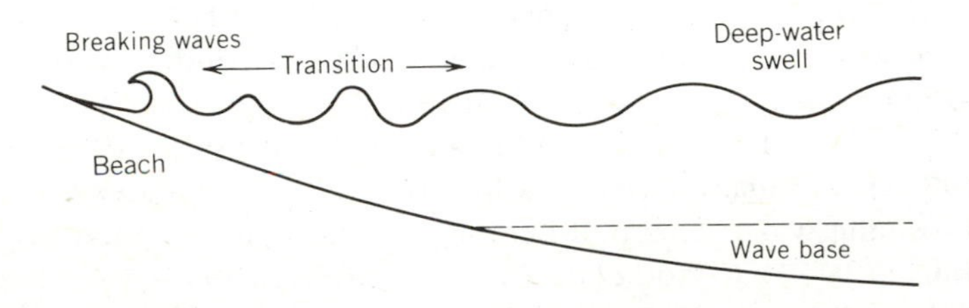

FIGURE 7.1 Shapes of water waves in deep and shallow water. Note the increase in height and decrease in wavelength in shallow water as waves "bunch up" on the shore.

FIGURE 7.2 Complex spectrum of a combination of simple waves.

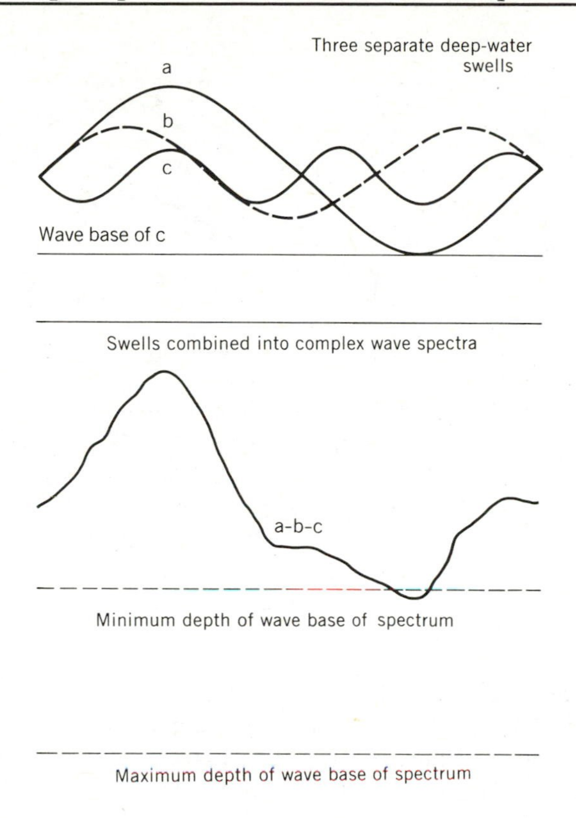

the wave moving shoreward and the lower part returning away from shore under the breaking wave. To understand the processes and deposits generated by waves, we must examine the processes acting where waves begin to "feel bottom" (the transition from deep-water swells to breaking waves) and those generated by breaking waves.

Deep-water swells are composed of many wave trains and form complex wave spectra that deviate significantly from the ideal sinusoidal shape of an isolated, theoretical wave train. Because the orbital motion of water within the wave extends to a depth of one-fourth the wavelength of a single wave train, for any particular spectra of waves there is a complex arrangement of depths where the bottom will be affected by the waves (Fig. 7.2). This generally assures that when real waves move into shallow water, the water movements are very complex and the wave base is actually a zone that changes position through time depending on the wavelength of the wave trains. Just as we used simplified systems to develop the concepts of unidirectional flow, we will first ignore this complexity of real waves and concentrate on fairly simple flow systems.

FIGURE 7.3 Terminology used in defining oscillatory flow.

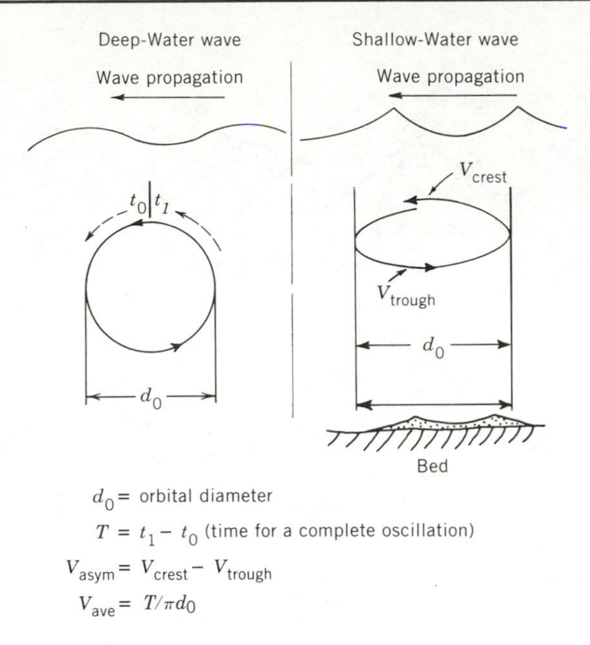

Wave-generated oscillatory flow in deep water can be described by several parameters (Fig. 7.3). The time it takes for a complete orbit is the **period of oscillation** (T) and the difference between extreme positions within the orbit is termed the **orbital diameter** (d_o). Velocity of flow within the orbit can be defined by the maximum orbital velocity, or the average **orbital velocity,** which is the circumference of the orbit (πd_0) divided by the period:

$$v_{ave} = \pi d_o / T.$$

When the waves move into shallow water and the circular orbit is deformed, this terminology must be modified. Above the bed, the water moves in an elliptical path (Fig. 7.3) and the orbital diameter is the maximum diameter of the ellipse. However, because the wave is steepening and leaning shoreward as it begins to build toward breaking, the movement does not precisely close the ellipse (Fig. 7.3). The average velocity can still be defined as for deep-water waves, but a distinct velocity asymmetry has formed. In fact, on the bed, the elliptical motion is replaced by alternating shoreward and seaward horizontal motion (Fig. 7.3). In this case the velocity asymmetry is defined as the difference between the velocity under the crest (directed shoreward) and that under the trough (directed seaward). This asymmetry develops because as the wave feels bottom it changes shape. The deep-water sinusoidal form is replaced by one of peaked crests and

FIGURE 7.4 Shallow-water wave shape and velocity asymmetry.

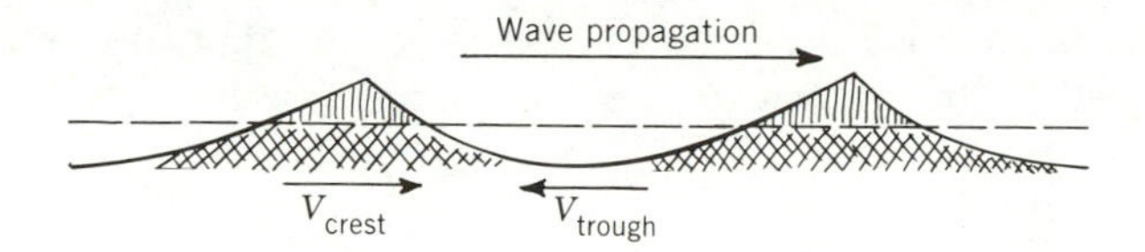

broad, rounded troughs (Fig. 7.4). If we pass a line through the wave that represents the mean water level, halfway between the crest and trough, we see that more water lies below the line than above. In waves, the volume of water that moves toward the shore approximately equals the volume moving seaward. As the crest moves over a point it must move faster to balance the larger, and slower, flow under the trough. This velocity difference between the crest and trough can be expressed as a velocity asymmetry, which is the velocity under the crest (v_{crest}) less the velocity under the trough (v_{trough}). The measured difference between these two velocities is the velocity asymmetry:

$$v_{asym.} = v_{crest} - v_{trough}.$$

These various parameters of waves, that is, period of the oscillation, orbital diameter, and orbital velocities, are determined by the height, length, and period of the wave and the depth of water. Period is directly related to wave length: the larger the wavelength, the longer the period. The velocity asymmetry is controlled by the asymmetry of the wave resulting from building and spilling during breaking or a mixture of waves forming complex wave spectra. When a sinusoidal, shoaling wave first feels bottom, the orbits are nearly circular, so there is little velocity asymmetry. As the wave moves into shallower water the orbit flattens and the difference between v_{crest} and v_{trough} increases. When the wave actually breaks, the already flattened orbit is transformed into pulsating flow (Fig. 7.5), with water in the upper part of the wave moving shoreward and that in the lower part

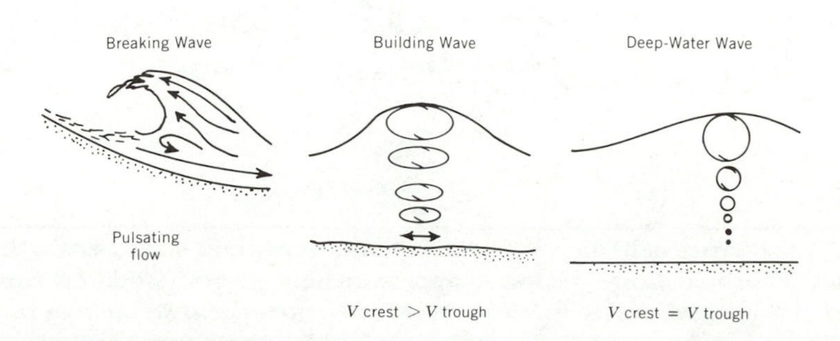

FIGURE 7.5 Transformation of oscillatory to pulsating flow as a wave shoals.

moving seaward. These different types of flow affect the bed in very different ways that lead to a significant variety of bedforms that are characteristic of these processes. We will examine the deeper-water processes first and then turn to the processes formed in the shallow areas where waves break.

BED CONFIGURATIONS GENERATED BY WAVES

Let us start our examination of wave-formed structures by examining a velocity versus grain size plot similar to that used in unidirectional flow (refer to Fig. 5.2 for comparison). If we start with a smooth bed and then initiate a symmetrical, oscillatory flow of a particular period over the bed (pass a wave of a particular wavelength over the surface), grains start to move at a threshold maximum velocity dependent on grain size (Fig. 7.6). At initiation of movement, ripples form on the bed. These first ripples are very low relief small forms called **rolling grain ripples** because they form by the action of grains rolling to-and-fro on the bed. These forms have been seen only in experiments and are not observed in

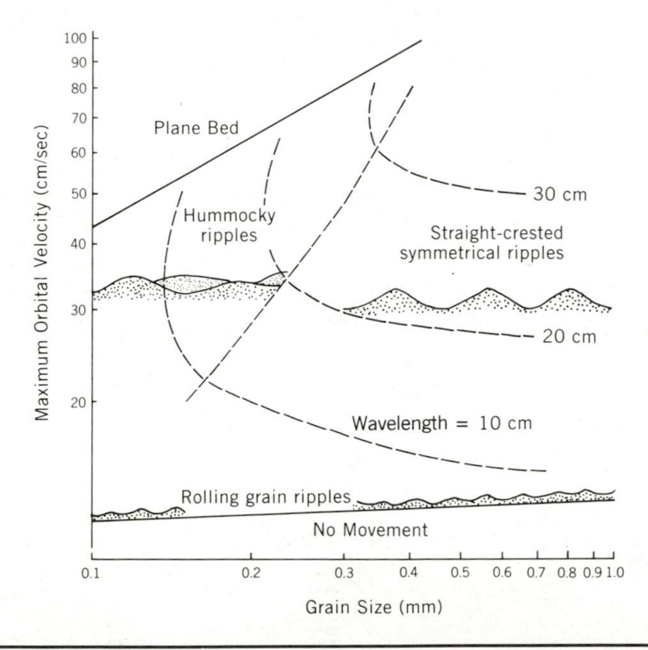

FIGURE 7.6 Types of bedforms formed on different grain-sized beds at different velocities of orbital flow (period is approximately 3 sec). Modified from J. C. Harms, J. B. Southard, and R. B. Walker, 1982, *Structures and Sequences in Clastic Rocks,* Society of Economic Paleontologists and Mineralogists Lecture Notes for Short Course No. 9, Fig. 2–34, p. 2–12. Reprinted by permission of SEPM.

natural wave environments. Even in experiments they are metastable and transform into steeper, symmetrical ripples **(vortex ripples)** that form as sand is transported by vortices or eddies over the growing ripple crests (Fig. 7.7). When higher velocities are reached, the ripples wash out and are replaced by a plane bed (Fig. 7.6). By comparing maximum flow velocity to grain size we see that two fields develop as a response to symmetrical, oscillatory flow: a wave ripple field and a plane bed field. The ripples formed in the lower part of this field are symmetrical and generally have long and straight to slightly sinuous crests (Fig. 7.8), but those created by higher velocities have discontinuous crests and form broad, low-relief hummocks (Fig. 7.8). Higher velocities also result in ripples with longer wavelengths, as does increased grain size (Fig. 7.6).

If the period of oscillation changes, but the flow velocity remains symmetrical, there is only minor modification of the scheme presented above. The ripples formed in symmetrical oscillatory flow will be symmetrical and contain opposing sets of small cross laminae. The maximum length for oscillatory, symmetrical ripples formed under these conditions has not been determined, but no ripples

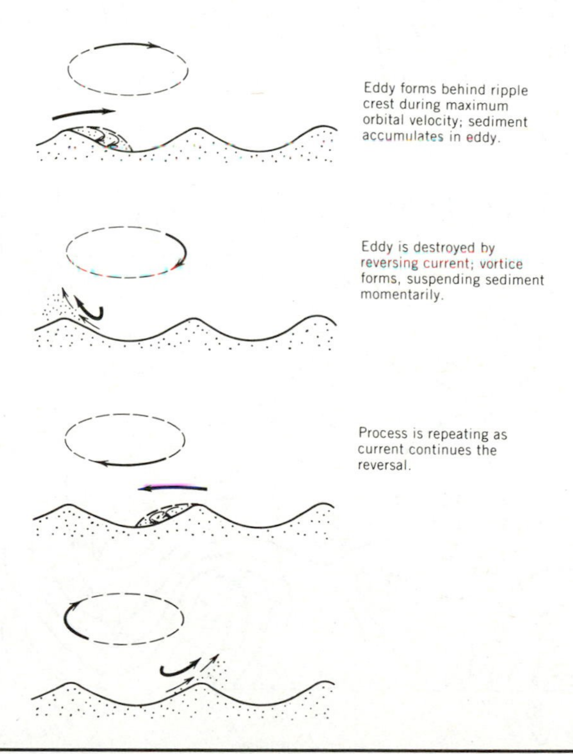

Eddy forms behind ripple crest during maximum orbital velocity; sediment accumulates in eddy.
Eddy is destroyed by reversing current; vortice forms, suspending sediment momentarily.
Process is repeating as current continues the reversal.

FIGURE 7.7 Formation of vortex ripples. From D. L. Inman and A. J. Bowen, 1963, *Proceedings of the 8th Conference on Coastal Engineering, Berkeley, California.*

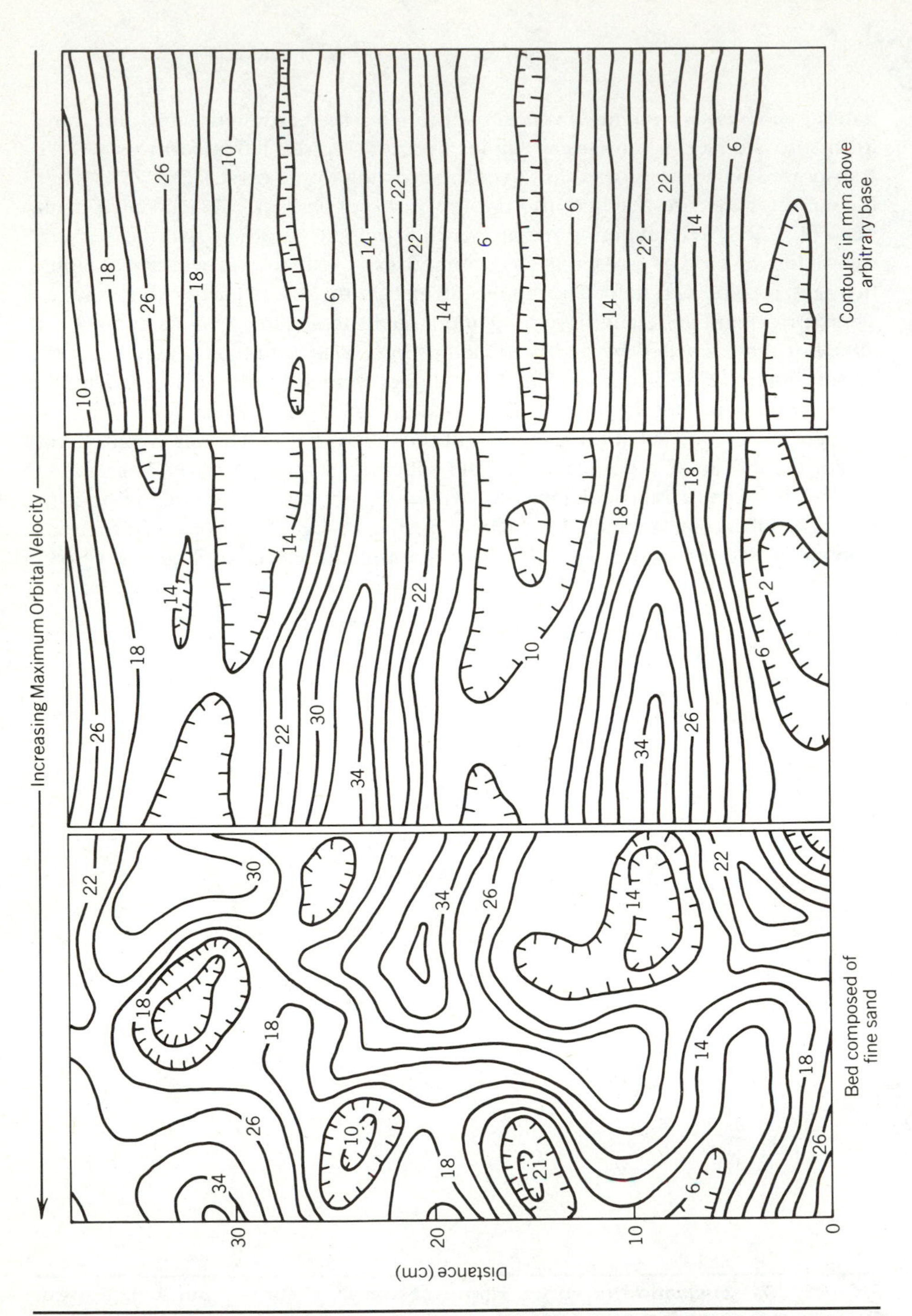

FIGURE 7.8 Contour maps of wave ripples formed under different velocities of orbital flow.

produced experimentally or identified in recent environments exceed approximately 60 cm in wavelength, and most symmetrical ripples have wavelengths of less than 25 cm. Because these observations have been limited in both depth, period, and velocity, large oscillatory ripples may form but have yet to be found because they require very high orbital velocities and long periods that occur only in relatively deep water during large storms. However, in the rock record, very large symmetrical ripples have been described in Ordovician rocks of northern Wales. These straight-crested, symmetrical megaripples have wavelengths of approximately 0.6 to 1.0 m (Fig. 7.9). These bedforms are composed of very coarse sand to pebbles and probably formed in water approximately 20 m deep by waves with periods of 8 to 10 sec. Such waves are common on the Pacific shelf of North America today, but very large symmetrical ripples have not been observed. This is probably because the shelf in most places is covered by finer-grained sediment and, as with unidirectional flow, grain size plays a major role in determining the bedform under a specific set of current conditions. In the case of the ripples in Wales, the grain size was large because volcanic eruptions had flooded the shelf with coarse-grained, poorly sorted debris, which was available to form wave-generated structures.

Grain-size control is similar to that in unidirectional flow but the wave period also has a major effect. Figure 7.10 is a plot of data generated by many authors. From this plot you can see that as grain size increases, the velocity needed to move the sediment increases (similar to the right side of the Hjulstrom plot, see Fig. 4.13). However, the period of the wave generating the flow is also important.

FIGURE 7.9 Large, straight-crested wave ripples in the Ordovician Capel Curig Volcanic Formation, north Wales. (Photograph by W. J. Fritz, 1984).

FIGURE 7.10 Grain size versus velocity plots. From H. E. Clifton and J. R. Dingler, 1984, *Marine Geology,* 60, Fig. 4, p. 174. Reprinted by permission of Elsevier Science Publishers, B.V.

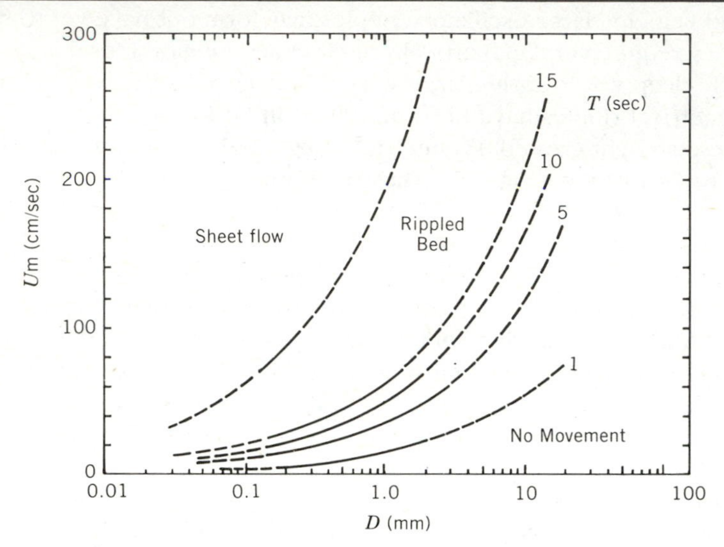

At the same velocity, shorter-period waves will move larger grain sizes of sediment. For example, at a velocity of 100 cm/sec, a wave with a 1-sec period would move grains on the order of 20 mm in diameter, whereas a 15-sec-period wave would move grains less than 8 mm and possibly no larger than 2 mm.

Grain size, along with orbital diameter, also controls ripple size. These relationships have been summarized by Clifton (1976) and Clifton and Dingler (1984) using a combination of field and laboratory data. They suggest that three separate types of symmetric ripples form under three different conditions. The first of these, **orbital ripples,** have wavelengths directly proportional to orbital diameter of the flow and not controlled by grain size at all (Fig. 7.11). As the orbital diameter of the flow increases so does the wavelength of the ripples. The grain size increases in the ripple because of the increased velocity, but it does not control the wavelength of the ripple. **Suborbital ripples** have the opposite relationship. As the orbital diameter increases, the wavelength of the ripples decreases. In these forms the grain size and the orbital diameter are important in controlling the wavelength of the ripple. Finally, **anorbital ripples** are independent of orbital diameter and are controlled instead by grain size. The distinction between these ripple types depends on the ratio between orbital diameter and grain size (d_o/D). Orbital ripples form at d_o/D values of from 100 to 3000. This requires relatively short orbital diameters, so orbital ripples are generally formed by short-period waves in shallow water. When the d_o/D ratios reach 1000 to 3000, suborbital ripples form, and at values above 5000, anorbital ripples form.

FIGURE 7.11 Ripple types. From H. E. Clifton and J. R. Dingler, 1984, *Marine Geology*, 60, Fig. 5, p. 177. Reprinted by permission of Elsevier Science Publishers, B.V.

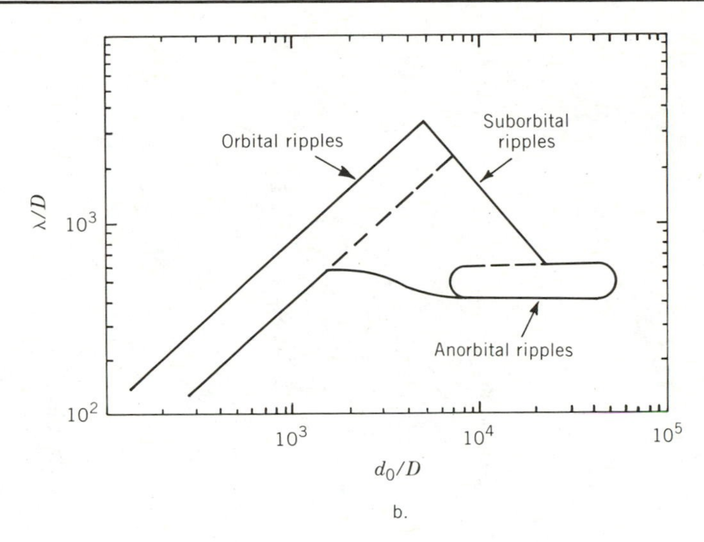

These relationships are extremely useful in interpreting past processes. For each type of ripple the orbital velocity/grain size ratios have corresponding values of wavelength/grain size. For example, from Fig. 7.11 we see that a d_o/D of 1000 corresponds to a wavelength/grain size ratio of approximately 250 to 800. Because we can measure grain size and ripple wavelength in the field, we can use these relationships to determine the range of orbital diameters that formed a particular set of ripples found in the rock record. We can then calculate the depth of water where the ripples formed from the orbital diameter, knowing something about the likely ranges in wave periods. The use of this extremely powerful tool in deciphering ancient depositional systems is explained in detail by Clifton (1976) and Clifton and Dingler (1984). Figure 7.12 shows a sedimentary structure formed by oscillation ripples. These forms are recognized by the abundant cross bed angles of less than 10 degrees.

Another type of large bedform, with cross bedding of less than 10 degrees, is significantly different from symmetrical ripples and possibly forms from long-period waves. These structures have been described from outcrops but have not yet been seen in recent environments. These forms, called hummocky beds (Fig. 7.13), are much larger than the hummocks produced experimentally but have a very similar morphology and have been attributed to strong oscillating currents with long periods that result from storm waves and possibly with associated underflow currents. However, they have also been described in sediments deposited in the surf zone by Greenwood and Sherman (1986), who found that

FIGURE 7.12 Small-scale cross lamination formed by oscillation ripples. Ordovician Lower Rhyolitic Tuff Formation, Moel Siabod, north Wales. (Photograph by W. J. Fritz, 1985).

FIGURE 7.13 Block diagram showing hummocky cross-stratification. Current direction unknown. From J. C. Harms, J. B. Southard, and R. G. Walker, 1982, *Structures and Sequences in Clastic Rocks*, Society of Economic Paleontologists and Mineralogists Lecture Notes for Short Course No. 9, Fig. 3–15, p. 3–31. Reprinted by permission of SEPM.

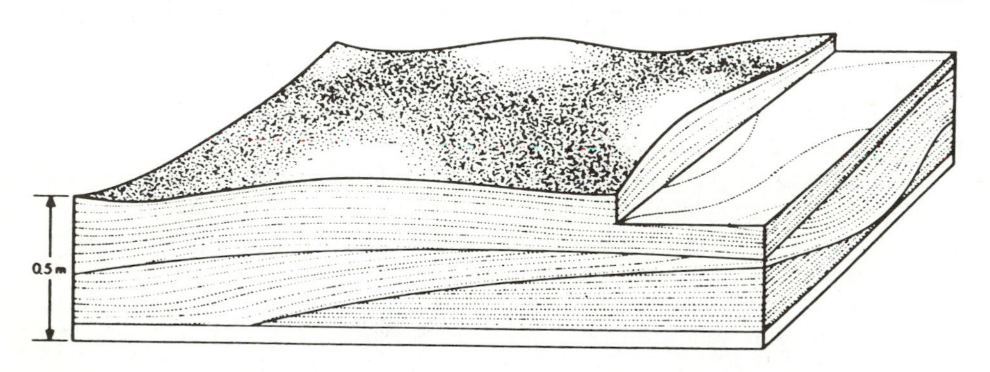

they formed from a combination of unidirectional and oscillatory flow with the latter dominant. Continuing work on hummocky bedding indicates that it can form at depths ranging from less than 2 m to tens of meters and under many wave conditions.

Even though there are few data on the details of large wave-generated bedforms, such as hummocks, and the processes that form them, a generalized scheme has been developed that encompasses the ideas published to date on oscillatory flow features. In that scheme (Fig. 7.14), ripples become larger, broader, and

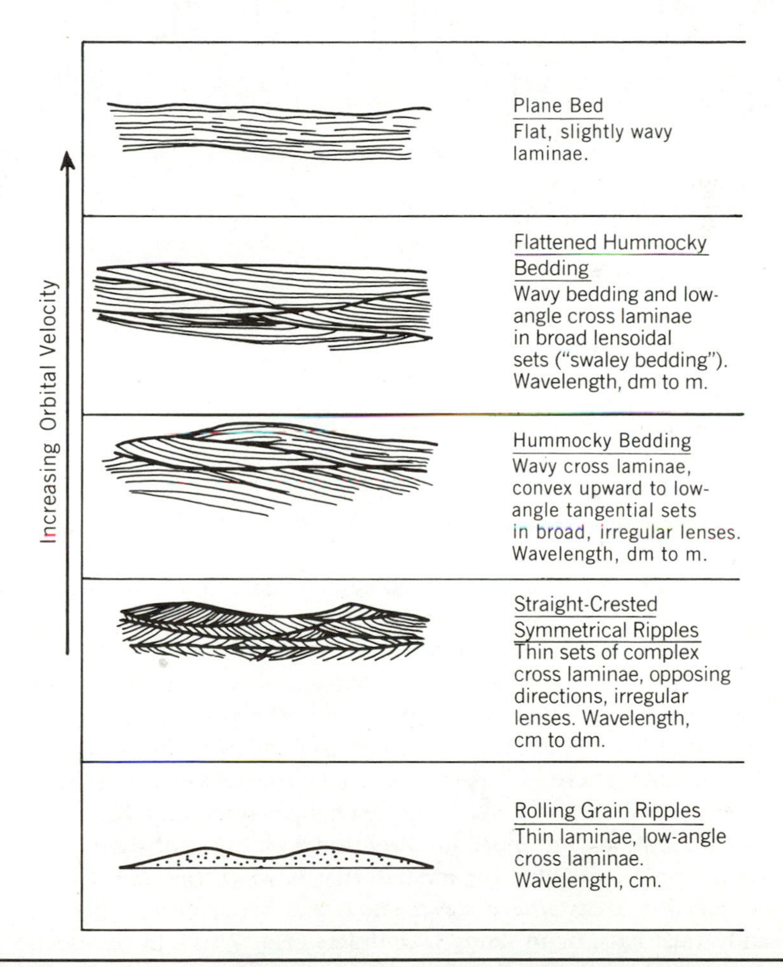

FIGURE 7.14 Types of bedforms and resulting bedding produced under different orbital velocities. Modified from J. C. Harms, J. B. Southard, and R. B. Walker, 1982, *Structures and Sequences in Clastic Rocks,* Society of Economic Paleontologists and Mineralogists Lecture Notes for Short Course No. 9, Fig. 3–16, p. 3–33. Reprinted by permission of SEPM.

FIGURE 7.15 Types of ripples formed under combined oscillatory and unidirectional flow. From J. C. Harms, J. B. Southard, and R. B. Walker, 1982, *Structures and Sequences in Clastic Rocks,* Society of Economic Paleontologists and Mineralogists Lecture Notes for Short Course No. 9, Fig. 3–12, p. 3–24. Reprinted by permission of SEPM.

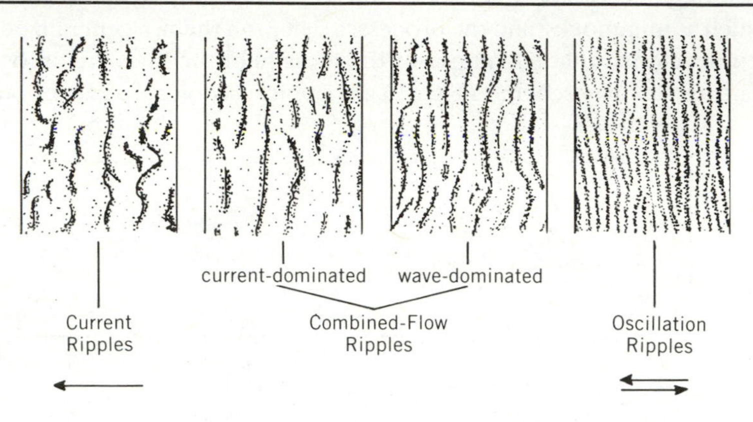

more complex with increased orbital velocity and period. At higher velocities they flatten into plane beds. This very tentative model must await more data to substantiate the larger forms, but both laboratory studies and field measurements have detailed the shapes of smaller ripples and the processes that form them. It seems likely that the complex, three-dimensional forms like hummocks form under energetic conditions with superimposed oscillatory and unidirectional currents.

Given all this complexity, it is well established that straight-crested symmetrical ripples form under symmetrical oscillating currents. The currents transport sand back and forth over the crest of the ripple and form opposing sets of cross laminae. If the current becomes asymmetrical, as when a wave approaches shore, one direction of migration dominates over the other. This preferential migration will result in asymmetrical ripples, the amount of asymmetry being dependent on the velocity asymmetry. There is a complete gradation between oscillation ripples and current ripples (Fig. 7.15). As the ripple changes symmetry, the cross laminae within it become more and more unidirectional and do not differ significantly from current ripple cross laminae; indeed, that is what they are. This sequence can be observed in areas where waves shoal and break into shallow water, on broad, sandy tidal flats, or in sandy lake deltas (Fig. 7.16). In the deeper areas where symmetrical oscillatory currents move sediment, symmetrical ripples form. As the waves shoal and break, the ripples become asymmetrical with the steep sides toward the shore. On the broad, shallow flat, where the pulses of breaking waves form an asymmetrical pulsating current, asymmetrical ripples form. These form in very shallow water and have a distinct shape unlike ripples forming in

FIGURE 7.16 Different shapes of ripples formed by oscillatory and pulsating currents produced by waves.

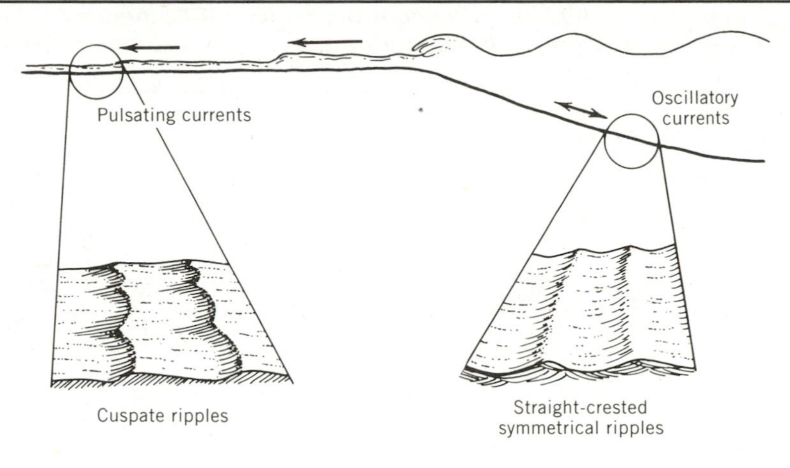

FIGURE 7.17 Cuspate ripple marks generated by pulsating currents formed by breaking waves on a tidal beach (Washington, left) and similar forms in the Precambrian Belt Supergroup, western Montana (right). (Photograph by J. N. Moore).

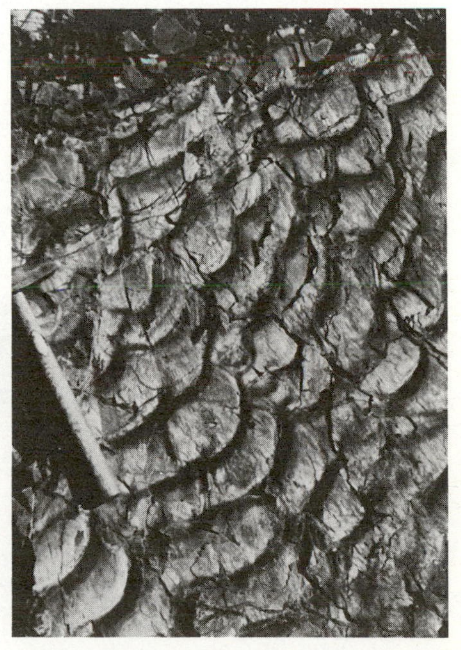

FIGURE 7.18 Ripple shape changes. From H. E. Clifton, 1976, Wave-formed sedimentary structures: A conceptual model. In R. A. Davis, Jr., and R. L. Ethington (Eds.), *Beach and Nearshore Sedimentation,* Society of Economic Paleontologists and Mineralogists Special Publication No. 24, Fig. 15, p. 136. Reprinted by permission of SEPM.

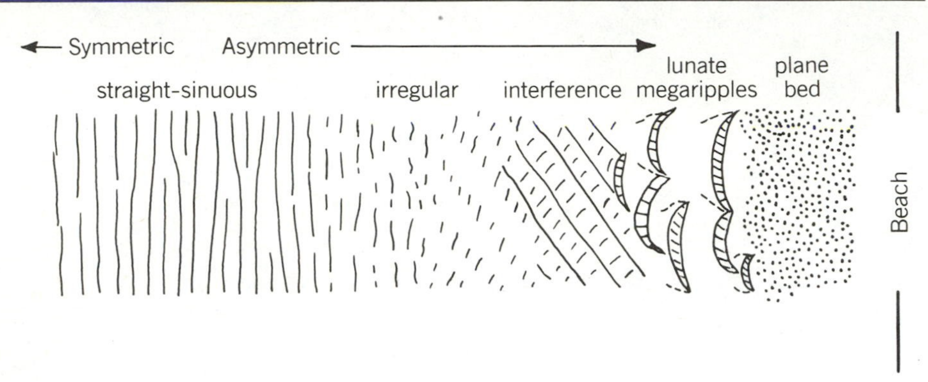

FIGURE 7.19 Shoreward-dipping cross lamination in a runnel system, Sapelo Island, Georgia. The shore is to the right and the sea to the left. (Photograph by J. H. Hoyt. Courtesy of V. J. Henry).

deeper flows. Their crests are cuspate and the horns of the cusps are aligned (Fig. 7.17).

Orbital–velocity asymmetry also forms asymmetrical ripples as waves move shoreward and build in height and break. Clifton (1976) identified this transition on some U.S. Pacific Northwest and Spanish beaches (Fig. 7.18). Wave-generated bedforms represent complex mixtures of processes that result in a mixture of asymmetrical and symmetrical structures, and on beaches these features become very pronounced. As waves break and move onto the beach, the oscillatory flow within the wave is submerged by the pulsating flow of the breakers. On beaches with well-developed longshore bars, these currents surge through the relatively deep water below the beach face. This situation generates pulsating unidirectional flow and large lunate megaripples form in the lower flow regime and migrate shoreward, leaving a record of shoreward-dipping cross laminae (Fig. 7.19). Flow in **runnels,** which are broad troughs parallel to the shoreline, can also form unidirectional flow bedforms (Fig. 7.20). Such transitions from oscillatory flow to unidirectional flow result in complex mixtures of large and small, symmetrical and asymmetrical bedforms and, hence, bedding types (Fig. 7.21).

The breaking-wave system simplifies somewhat as the wave continues onto

FIGURE 7.20 Complex assemblage of megaripples indicating longshore pulsating flow in a runnel on Sapelo Island, Georgia. (Photograph by W. J. Fritz, 1983).

FIGURE 7.21 Complex bedding produced by nearshore oscillatory currents. From R. G. D. Davidson-Arnott and B. Greenwood, 1976, Facies relationships on a barred coast, Kouchibouguac Bay, New Brunswick, Canada. In R. A. Davis and R. L. Ethington (Eds.), *Beach and Nearshore Sedimentation,* Society of Economic Paleontologists and Mineralogists Special Publication No. 24, Fig. 4, p. 154. Reprinted by permission of SEPM.

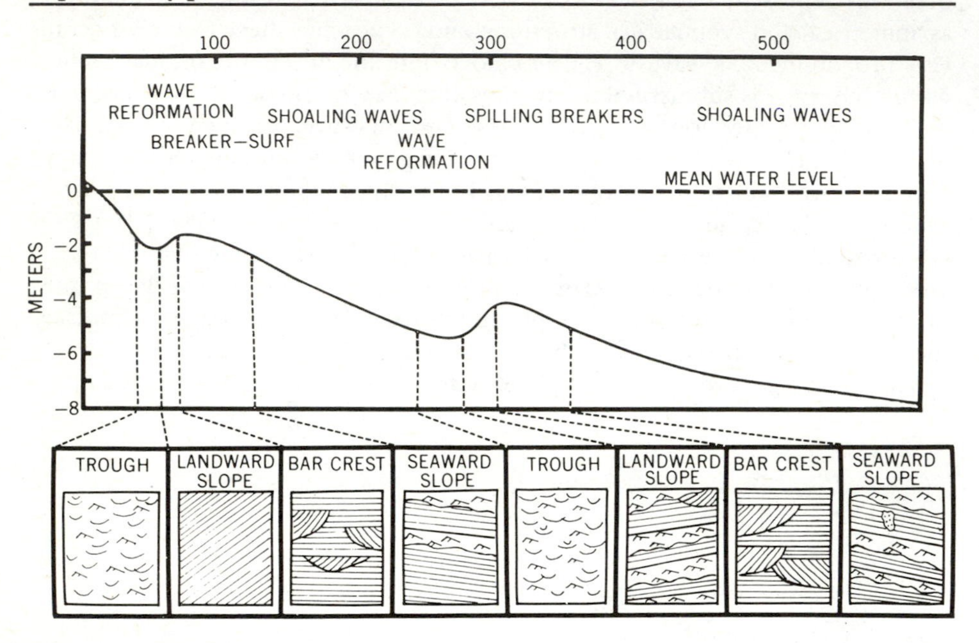

the beach face. The last swash of the wave forms reversing, high-velocity, shallow flow that produces upper flow regime bedforms. Beach faces are commonly dominated by plane beds and antidunes are common features (Fig. 7.22), but they should not be interpreted as the only features indicative of breaking waves. Wave-generated sedimentary structures are a complex mixture of bedforms that are strongly dependent on the depth of water, size and period of the waves, and position within the shoaling wave zone. To decipher the details of these processes you must carefully examine the internal and external shape of bedforms rather than rely on one specific feature.

TIDAL-GENERATED CURRENTS

Tidal currents form oscillating flows with significantly longer periods of reversal than wave-generated currents. The velocity of tidal currents is often extremely asymmetrical and remains near zero for hours. This long-term asymmetry results

FIGURE 7.22 Plane beds (right of photo) and antidunes (center) on a ridge in the lower foreshore, Sapelo Island, Georgia. Note runnel on left filled with small ripples (ocean to right and land to left). (Photograph by W. J. Fritz, 1983).

from the periodic change in sea level caused by the gravitational attraction of the moon and the sun on the ocean. The gravitational effects of the moon on the earth's oceans are approximately twice that of the sun, so the relative positions of the moon and sun cause variations in the height of the global tidal bulge. In the open ocean, the tidal bulge would be approximately 50 cm in height, and because the earth rotates beneath this bulge, we would expect to find two high and two low tides of approximately the same height each day **(semidiurnal tides).** This simple equilibrium theory of tides is complicated by many factors, including the tilt of the earth, size and shape of ocean basins, the extreme length of the tidal standing wave and the irregular shape of the shoreline where the tide moves onto the shelf. The shallow-water effects cause the midocean tidal range to be modified dramatically. On open coasts and highly restricted seas (like the Mediterranean Sea), tidal ranges generally fall within 2 to 4 m. However, along coasts crenulated by embayments, tidal ranges can reach well over 4 m. The Gulf of California has a tidal range of over 9 m, and the Bay of Fundy, in Nova Scotia, is famous for tides with ranges of nearly 16 m. Currents generated by the tides in such tidally dominated bays can reach velocities of up to 10 m/sec.

The different depths and sizes of ocean basins can result in large departures from the expected semidiurnal tides. However, the Atlantic Coast of North America generally exhibits tidal currents that are very nearly semidiurnal (Fig. 7.23). The Pacific Coast of North America has **mixed tides** that are extremely asymmetrical and complicated (Fig. 7.23). Some coasts have **diurnal tides,** with only one high and one low tide each day. These complex changes in sea level along the coast and the associated tidally generated currents result in fairly characteristic sedimentary structures. At least once a day (diurnal tides) or twice a day (semidiurnal tides), tidally dominated bays and estuaries are covered and uncovered by the tides. Let us examine a semidiurnal tidal flat over a complete tidal cycle.

If we wander out onto the tidal flat at low tide we would see a surface exposed to the air (subaerially exposed). As the tide rises, the surface will be covered by the inflowing tide. Water will flow shoreward and the depth will increase until high tide is reached and the current stops. At high tide the current will stop and eventually reverse and begin to flow seaward. At the second low tide, the flat will be exposed again. This process would be repeated for the second high tide and continue throughout the year with modifications in tidal range and strength of the current.

BEDFORMS CREATED BY TIDAL PROCESSES

The changing tides have major effects on the processes of sedimentation. Strong, unidirectional currents will flow over the tidal flat in relatively deep water (de-

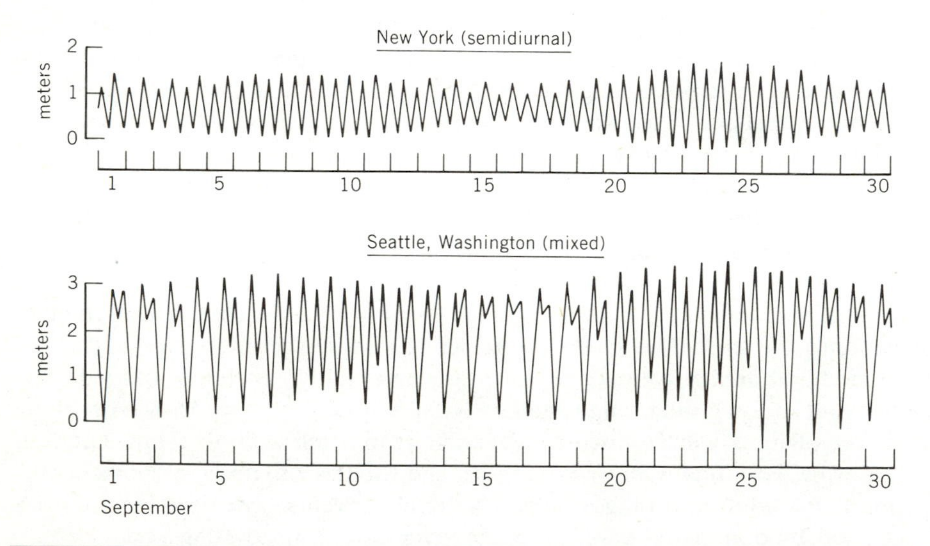

FIGURE 7.23 Tidal charts showing tidal changes at New York and Seattle. From U.S. Oceanographic office.

pending on the tidal range) during the flood tide. These processes can transport bedload sediment and create large and small ripples in the lower flow regime. At and near the high-tide stand, the water would nearly stop and remain relatively deep. Any suspended load carried by the tidal current would then settle out of the water column. During the ebb flow, the currents could again form lower flow regime bedforms but they would migrate in the opposite (or at least different) direction from the flood-tide bedforms. As the tide drained off the flat, the surface would emerge and be subjected to various emergent, subaerial processes including desiccation and other penecontemporaneous modifications (see Chapter 3). Let us look at the details of some of these processes that form very distinct sedimentary structures.

If we start our examination at the beginning of flood tide and continue through a tidal cycle we can get a good idea of the mechanisms that create some of these particular structures. We will consider two examples that delineate tidal features involving large ripples and small ripples.

First, let us look at a situation where the tidal currents are symmetrical and strong enough and deep enough to form ripples (see Chapter 4). During flood tide, the ripples on the bed would migrate landward, and a train of ripple cross laminae would form (Fig. 7.24*a*). As the current stopped, the ripples would remain on the surface because the current decreased in velocity as the depth increased. When the velocity reached a low enough value such that sand could no longer be transported, the ripples would stop migrating. If suspension sediment was

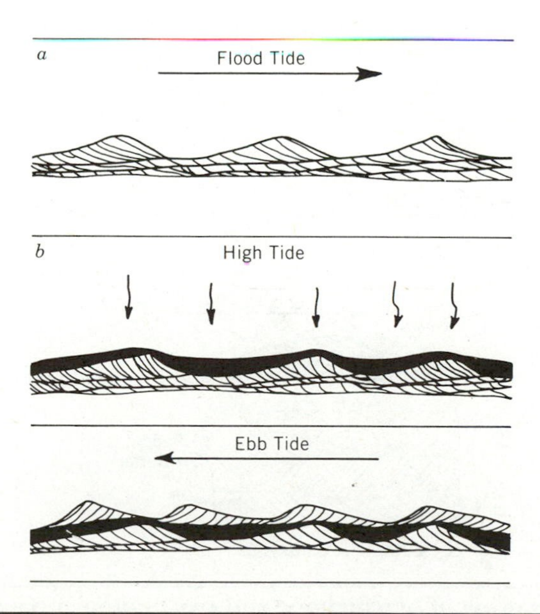

FIGURE 7.24 Ripple migration during tidal changes.

available in the overlying water, mud could then settle out onto the ripple surface (Fig. 7.24*b*). As the tide ebbed and flowed seaward, sand could then move over the mud surface and form a new set of ripples (Fig. 7.24*c*). Mud would not normally be resuspended because the cohesive, fine-grained mud takes considerably more energy to move than fine sand (Chapter 4). Repetition of these processes would result in mixed layers of mud and sand. If the ripples were "starved" so that they were only isolated ripples on the surface, flaser bedding would result (see Fig. 3.5); if a continuous sand surface alternated with mud, which filled in only the ripple troughs, then lenticular bedding would form; and if mud and sand were deposited about equally, then wavy bedding would form. If we were to look carefully at the direction of ripple foresets, we would see that different levels of ripples would have opposing directions because one was made from the ebb tide and one from the flood tide. The shapes of the resulting bedding would depend on the morphology of the ripples, which is determined by the flow characteristics.

If the tidal currents are not completely symmetrical then the ripple foresets do not necessarily directly oppose one another in different layers. Currents that form ripples may be associated with only one of the flows, either ebb or flood, or the two tides may not flow in perfectly opposite directions. In studies of modern tidal estuaries and bays, it is quite common for flood tides to have very different magnitudes and directions than ebb tides in different areas of the bay. This results in extremely asymmetric currents that often move and deposit sediment in only one dominant direction. In areas where the currents are equal and opposite, very distinctive tidal structures can form. If the currents are strong enough, large ripples form in the flood tide, leaving behind a trail of cross laminae.

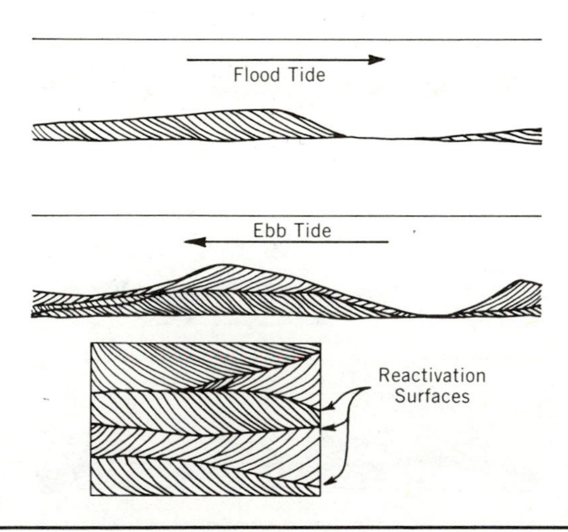

FIGURE 7.25 Drawing of herringbone and reactivation surfaces.

FIGURE 7.26 Herringbone bedding (best seen in the lower center part of the photo) from deposition on incoming and outgoing tides. Vertical V-shaped wedge structure in upper right corner of photo is probably an escape structure. Cretaceous Eutaw Formation near Columbus, Georgia. (Photograph by W. J. Frazier).

FIGURE 7.27 Drawing of a reactivation surface where cross beds dip in the same direction as the surface. From J. C. Harms, J. B. Southard, D. R. Speargin, and R. G. Walker, 1975, *Depositional Environments as Interpreted from Primary Sedimentary Structures and Stratification Sequences,* Society of Economic Paleontologists and Mineralogists Short Course No. 2, Fig. 3–4, p. 51. Reprinted by permission of SEPM.

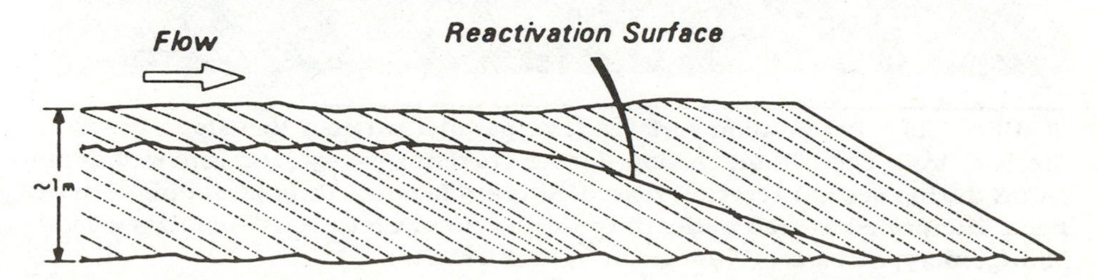

As the current reverses. The front of the large ripple is eroded and a new form is created that migrates in the opposite direction (Fig. 7.25). Two sets of opposing cross laminae that form separated by a surface of erosion are called **herringbone cross laminae** because they resemble the suit fabric of the same name (Figs. 7.25, and 7.26). Remember, though, that true herringbone cross laminae have directly opposing foresets. They form best when straight-crested large ripples migrate in opposite directions; the relationships become more complex when complex bedforms are involved. Also, cross laminae that look like herringbone can form when two lunate forms cut through one another and produce overlapping trough cross beds. Erosion during tidal changes also produces another type of **reactivation surface.** Such structures form as bedforms migrate during one part of the tidal cycle and become eroded during the opposite-flow tide. Then with the succeeding tide, identical flow conditions return and the bedform migrates again, that is, it is reactivated and produces almost the same kind of cross laminae as the bedform migration during the previous tidal cycles (Figs. 7.27,

FIGURE 7.28 Reactivation surfaces (two parallel surfaces that slope gently to the left) from deposition, erosion during current reversal, and renewed deposition during second reversal. Disrupted cross laminae throughout photo have been interpreted as crab tracks in vertical section. Cretaceous Tombigbee Sand, Montgomery, Alabama. (Photograph by W. J. Frazier).

FIGURE 7.29 Reactivation surfaces in the Cretaceous Eutaw Formation near Columbus, Georgia. (Photograph by W. J. Frazier).

7.28, and 7.29). Thus, reactivation surfaces yield a kind of compound cross bedding.

Tidal processes are also responsible for a number of other surface features. Because the tides periodically leave large areas of the tidal flat emergent, ripple marks and other features form that are indicative of the periodic water-level changes. Ripples forming at high tide will be controlled by the flow conditions and sediment parameters. But as the tide falls, those ripples will be modified by changing current conditions. Common structures on tidal flats are flat-topped ripple marks, sometimes called "scuffed" ripples (Fig. 7.30). These form as the ripples emerge when the tide falls and the crests are reworked by the shallower currents (Fig. 7.31). Interference ripple marks are very common on tidal flats because of the complex interaction of tidal currents and wave-generated currents. The last stages of low tide often produce Runzelmarken as small currents move sediment in extremely shallow water. Also, myriad late-stage emergence erosional features form.

The structures formed by tidal processes are distinct because the depths, magnitudes, and directions of currents change frequently and repetitively. The settling of mud from suspension **(mud drapes)** is also a common feature because

FIGURE 7.30 Flat-topped ripples, Capital Reef National Monument, Utah. (Photograph by Richard Weymouth Brooks). From Flint and Skinner, Fig. 12.13, *Physical Geology*, 2nd ed.

of the stagnant conditions at high tide. However, such features can also form in any environment where the currents change rapidly, for example, in large river systems, so be cautious when using these features to delineate depositional processes. You must remember that none of these features is exclusively tidal but represents periodic flooding and emergence of the bed during deposition.

EPILOGUE

A wrath of water—it bulged a full three feet over all other froth in the channel, as if some great-headed creature was seeking surface—careened

FIGURE 7.31 Drawing illustrating the formation of reworked flat-topped wave ripples.

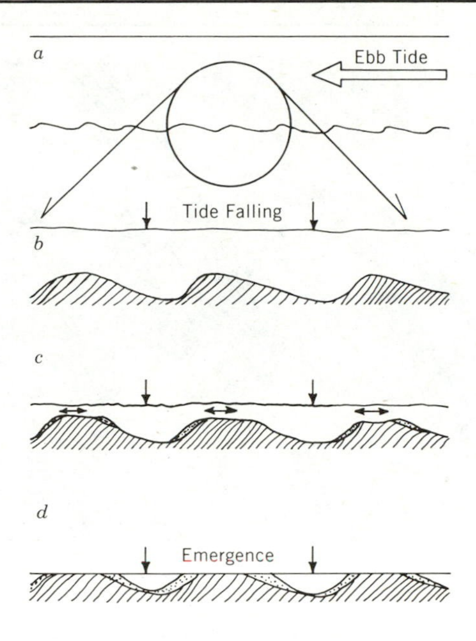

in. Surf spewed over Karlsson and Wennberg, both of them clenching eyes tight against the salt sting.

Ivan Doig,* The Sea Runners, *1982

Tides and waves generate powerful currents that move prodigious amounts of sediment. Although both these currents reverse direction, they have very different results. The long duration of tidal currents is just a special case of unidirectional flow, in which the currents change direction and magnitude periodically. Geologists have tied interpretations of sedimentary rock packages directly to sets of structures (or even individual structures). Thus, herringbone cross laminae, or flaser bedding, were considered *a priori* evidence of tidal deposition. This is not necessarily the case! They represent alternations in current direction and velocity that can occur in many environments. This is emphasized by the recent description of herringbone cross laminae in fluvial sediments by Alam *et al.* (1985). Reactivation surfaces, flaser bedding, Runzelmarken, flat-topped ripple marks, and mudcracks can suggest tidal processes but must be dealt with in the context of the entire package of sedimentary structures and processes.

The interpretation of wave-generated sedimentary structures (Fig. 7.32) also demands diligence to the process model. In the past, all wave ripples were considered symmetric, and in fact the terms were synonymous. It is now quite

FIGURE 7.32 Bedding plane exposure of straight-crested to slightly sinuous wave ripples on Dakota sandstone. (Photograph by J. R. Stacy, #588, U.S. Geological Survey).

clear that most wave-produced ripples are asymmetrical and are formed as they migrate in the direction of wave propagation. Different bedforms are created under different flow situations and are controlled by wave and sediment parameters similar to those operating under unidirectional flow. You must observe carefully to distinguish wave and tidal processes recorded in the sediments and rocks and draw on the large amount of laboratory and field data published in the literature. Do not rely on simplifications or you will be blinded by the salt spray.

OUTSIDE READING

Experimental Work on Oscillating Currents
Bliven *et al.* (1977); Carstens *et al.* (1969); Dingler (1974); Dingler and Inman (1977); Harms (1969); Inman and Bowen (1963); Komar (1974, 1976); Lofquist (1978); Miller and Komar (1980a).

Field Applications to Wave-Generated Bedforms
Alam *et al.* (1985); J. R. L. Allen (1979); P. A. Allen (1981); Bagnold (1946); Clifton (1969, 1976); Clifton and Dingler (1984); Clifton *et al.* (1971); Dott and Bourgeois (1982); Greenwood and Sherman (1986); Harms *et al.* (1982); Inman (1957); Miller and Komar (1980b); Moore *et al.* (1984); Nøttvedt and Kreisa (1987).

Tidal Processes and Structures

Clifton *et al.* (1971); Dabrio (1982); Dalrymple *et al.* (1978); Davidson-Arnott and Greenwood (1976); Davis (1985); Davis and Ethington (1976); Davis *et al.* (1972); Evans (1965); Ginsburg (1975); Klein (1977); Knight and Dalrymple (1975); Komar (1976); Orford and Wright (1978); Reineck (1972); Reineck and Wunderlich (1969); Reineck and Singh (1980); van Straaten (1954); Zarillo (1982).

Chapter 8

Transport and Deposition by Sediment Flows

DEFINITIONS AND PROCESSES

So far we have discussed the role of water and air in moving and depositing sediment. In those discussions, sediment moved because of the drag of relatively sediment-free fluid. The load of sediment was minimal within the fluid so that the properties of flow were determined by fluid properties. Increasing the load of sediment increases the viscosity and density significantly, which drastically changes the type of flow. Let us now consider a fluid composed of large amounts of sediment, ranging from 20 to 70%, so that the flow is controlled by the properties of the sediment–water mixture. Such flows are termed **sediment gravity flows** and are dependent on the sediment for movement. The sediment, in some way, maintains movement such that the flow would not exist without the sediment. Sediment flows range from those moving under completely dry conditions (grain flows) to those that are very low density fluids, not much denser than water. The common feature of all these flows is that they move as a result of the gravitational attraction on the sediment grains themselves. In the case of grain flows, the flow can operate under dry or even vacuum conditions because the mass of solid grains itself behaves as a fluid as a result of dispersive pressure caused by grain interaction.

The recognition of sediment gravity flows is fairly recent. Although subaerial sediment flows have been discussed for many decades (landslides, slumps, and mudflows), the observation that sediment can flow as a mass subaqueously was not considered until well into the twentieth century. In the last 30 years there has been an explosive increase in the work on sediment gravity flows. Both experimental and field work have led to a myriad of interpretations and classification schemes related to sediment gravity flows and their deposits. The details of this work are well documented in summaries by Middleton (1969), Middleton and Hampton (1973, 1976), and Nardin *et al.* (1979) and by several papers by Lowe (1976a,b, 1979a,b, 1982). We will use the conceptual frameworks constructed by these authors and others to develop the details of sediment gravity flow processes and their deposits. Unfortunately, different authors have used different terms for the same features and processes. For simplicity we will combine this terminology into a compilation of sediment gravity flow types and principles.

Sediment flows are mixtures of sediment and fluid (water or air). We can classify sediment flows by the way they move, as either a fluid (Newtonian flow) or a plastic (non-Newtonian behavior). Sediment gravity flows that act like fluids

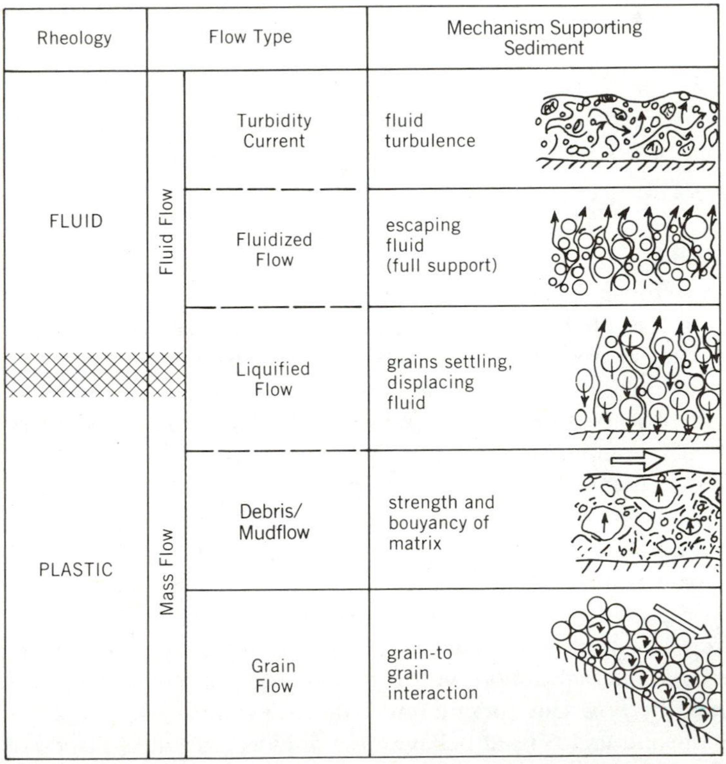

Rheology	Flow Type		Mechanism Supporting Sediment
FLUID	Fluid Flow	Turbidity Current	fluid turbulence
		Fluidized Flow	escaping fluid (full support)
		Liquified Flow	grains settling, displacing fluid
PLASTIC	Mass Flow	Debris/ Mudflow	strength and bouyancy of matrix
		Grain Flow	grain-to grain interaction

FIGURE 8.1 Type of flow and mechanisms of sediment transport for different types of sediment flows. Modified and compiled from D. R. Lowe, 1982, *Journal of Sedimentary Petrology*, 52, Fig. 1, p. 280, and Nardin *et al.* (1979). Reprinted by permission of SEPM.

are called **fluidal flows** and those with a plastic behavior are **mass flows** (Fig. 8.1). To subdivide sediment gravity flows further we must examine the mechanism that keeps the sediment supported during the flow. Mass flows are of two types: mudflow/debris flow and grain flow. Fluidal flows can move by liquified flow, fluidized flow, and turbidity current. The mechanism supporting sediment in each of these flow types differs depending on the sediment concentration and turbulence of the flow.

Actually, fluidal and plastic sediment flows may be thought of as end-members in a continuum of flow types. The concentration of sediment and the mechanism of flow control the type of flow that forms in a particular circumstance. Some of the mechanisms operate in both air and water and in differing sediment–fluid mixtures. Others operate under fixed conditions of sediment concentration and fluid type. The specific mechanism of flow can change during the progress of a particular flow and most large masses of sediment moving as sediment gravity flows are probably moving by more than one mechanism. In natural systems these types of flow can grade into one another and are not strictly confined to one mechanism (Fig. 8.2).

Sediment accumulates from fluidal flows by deposition from bedload (traction sedimentation) or from suspended load (suspended sedimentation). Because these mechanisms are similar to those in flowing water, the deposit forms by

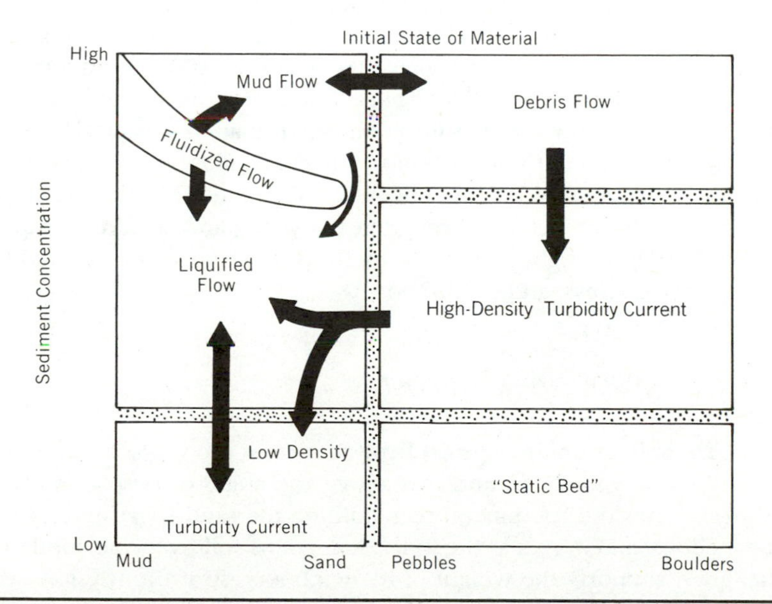

FIGURE 8.2 Types of sediment flow mechanisms under different sediment concentrations and sediment grain sizes. Modified and compiled from data presented in Lowe (1982) and Middleton and Hampton (1973).

FIGURE 8.3 Rheological classification for flows of sediment–water mixtures. NFT is the Newtonian fluid threshold and LT is the liquefaction threshold. From Pierson and Costa (1984).

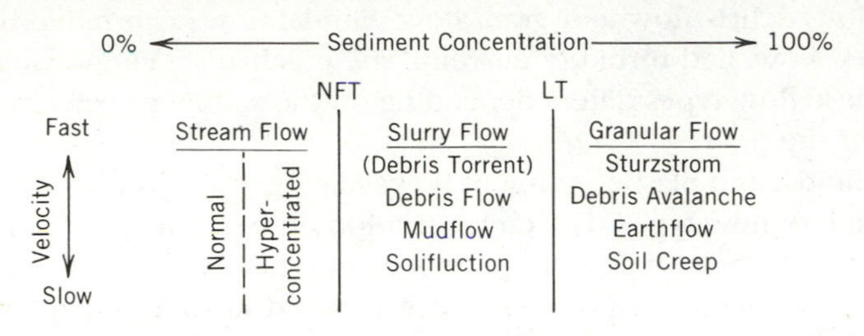

accretion of the bed as material is deposited at the base of the flow. This mechanism of deposition is very different from that in mudflows/debris flows, which essentially solidify *en masse* when they stop flowing. Let us now look at each of the various types of flow and examine the details of flow, their restrictions, and resulting deposits. Because most of the processes of sediment flow are difficult to model and observe, these discussions are somewhat tentative. Many of the ideas and comments are conceptual models built to explain what we see in ancient and modern deposits and are not based on experiments or physical models.

Pierson and Costa (1984) present a rheological classification of subaerial sediment–water flows (Fig. 8.3). This classification considers a continuum from clear water with no sediment to dry granular flow made of 100% sediment. At low sediment concentrations, flows behaving in a liquid Newtonian fashion are termed "stream flow." Stream flow changes into non-Newtonian "slurry flow" and finally "granular flow" as sediment concentrations increase. Flow type is classified not only by the percentage of sediment concentration but also by velocity (Fig. 8.3). Low-velocity flow is characterized by grain buoyancy and structural support of the grains by the flow. Higher-energy/velocity flow becomes dominated by turbulence, dispersive stress, and fluidization.

GRAIN FLOWS AND THEIR DEPOSITS

An easily observable example of **grain flow** occurs on the slipfaces of sand dunes (Fig. 8.4). When the sand accumulates above the angle of repose, the mass of the sand overcomes the frictional forces holding the sand together and it flows down the incline. Sand moves as the individual grains roll over one another. Each individual grain supports the weight of its neighbors, so if the friction between moving grains overcomes the inertia the flow will stop. In air, grain flows need slopes of approximately 30° to flow, although the exact angle does vary depending on grain size, shape, sorting, angularity, and water content. The slope is less in

FIGURE 8.4 Grain flow on dry, medium-grained sand on a slipface (ca. 40° slope) of a modern eolian dune on the backshore of Sapelo Island, Georgia. Flow on left is approximately 10 cm wide. (Photograph by W. J. Fritz).

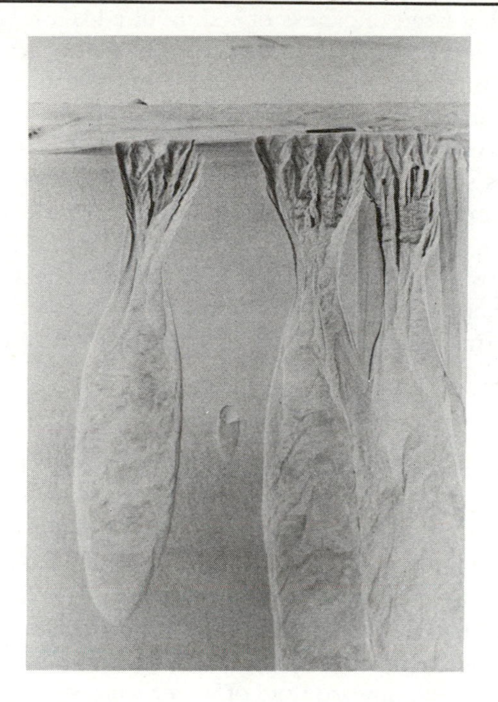

water because the higher density of the water adds a little buoyancy to the grains and reduces internal friction, so that grain flows can form on subaqueous slopes steeper than approximately 18°.

Both subaqueous and subaerial grain flows move by the interaction of grains, rather like a mass of marbles rolling down an incline. As the grains interact with one another the flow is perpetrated. Such direct grain-to-grain interaction is called **dispersive pressure** and forms only in highly concentrated flows. The only fluid present in grain flows is air or water in the pore space between grains. Because grain flows are dependent on the interaction of individual grains, they form best in well-sorted sediments. If there is a large component of clay and silt in the mixture, the strength of the fine material keeps the grains from rolling, and the dispersive pressure is minimal. When dispersive pressure is high, large grains sometimes migrate upward into zones of lower shear stress. This effect produces one type of **inverse grading** characterized in general by poorly sorted layers with the greatest concentration of large clasts near the top. Inverse grading can also form in grain flows by a kinetic sieve mechanism whereby small grains settle through pore spaces between large grains and into the lower part of the flow.

Grain flows move by the dispersive pressure between grains as they collide with one another during movement. Because the grains that move must support the entire weight of the grains above the flow, grain flows have a very limited thickness, because a large thickness of sediment adds a large frictional force to the grains, which freezes the flow. It is very unlikely that grain flows greater than 5 cm thick can form at all, and in cases observed to date, in both subaerial and subaqueous environments, grain flows never exceed a few centimeters in thickness. So, it is extremely unlikely that thick grain flow deposits form in nature. This conclusion is strongly supported by experimental and theoretical work by Middleton and Hampton (1973, 1976) and Lowe (1976a). The reexamination by Van der Kamp *et al.* (1973) and Link (1975) of thick sandstone beds interpreted as grain flow deposits by Stauffer in 1967 contradicts the supposed classic grain flow sequence, thereby limiting grain flows to the steep slipfaces of dunes and other large bedforms. They apparently did not form thick, extensive deposits in the recent or in the past.

FLUIDIZED AND LIQUIFIED FLOWS AND THEIR DEPOSITS

If the dominant mechanism keeping sediment entrained in a flow results from the movement of fluid through the sediment, the flow is termed liquidized flow or a fluidized flow. In the case of a **fluidized flow,** fluid moves upward through the sediment, thus buoying the grains. Such a mechanism forms where unstable masses of sediment are shaken by earthquakes and the fluid, while escaping, suspends the grains. The upward flow of water causes a tremendous decrease in strength of the mass because the grains are no longer supported by grain-to-grain contact, so that fluidized flows will move on very low slopes. They occur both subaerially and subaqueously. Once a sediment is fluidized it can flow down very gentle slopes as the fluid moves upward through the sediment. Probably slopes of from 3 to 10 degrees are required for the mass to move. Some very spectacular fluidized flows have formed when large masses of rock have slid from mountainsides in the Andes and entrained air. The trapped air fluidizes the mass of fragmented rock and a sediment flow is created. These gigantic flows moved tens of kilometers at tremendous speeds and caused the destruction of entire villages. Subaqueous fluidized flows are less devastating but they may cover huge areas. Because they occur beneath the oceans or lakes, they are generally not observable.

Fluidized flows represent sediment gravity flows that depend on fluid escape for their movement; however, if the grains in a flow are only partially supported by escaping fluid, then the flow is termed a **liquified flow.** In liquified flows, high concentrations of grains settle through the fluid instead of the fluid flowing upward through the sediment. Liquified flows probably form during the last stages of high-density turbidite deposition. Such flows may also form where sediment is fluidized by an outside event, say an earthquake, and the sediment then moves downslope with the water escaping upward as the grains settle downward through the moving mass. If the slope is steep enough, such flows may turn into turbidity currents as the turbulence increases. If they do, when they again slow they may return to liquified flow and the resulting deposit will not record the effects of the turbidity current. As the sediment is deposited, water escape structures will form. The deposit may be completely massive or show some vague planar laminae resulting from laminar flow. Because these are slow-moving flows, and very concentrated, they do not erode the underlying sediment, thus they tend to have sharp depositional bases. They often form the upper part of the sequence generated by sandy turbidites (see the following) and should not be thought of as forming exclusively and separately from other types of deposits.

Coarse-grained sediment (sands or gravels) does not fluidize easily because of the large amount of thoroughly connected pore space and the tight packing of the grains, not to mention the mass of the grains themselves. Fluidization occurs when the pore space is expanded by water moving through the sediment. If the sediment contains a large percentage of poorly consolidated clays, when the mass is shaken or otherwise disturbed, the fabric can collapse to squeeze water into the pore space and the sediment will liquify. This happens commonly in clays that have a distinct packing that collapses during applied stress; such clays are termed **thixotropic.** Note that thixotropic behavior can occur once or twice in a given deposit but not repeatedly, because the packing becomes tighter as a result of this behavior.

FIGURE 8.5 Dish structures and pillars. From D. R. Lowe, 1982, *Journal of Sedimentary Petrology,* 52, Fig. 3B, p. 281. Reprinted by permission of SEPM.

Once the sediment is flowing, the duration of flow depends on the amount of water available. Because of the limited amount of pore water in a mass of sediment, even large flows would stop within a matter of hours if only liquified flow is the transport mechanism. Such flows would only be important very near the source area, because they would not have time to move great distances. When the flow slows as the water finally escapes, the sediment freezes from the frictional interaction of the grains themselves. The flow will stop from the bottom up. The water moves upward as the sediment at the base freezes, forming **dish structures** (Fig. 8.5). Dish structures are often associated with **pillar structures,** which form as columns of water move through the sediment during dewatering.

TURBIDITY CURRENTS AND THEIR DEPOSITS

If turbulence within the flow is the primary mechanism that keeps grains supported, the flow is called a **turbidity current.** For a sediment flow to be dominated by turbulence it must be quite fluid (remember R_e). Such flows form in areas where sediment is accumulating subaqueously above or near a slope, so that as the sediment moves downslope it can mix with water. As mixing proceeds, the result is a fluid that flows turbulently down the slope with tremendous speed. Turbidity flows can form in any situation where sediment and water mix to form dense fluid and can be composed of any sediment as long as the water content is high. The composition of turbidity currents is quite varied and they deposit a wide range of sediment types and sedimentary structures, depending on the concentration and grain size of the sediment. The turbulence of turbidity currents affects different grain sizes of sediment in different ways and we can best describe their mechanisms of flow and their resulting deposits by examining three populations of sediment sizes: (1) mud and fine- to medium-grained sand; (2) coarse-grained sand to small pebbles; and (3) pebbles and cobbles.

Sediment in the mud to sand size range will be suspended by the turbulence within the flow regardless of the concentration of sediment. Even very low density flows, if they maintain a high enough velocity, will keep mud and sand in suspension, including sediment-free water (remember the Hjulstrom diagram from Chapter 4). This means that **low-density turbidity currents** with sediment concentration less than approximately 20% can transport and deposit clay, silt, and sand. As in flowing water, when the velocity of flow decreases, sand will be deposited from bedload and the mud from suspension load. For turbidity flows containing coarse-grained sand to small pebbles, the turbulence is not great enough to keep these grains in suspension unless the density of the flow is fairly high. If the flow contains a large percentage of sediment, then grains are supported by turbulence and impeded from settling by the high concentration of grains and the buoyancy of the finer-grained matrix (population 1 above). Low-density turbidity flows cannot transport this coarser-grained material (populations 2 and 3 above).

Low-density turbidity currents deposit sediment carried by both suspension and bedload transport. As a flow decelerates, the coarser-grained material carried in suspension by the turbulence will move onto the bed and be deposited by traction. Traction sedimentation occurs at high velocities and mostly in relatively thin flows, so that upper flow regime plane beds are formed (Tt in Fig. 8.6, for turbidite traction unit). As the velocity continues to decrease, lower flow regime bedforms create ripples and/or large ripples (upper part of Tt in Fig. 8.6) and so form cross laminae. Because the velocity is decreasing rapidly at this stage, sediment may fall from suspension at a high rate to form climbing ripples. With continued deceleration, the silt in suspension settles onto the bed, where it may be modified by the waning traction currents to form laminae characteristic of the lower flow regime plane bed (Ts in Fig. 8.6, for turbidite suspension unit). Finally, as the flow comes to a stop, the remaining suspended sediment rains from the overlying water to form a deposit of silt and clay, which often grades into the nonturbidity current **hemipelagic** suspension sedimentation composed of a mixture of terrigenous clay and pelagic organic sediment (iTp in Fig. 8.6, for

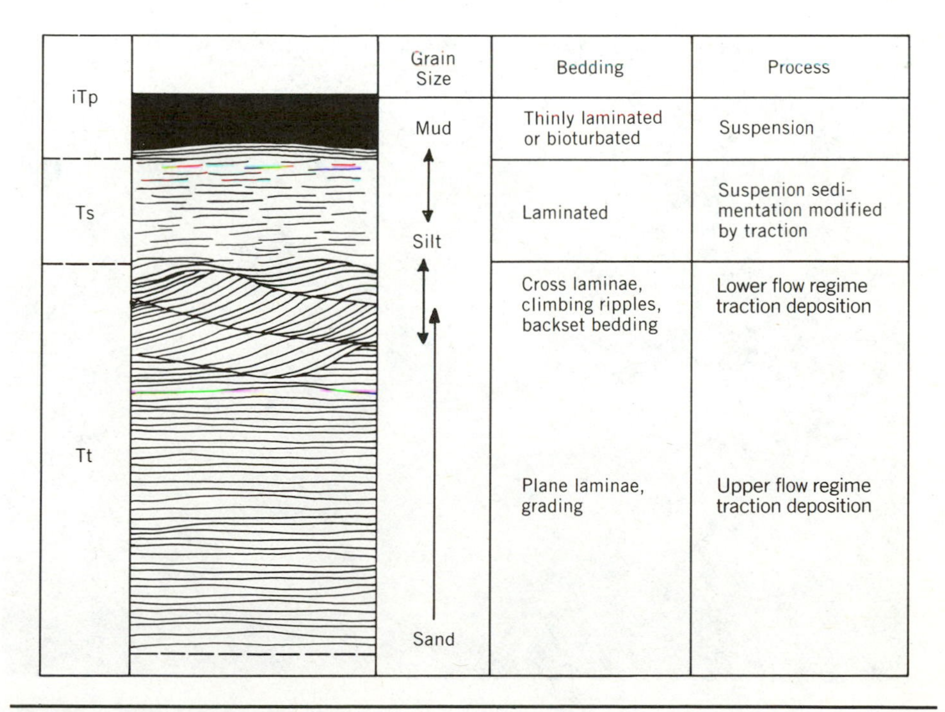

iTp
Ts
Tt
Grain Size
Mud
Silt
Sand
Bedding
Thinly laminated or bioturbated
Laminated
Cross laminae, climbing ripples, backset bedding
Plane laminae, grading
Process
Suspension
Suspenion sedi- mentation modified by traction
Lower flow regime traction deposition
Upper flow regime traction deposition

FIGURE 8.6 Sedimentary sequence generated by a low-density turbidity current.

interturbidite pelagic unit). This sequence represents the ideal deposit from a low-density turbidity current (a **turbidite**) and was first described by Bouma (1962) in flysch deposits of the Alps. Bouma used the units Ta, Tb, Tc (Tt above), Td (Ts above), and Te (iTp above) as well. The classic **Bouma sequence** has been very useful in describing turbidites but has been replaced by models that incorporate a greater understanding of the restrictions on processes governing sedimentary sequences produced by turbidity currents.

The high-velocity, thin flows of low-density turbidity currents can form antidunes as well as plane beds (within unit Tt above). Antidunes have been reported from several turbidites and form distinct **backset bedding** within some turbidites. If the turbidity flows are thick enough, large-scale cross bedding can also form as large ripples migrate along the bed. Because the flow is very turbulent and unsteady, these features tend to mix with the plane beds, backset bedding, and ripple cross laminae to create a hodgepodge of complex sedimentary structures (Fig. 8.7). Even hummocky bedding has been described from turbidite layers. A complete range of these structures can form by low-density turbidity currents to produce sequences that nearly always deviate from the ideal sequence described by Bouma.

Among the fluid flows, only a **high-density turbidity current** can by itself

FIGURE 8.7 Backset bedding in a turbidite sequence. (Photo courtesy of Dr. Richard N. Hiscott).

support and transport grains larger than sand and, therefore, form very different sequences than those produced by low-density turbidity currents. In high-density flows the matrix is important in adding a buoyant lift to the grains, and turbulence, combined with collisions of grains during the flow, keeps the sediment entrained. When sediment load reaches approximately 20 to 30%, the grains are probably significantly hindered from settling by grain-to-grain collisions and resulting dispersive pressure. If the concentration of grains falls below this level, the larger grains will be deposited and the flow stops. Only very high turbulence could keep the coarser fraction entrained at low sediment concentrations. The mud and sand population will be transported at all sediment concentrations and turbulences, so that there is a complete gradation between sediment transported by low-density flows and that by high-density flows.

However, structures formed by high-density turbidity flows differ considerably from those found in low-density turbidites. Within high-density flows, the grain size of the material carried controls the resulting deposits. The grains are supported by turbulence and settling is hindered by the high concentration of finer sediment in the matrix of the flow. As velocity of the flow decreases, the sediment accumulates on the bed by three mechanisms: (1) traction load, (2) traction carpet accretion, and (3) suspension sedimentation. During the first stages of slowing, plane bed laminae form from high-velocity flows as do large-ripple cross laminae. Because the current is very turbulent and unsteady, the bedforms are strongly mixed together and form complex bedding packages (Fig. 8.8).

Because of the large variation in velocity, the flow may alternately erode its bed and deposit, so that erosional surfaces may separate complex packages of bedding structures. With continued velocity decrease, sediment concentrates at the base of the flow. Grains collide and are incorporated into the traction load as sediment falls from suspension, forming a **traction carpet.** The traction carpet moves as a result of the energy from grain collisions as they accumulate on the bed, similar to the traction carpet formed by saltating grains in wind transport. When the traction carpet becomes too densely packed to move, sediment freezes into a lamina or layer. Another carpet forms above as more grains accumulate from the suspension, thus forming a sequence of layers or laminae with different grain sizes (Fig. 8.9). The grain size of these layers generally controls the thickness, with the coarser ones being thicker. Because of the strong velocity gradient within the traction carpet, inverse grading can form. (These types of deposits are very common in the Ta unit of the Bouma sequence and proximal turbidites of other authors.)

If the velocity drop is very rapid, a dense liquified bed forms that is devoid of traction structures. As water moves upward in escaping the flow, sediment is liquified. The water flowing from the sediment forms dish structures and pillar structures. This process allows nearly instantaneous settling of suspended sediment from high-density, sandy turbidity currents and forms very thick deposits of sand. These deposits are massive or show poor grading and primary water

FIGURE 8.8 Sedimentary sequence generated by high-density turbidity current. Modified from D. R. Lowe, 1982, *Journal of Sedimentary Petrology*, 52, Fig. 8, p. 288. Reprinted by permission of SEPM.

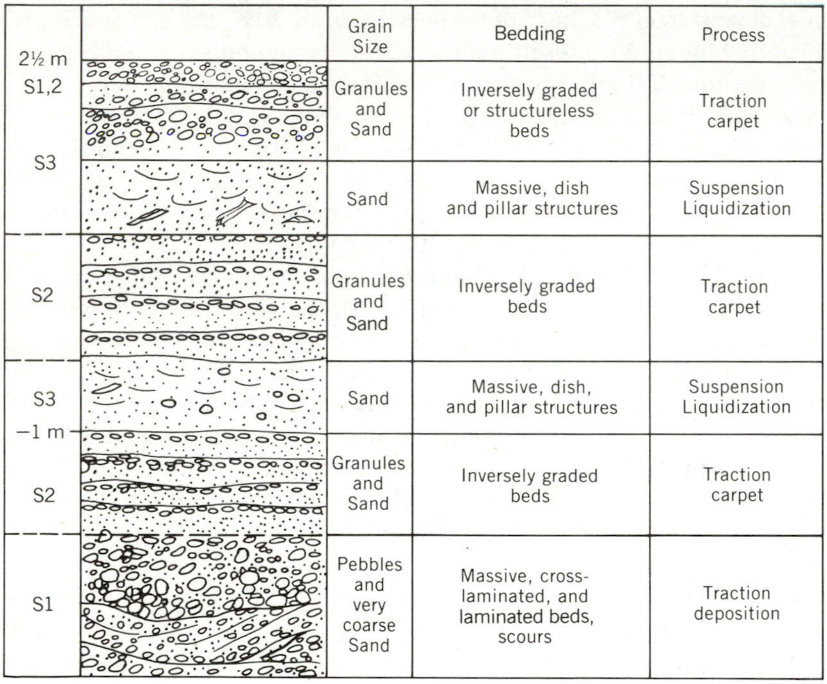

		Grain Size	Bedding	Process
2½ m S1,2		Granules and Sand	Inversely graded or structureless beds	Traction carpet
S3		Sand	Massive, dish and pillar structures	Suspension Liquidization
S2		Granules and Sand	Inversely graded beds	Traction carpet
S3		Sand	Massive, dish, and pillar structures	Suspension Liquidization
−1 m S2		Granules and Sand	Inversely graded beds	Traction carpet
S1		Pebbles and very coarse Sand	Massive, cross-laminated, and laminated beds, scours	Traction deposition

escape structures and have been misinterpreted in the past as grain-flow deposits. The ideal sequence of a sandy, high-density turbidite changes from traction load at the base to a sequence of traction carpet deposits and an uppermost, thick, liquified-flow deposit (Fig. 8.10). The different packages can mix to form a complex assemblage of sedimentary structures.

The above-mentioned structures will form in turbidity flows that do not carry a large percentage of very coarse grained material. If a large range in grain sizes is present and pebbles and cobbles are included, the resulting sedimentary structures are somewhat different. When these gravelly, high-density turbidity flows decelerate, the coarser grains concentrate near the base of the flow because of their large mass. Grain dispersive pressure is the main mechanism that keeps grains supported. Because pebbles and cobbles are too large to move by traction, no traction structures are present (i.e., cross laminae, ripple marks, etc.). Instead, traction carpets deposit layers of coarse-grained sediment so that the lower part of the bed may be inversely or normally graded. Inverse grading forms during the traction carpet deposition and is caused by the dispersive pressure of the grains whereby large grains move upward in the flow into zones of least shear stress. Normally graded beds form from suspension deposition as the velocity drops. After the coarser material has settled out, the remaining sandy fraction may continue on to form high-density, sandy turbidites farther downslope (Fig. 8.10).

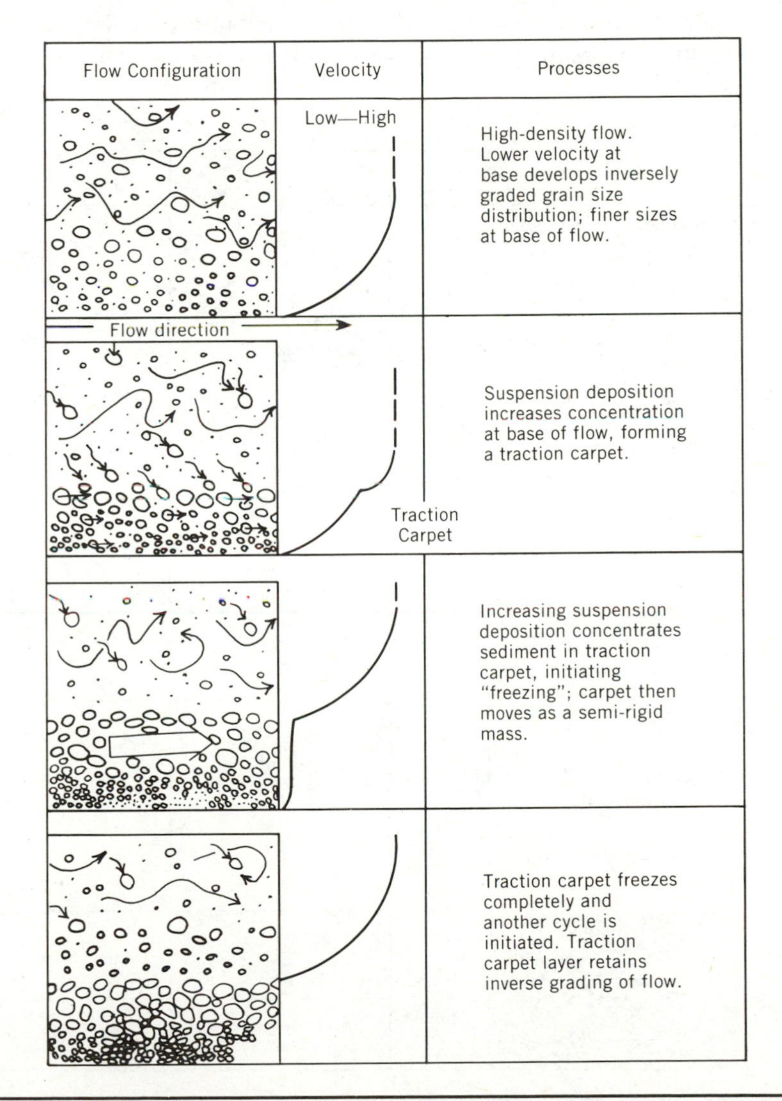

FIGURE 8.9 Diagram depicting the formation of traction carpet deposits. Modified from D. R. Lowe, 1982, *Journal of Sedimentary Petrology*, 52, Fig. 7, p. 287. Reprinted by permission of SEPM.

Turbidites are extremely complex and can contain a variety of sedimentary structures and sequences. The type of turbidite and the resulting sequence depends on the concentration of sediment and the grain sizes deposited. While keeping all the foregoing discussion in mind, let us discuss the changes we might expect as a turbidity current flows away from its source. We shall assume that the current contains all possible ranges of grain sizes so we can build an ideal sequence. Figure 8.11 shows such an idealized distribution. At the base of the figure we see the dominant type of turbidite versus distance from the start of

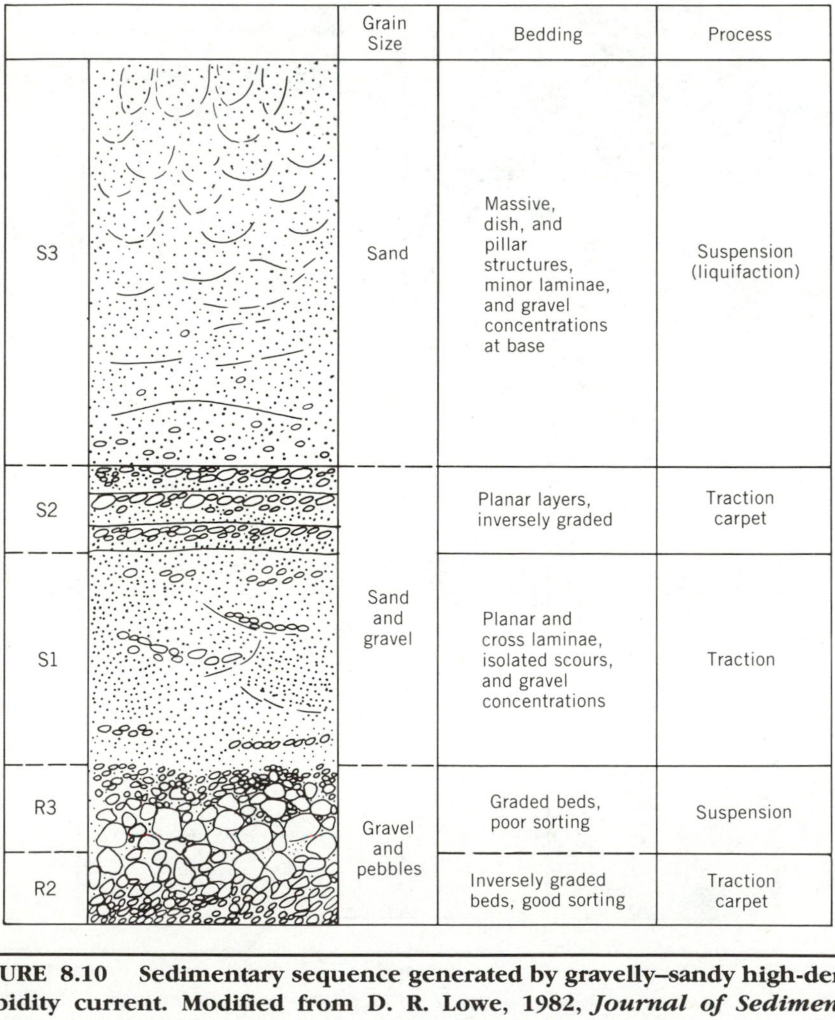

	Grain Size	Bedding	Process
S3	Sand	Massive, dish, and pillar structures, minor laminae, and gravel concentrations at base	Suspension (liquifaction)
S2	Sand and gravel	Planar layers, inversely graded	Traction carpet
S1		Planar and cross laminae, isolated scours, and gravel concentrations	Traction
R3	Gravel and pebbles	Graded beds, poor sorting	Suspension
R2		Inversely graded beds, good sorting	Traction carpet

FIGURE 8.10 Sedimentary sequence generated by gravelly–sandy high-density turbidity current. Modified from D. R. Lowe, 1982, *Journal of Sedimentary Petrology*, 52, Fig. 11, p. 291. Reprinted by permission of SEPM.

FIGURE 8.11 Types of sediment flow processes expected at relative distances from the source. Modified from D. R. Lowe, 1982, *Journal of Sedimentary Petrology*, 52, Fig. 10, p. 290. Reprinted by permission of SEPM.

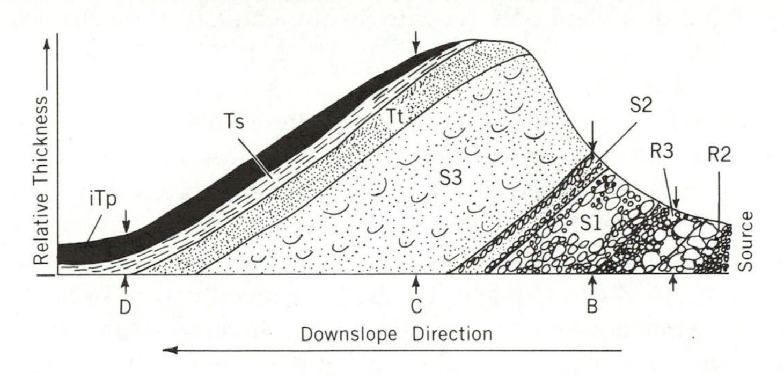

the flow. We can predict the possible vertical sequence by passing a line vertically through the figure at any particular point along the base. If we want to predict the vertical sequence very near the source we pass a vertical line through the figure at that point. Such a line near the right side of the figure (point A) shows that only gravel from high-density turbidity currents is deposited (units R2 and R3). Moving farther downslope (point B), basal gravels (S1) underlie traction carpet and traction deposits from sandy flows (S2). Farther away still (point C), we see that the vertical sequence could contain a thick, basal, suspension deposit (liquified flow, S3) and low-density turbidites above (Tt, Ts, iTP). Only suspension deposits (Ts, iTP) of low-density turbidites can form at relatively large distances from the source (point D). This diagram predicts that only certain packages of sedimentary structures and grain sizes can occur together and the association is dependent on the distance from the source of the flow. Even though this is an idealized model, it is a powerful tool in determining the relative position of real sequences deposited as turbidites.

DEBRIS FLOWS, MUDFLOWS AND THEIR DEPOSITS

Debris flows and mudflows depend on the cohesive strength of the matrix to keep grains supported during flow. Enough water mixes with the sediment so that a viscous, plastic material forms, but not enough to form a more liquid, turbulent flow. Mudflows and debris flows form either subaerially (although they may be under water in streams and water saturated) or subaqueously. The flow in both cases depends on the strength of the matrix, and the larger grains are supported by the high viscosity and density of the matrix–water mixture. Although these processes have been observed by many people in subaerial environments, there is very little known about their subaqueous counterparts. Work by Blackwelder in 1928 stands as a basis for the more modern approaches to mechanisms of movement and dynamics of mudflows and debris flows by Hooke (1967) and Johnson (1965, 1970). Subaqueous mudflows and debris flows have been studied and described only recently in the context of mechanism of flow by Bagnold (1954, 1956), Hampton (1970, 1972, 1975, 1979), and Middleton and Hampton (1973, 1976).

The textures of the deposits resulting from these flows can vary significantly depending on the concentration of coarse grains. In flows carrying boulders but with a large component of fine material, a matrix-supported deposit is formed with the grains separated by the matrix. These are generally, but not exclusively called **mudflows.** The separation of the clasts occurs as the flow solidifies and the larger grains are frozen in place. The bed is generally massive because there is little or no turbulence and traction deposition. Inverse grading is a common feature of these deposits. On the other hand, if the concentration of large grains is great enough, the grains will be in contact with one another but the matrix

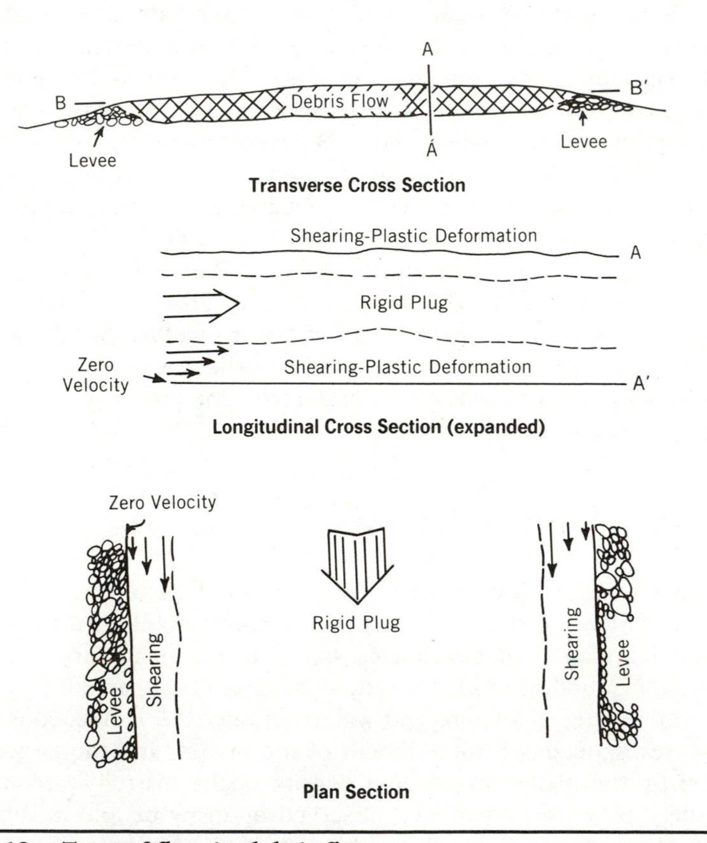

FIGURE 8.12 Type of flow in debris flow.

still acts as the transport agent. These flows result in a massive deposit that has a matrix between grains but is somewhat or entirely grain supported. These deposits are generally called **debris flows.**

Debris flows and mudflows are episodic events that, on land, result from torrential rainfall in regions of little vegetation. In desert regions they commonly form as a result of heavy rains that wash material together in streambeds. The torrential mixture of mud and coarser-grained material can travel many kilometers before the sediment dewaters and the flow stops. The dense and viscous matrix supports larger sediment that can be transported many kilometers. Only very low percentages of clay are required for the matrix to form a cohesive mass that will support large clasts. In a debris flow described by Curry (1966), the matrix contained 1% clay but the density of the matrix was 2.5 g/cc, approximating that of most sediment grains. This resulted from the very low water content of only 9%. Such flows may transport very large boulders, sometimes up to 2 to 3 m in diameter. This is possible because the density of the matrix approximates that of the boulders and the matrix has a cohesive strength, so boulders are actually carried on the matrix. As grains bump into one another the dispersive

FIGURE 8.13 Photograph of amalgamated traction deposits in a volcanic ash sediment flow. Eocene Lamar River Formation, Mount Hornaday, Yellowstone National Park. (Photograph by W. J. Fritz, 1982).

FIGURE 8.14 Types of bedding and associated processes of sedimentation for different types of debris flows and mudflows. With additions from D. R. Lowe, 1982, *Journal of Sedimentary Petrology*, 52, Fig. 13, p. 294. Reprinted by permission of SEPM.

Bedding	Processes
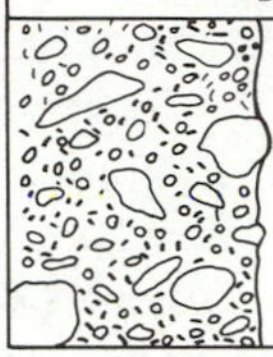 Massive, matrix-supported, clast suspended in matrix, possibly inversely graded	Mudflow with high percentage of fine-grained material; subaerial/subaqueous
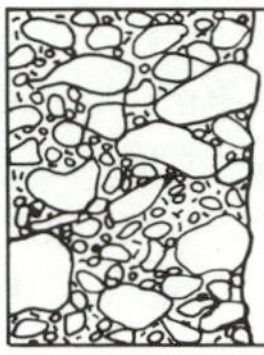 Massive, clasts supported, muddy matrix fills space between clasts	Debris flow with low percentage of fine-grained material; subaerial/subaqueous
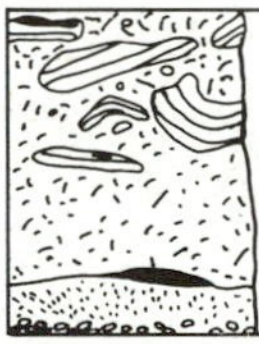 Stratified, basal layer containing laminae overlain by matrix-supported, inversely graded sequence; soft-sediment deformation	Suspension Sedimentation at base from turbulent mudflow, freezing of remaining mass by dewatering; subaqueous
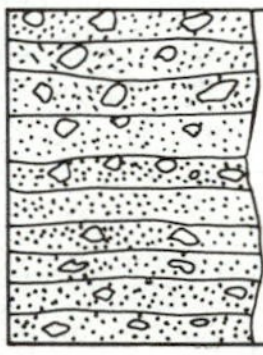 Rhythmic layering of poorly sorted sediment, matrix-supported larger clasts, monolithic (volcanic)	Amalgamated traction carpets from fluid, turbulent mudflows; subaerial (volcanic ash)

pressure of grain interaction probably helps keep them in transportation and often produces flows that are weakly inversely graded. These flows can move on slopes as low as 1 to 2 degrees on land.

Deposition from mudflows and debris flows occurs when the internal stress is overcome by the cohesion of the matrix. The flow then freezes as the matrix solidifies. The flow is mass dependent so that on the edges of a flow where there is little mass and the flow thins, material is deposited to form plugs of material or levees (Fig. 8.12). Sediment may also solidify on the top or bottom of the flow, where the velocity is lower, and often a plug of solidified material is carried along within the flow. This mass solidification forms very distinct textures within mudflow and debris flow deposits.

In subaqueous mudflows and debris flows, where the matrix may be quite

FIGURE 8.15 Massive, fine-grained sediment flow (hyperconcentrated stream flow) deposit from Mount St. Helens, Washington, on the (top) lower Cowlitz River and (bottom) North Fork Toutle River. (Photographs by W. J. Fritz, 1982).

FIGURE 8.16 Conceptual diagram relating various types of mass flow deposits. Arrows represent processes that evolve into another type and bars represent gradational processes that do not lead to another type of process. Modified from D. R. Lowe, 1982, *Journal of Sedimentary Petrology*, 52, Fig. 12, p. 292. Reprinted by permission of SEPM.

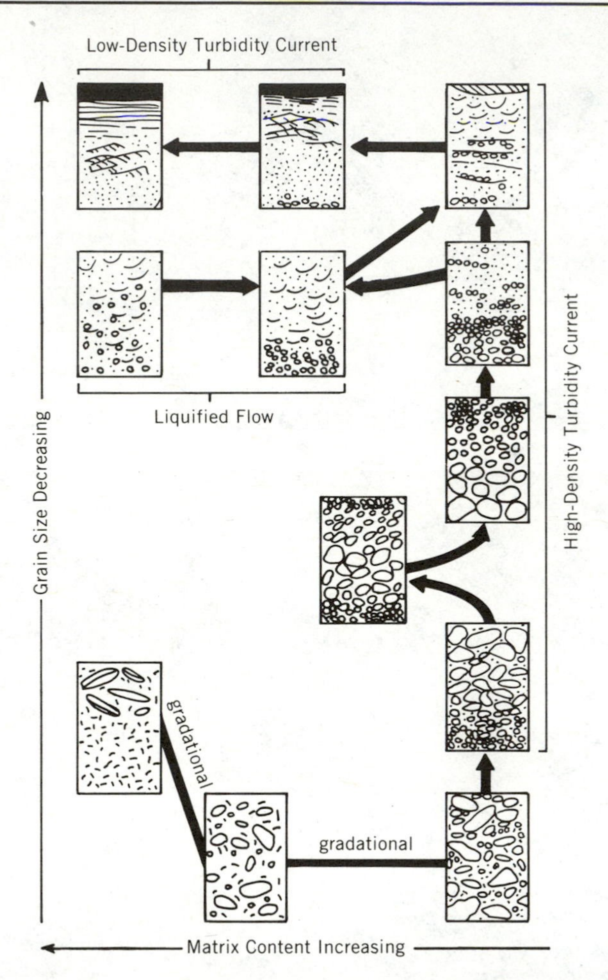

fluid, coarser material concentrates at the bottom of the flow. These flows may move with some turbulence, which helps keep grains supported, so that when the velocity decreases the largest grains settle out of suspension to form a basal grain-supported layer. The matrix-supported remainder of the flow freezes and forms a structureless, matrix-supported unit. Mudflow deposits vary depending on the type of material and fluidity of the flow (Fig. 8.13).

FIGURE 8.17 Fine-grained horizontally stratified deposit (probably hyperconcentrated stream flow) along the Cowlitz River, Washington, from the 1980 eruption of Mount St. Helens. (Photograph by W. J. Fritz, 1980).

In all cases, the features distinguishing a mudflow are massive bedding and matrix support of the clasts, although some exhibit weak inverse or symmetrical grading. For deposits of only fine material (mud and sand), the sedimentary textures are not clearly developed or are absent. These are common deposits in volcanic regions, where fine-grained ash mixes with water to form fine-grained mudflows or **lahars.** These flows are quite fluid and probably contain some turbulence as they flow. The resulting deposits show laminae that may form as amalgamated traction carpets (Figs. 8.13 and 8.14), but generally they are massive with coarser grains supported by the matrix (Fig. 8.15). Figures 8.16–8.23 illustrate additional features of mud and debris flows (lahars) from recent volcanic activity at Mount St. Helens and Nevado del Ruiz in Colombia.

FIGURE 8.18 Horizontally stratified sediment flow deposit along the North Fork of the Toutle River from recent mudflows at Mount St. Helens, Washington. Note basal graded gravel, middle unit of horizontally stratified sand, and weak inverse grading with the largest clasts on the surface of the flow. (Photograph by W. J. Fritz).

FIGURE 8.19 Graded bed of conglomerate along the North Fork of the Toutle River from a 1982 sediment flow at Mount St. Helens, Washington. (Photograph by W. J. Fritz, 1982).

FIGURE 8.20 Conglomerate from the 1985 lahar generated by the November eruption of Nevado del Ruiz, Colombia. Deposit along the headwaters of the Rio Guali. (Photograph by W. J. Fritz, 1985).

FIGURE 8.21 Lahar deposit along the Rio Guali at Mariquita, Colombia, from the November 1985 eruption of Nevado del Ruiz. Note the giant boulder (ca. 10 m in diameter) that was moved for a considerable distance by the flow. Flow velocity was about 28 km/hr at this locality. (Photograph by W. J. Fritz, 1985).

FIGURE 8.22 Thin lahar deposit less than 1 m deep at Armero, Colombia, along the Rio Lagunillas from the 13 November 1985 eruption of Nevado del Ruiz. This flow destroyed the city, resulting in the loss of more than 23,000 lives. The city was pulverized and swept downstream by the lahar, which was moving at approximately 38 km/hr with a 35–36% solids by volume content. (Photograph by W. J. Fritz, 1985).

EPILOGUE

And yonder, whenever the roads of the rain
come forth,
Torrents will rush forth,
Silt will rush forth,
Mountains will be washed out,
Logs will be washed down,
Yonder all the mossy mountains will drip with water.
The clay-lined hollows of our earth mother
Will overflow with water. . . .

Zuñi Indian Prayer

FIGURE 8.23 Massive, matrix-supported debris flow deposit along the North Fork of the Toutle River from the May 1980 eruption of Mount St. Helens, Washington. (Photograph by J. N. Moore, 1980).

The mechanisms and deposits discussed so far define a continuum and do not reside in distinct, unvarying classes. Each of the forms can move into other types of flow and few are restricted. The only flow type separated from the others is grain flow, which only forms thin, slipface deposits on large dunes. It is important to realize that all types of sediment flows grade into one another and one type may evolve into another during the lifetime of a flow. These concepts are presented in Fig. 8.14, which is modified from Lowe's compilation of sediment gravity flow mechanisms. Nearly all subaqueous mass flows are initiated by a slide or slump of competent material. When the mass mixes with water, a sediment flow is created that can take one of several paths. The deposits resulting from these different types of flow are distinct and represent the last stages of flow during deposition. They could have had, and probably did have, a varied history before deposition. When deciphering the mechanisms that formed sediment flow deposits, one must always keep these principles in mind or an unrealistically simple picture will emerge from the analysis.

OUTSIDE READING

Grain Flows

Link (1975); Lowe (1976a); Middleton and Hampton (1973, 1976); Stauffer (1967); Van der Kamp *et al.* (1973).

Turbidity Flows

Bouma (1962); Chan and Dott (1983, 1986); Middleton and Bouma (1973); Mutti and Ricci-Lucchi (1972); Nelson and Nielson (1984); Nielson and Abbott (1981); Skipper (1971); Stanley and Kelling (1978); Walker (1975b).

Debris Flows, Mudflows, and Lahars

Bagnold (1954, 1956); Blackwelder (1928); Curry (1966); Fisk (1974); Fritz and Harrison (1985a, b); Hampton (1970, 1972, 1975, 1979); Harrison and Fritz (1982); Hooke (1967); Janda *et al.* (1981); Johnson (1965, 1970); Lowe (1976a, b, 1979a, b, 1982); Middleton (1969); Middleton and Hampton (1973, 1976); Naranjo *et al.* (1986); Nardin *et al.* (1979); Naylor (1980); Pierson (1982, 1985); Pierson and Costa (1984); Pierson and Scott (1985); Sanders (1965); Smith (1986); Waitt *et al.* (1983).

Chapter 9

Facies and Modeling Environments of Deposition

INTRODUCTION

In the previous chapters, we have developed the tools that you will need to understand the processes that deposit sediment and form sedimentary structures. Let us now return to some of the material presented in Chapter 1 to organize these concepts and discuss the philosophy of models of depositional environments. Such models are essential for making final interpretations of earth history and paleoenvironments from the details of sedimentary structures and textures and stratigraphic relationships. One of the basic concepts of this organization is that of equivalency. Sedimentary packages can be equivalent in a number of ways (see Chapter 1). If we compare a sequence from different regions and consider the units as time-stratigraphic equivalents, we concede that they were deposited during the same (or nearly the same) interval of time. The establishment of this time equivalency is often quite difficult. We could use radiometric dating, if the appropriate rock types were included in the sequence, or we could use the construction of biostratigraphic zones from fossil assemblages. Whatever the procedure, the importance of such a correlation would be the fact that the two sequences represent the same interval of time. The most gross time correlation is the establishment of correlative sequences of the various geologic systems throughout the world. Rocks designated Ordovician in North America are considered time equivalent to Ordovician strata in Europe, Asia, or any place else on the globe. On the finest end of the time-correlation scale is the correlation of individual time-equivalent beds or events, for example, individual volcanic ash beds or flows or deposits from meteorite impacts. The middle ground is mostly constructed from biostratigraphic correlation using various fossil zones.

The establishment of time equivalency is a major goal of stratigraphy but commonly one that is unattainable. However, lithologic (or physical) correlation is another type of equivalency that commonly can be established but differs radically from time correlation. Lithostratigraphic units can be correlated from place to place using the basic constructs of physical stratigraphy that we developed in Chapter 1. An excellent example of lithologic correlation is the systematic establishment of formations and the mapping of those formations throughout a large region. In such work, descriptive parameters and stratigraphic sequence guide the correlation. Although a particular formation, for example, the Lower Cambrian Poleta Formation of the western United States, may be traced over many thousands of square kilometers, it is done so only on physical characteristics and has no inherent time equivalency. In fact, it is extremely unlikely that a lithostratigraphic unit is also a time-stratigraphic unit, except in the case of an episodically deposited unit such as a volcanic ash fall, a storm deposit, or a turbidite flow. This basic precept underpins the concepts of **facies** and its grandchild, the **facies model.**

Johannes Walther in 1893 was the first to thoroughly describe the relationship between time equivalency and lithologic equivalency (see Chapter 1). He de-

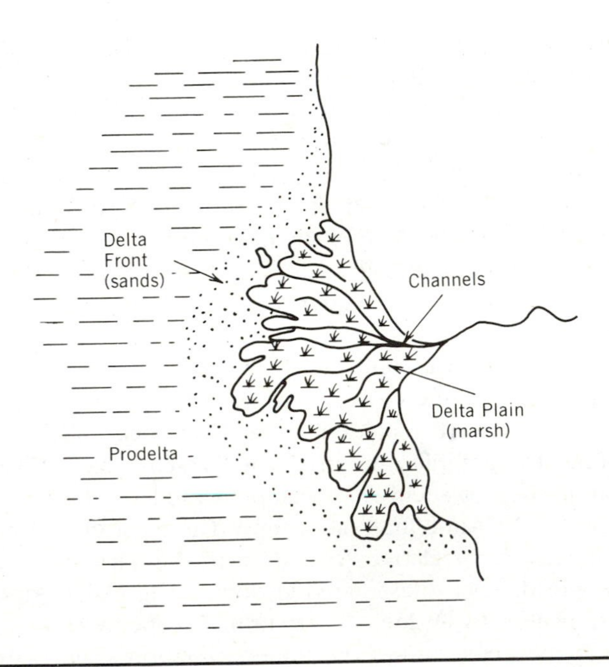

FIGURE 9.1 Facies map for a delta.

FIGURE 9.2 Lithologic units generated from the progradation of facies of the delta model shown in Fig. 9.1. Note that time lines transgress the lithologic units.

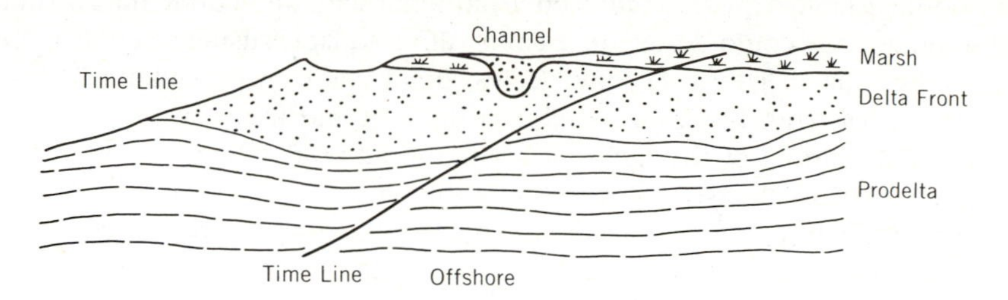

termined that a vertical succession of strata would be formed by the lateral migration of adjacent environments containing the original sediments that eventually became rocks. An excellent example of this process is the progradation of a delta (Fig. 9.1). A simple model of such a deposit shows that the system is composed of different environments of deposition. Each of these environments has different sets of processes affecting sedimentation, resulting in different types of sediments. On the delta plain surrounding the river channel that brings sediment to a delta, marsh sediments are deposited, whereas coarser-grained sediment is carried in the channel. As the channel enters the ocean, sediment is deposited as the velocity decreases. Bars of sediment derived from the channel develop near the river mouth and extend to shallow depths in the ocean. Finer-grained sediment carried in suspension by the river currents settles out father from the river mouth. This results in a lateral succession of sediment types deposited in different environments under different sets of processes. As the delta continues to grow it progrades into the sea, leaving behind a record of these different environments. It is extremely important to see that time lines in this sequence cut across stratigraphic units (Fig. 9.2). The units themselves are time transgressive, that is, the different, distinctive units are not time equivalent, only lithologically equivalent. If such deltas prograde long enough, a distinctive vertical sequence of stratigraphic units will be formed over a broad region as a result of the migration of the different depositional environments. Such deposits were extremely common throughout what is now the Rocky Mountains as the result of large deltas that prograded into the Cretaceous inland sea that covered the central United States.

In the rock record, such sequences record the paleogeography of the site of deposition. Different sedimentary processes dominate different environments and produce characteristic deposits. The concept of facies is intimately tied to these differences. The term facies has been used in many different ways to address the fact that different types of deposits form at approximately the same time. In the deltaic example given, the sequence could be divided into the sediment deposited in each of the separate environments: marsh, channel, delta bar, prodelta, and offshore. Each of these would have a particular set of sediment types and sedimentary structures associated with the different processes acting at each place. If we could establish these detailed environments from the sedimentary structures and sequence, we could designate them as different facies using an environmental designation: marsh facies, channel facies, delta bar facies, prodelta facies, and offshore facies. This, however, is often a difficult and imprecise task. If we did not know the origin of the different stratigraphic units when we examined the sequence (the usual case, for remember that stratigraphers earn their keep by deciphering such sequences), we would not be able to apply these names without considerable work and interpretation. To be cautious, we might rather apply some specifically lithologic definition to the different rock packages representing the major type of sediment in each of the facies.

If we examine the general processes that are actively depositing sediment in each of the environments, we can see that the resulting deposits would be relatively diagnostic. We would expect the marsh to contain very organic-rich muds because it would be covered by plants and receive sediment mainly by flooding from the channel. The channel itself would contain coarser-grained sediment, possibly sand moving as bedforms, and therefore be cross-bedded. The delta bar deposits also would likely be dominated by sand, but would probably contain somewhat different sedimentary structures because of the influence of waves and tides on the deposits. The prodelta and offshore deposits would be dominated by mud that settled from suspension, but because the prodelta deposits are adjacent to the river mouth, they would likely have more sand and contain more features produced by waves and tides. So, if we looked at the deposits carefully, we could probably distinguish them easily by physical (and biological) characteristics without referring to the actual environment. We then would have facies with lithologic designations: coal-organic mudstone facies (for the marsh deposits), trough-cross-bedded sandstone facies (for the channel deposits), tabular cross-bedded and rippled sandstone facies (for the delta bar deposits), rippled sandstone and bioturbated mudstone facies (for the prodelta deposits), and the bioturbated-laminated mudstone facies (for the offshore deposits). Although this terminology seems more cumbersome, it is considerably more precise because it does not depend on the interpretation of sedimentary structures and textures. Such facies designations, which rely mostly on the lithologic characteristics of the rocks, yield rock packages termed **lithofacies.** When lithofacies become complex and encompass several types of sedimentary structures and rocks, they are commonly given number or letter designations. A. D. Miall, in 1978, introduced such a scheme to describe the complex packages of sediments produced by braided rivers. His pioneering work has led to the adoption of similar descriptive lithofacies schemes to many sequences of strata resulting from many different processes in varied environments. These schemes are invaluable in determining depositional environments and establishing models that represent different facies models. By carefully establishing lithofacies based on detailed description, it is possible to describe facies without knowing or using preconceived ideas about facies. From the lithofacies designations the processes can be determined. Then a model describing the depositional environment can be formulated that encompasses the assemblage of processes in the different lithofacies. Examples of such lithofacies with codes used as a shorthand for each are shown in Figs. 9.3 and 9.4.

PROCESS–RESPONSE MODELS

Models are extremely useful in sedimentology and stratigraphy just as they are in many disciplines. Such simplifications allow complex systems to be distilled into major components that can be relatively easy to identify in the complexity of natural systems. But all models try to simulate real, complex systems and therefore have limitations. Models come in several types that are extremely different in their usefulness and application. For interpreting sedimentary rocks we tend to use a mixture of two types of models: those based on an actual system (actualistic models) and those based on sets of processes (process–response models). Models based on actual examples can be fraught with limitations. For example, we might carefully examine all the deposits forming in a playa lake. We could then build a model that would contain those descriptions and, when we found a sequence that matched it, make the assumption that the sequence was deposited in a playa lake. This uncritical, "cookbook" approach to facies modeling leads to many misinterpretations. There are many types of playas, some dominated by terrigenous sediments and some by chemical precipitates. The sedimentary processes are extremely different from playa to playa and the major features change over the years as the climate changes. To truly model deposition on playas and interpret deposits formed by ancient playas, we would have to distill information from all modern and ancient playa deposits, clearly an impossible task. But we can develop a precise and useful facies model for playa deposits by examining the processes that operate on modern playas and combine that information with processes determined from ancient playa deposits.

At first this may seem like a big order—it is! But it leads to models that more closely approximate real environments than do limited actualistic models. By monitoring the processes that act in different modern environments in the framework of hydrodynamic and other concepts, we can develop a model that relies on the basic aspects of a particular facies. Such modeling is not limited to modern environments because we can interpret sedimentary structures to determine the processes that formed them. Such models, based on physical, chemical, and biological processes, are inherently more versatile than actualistic models based on specific sites. Process–response models essentially start at the descriptive level and move through a hierarchy of interpretation. The first step in building the model relies on detailed and accurate description of sedimentary structures

Facies Code	Lithofacies	Sedimentary Structures	Interpretation
Gms	Massive, matrix-supported gravel	Massive bedded	Debris flow
Gmu	Massive, unsorted, clast-supported gravel	Normal-inverse graded crudely bedded, no fabric or inbrication	Debris flow
Gm	Massive or crudely bedded gravel	Horizontal bedding imbrication	Hyperconcentrated stream flow, sieve deposits, lag, bars, sediment flow
Gt	Gravel, stratified	Trough cross beds	Channel fill
Gp	Gravel, stratified	Planar cross beds	Linguoid bars or deltaic growth from older bar remnants
St	Medium to very coarse grained sand, may be pebbly, possible volcanic grains	Solitary or grouped trough cross beds	Lower-flow regime dunes
Sp	Medium to very coarse grained sand, may be pebbly	Solitary or grouped planar cross beds	Linguoid, transverse bars, sand waves, lower-flow regime
Sr	Sand, very fine to coarse	Ripple marks	Ripples
Sh	Very fine to coarse grained sand, pebbles	Horizontal lamination	Planar bed flow, upper and lower flow regime
Sl	Sand, fine	Low-angle cross beds (less than 10°)	Scour fills, crevasse splays, antidunes
Se	Erosional scours with intraclasts	Crude cross bedding	Scour fills
Ss	Sand, fine to coarse	Broad, shallow scours including eta cross-stratification	Scour fills
Sse, She, Spe	Sand	Analogous to Ss, Sh, Sp	Eolian deposits
Fh	Silt, mud, ash	Crude horizontal bedding	Planar bed flow, traction carpets
Fl(a)	Sand, silt, mud (ash)	Fine lamination, very small ripples	Overbank, waning flood deposits
Fsc	Silt, mud	Laminated to massive	Backswamp deposits

FIGURE 9.3 Lithofacies codes derived from sedimentary structures and interpretation of fluvial and volcaniclastic deposits. Fluvial codes from Miall (1978) with addition of various units for volcaniclastic environments from Fritz and Harrison (1985a).

Facies Code	Lithofacies	Sedimentary Structures	Interpretation
Fcf	Mud	Massive, with freshwater molluscs	Backswamp pond deposits
Fm	Mud, silt	Massive, desiccation cracks	Overbank or drape
Fr	Silt, mud	Rootlets	Seatearth
C	Coal, carbonaceous mud	Plants, mud films, paleosols, organic zones	Swamp, lacustrine overbank deposits
P	Carbonate	Soil features	Soil

FIGURE 9.4 Example of using lithofacies codes to interpret sedimentary facies, environments, and processes commonly associated with continental intermediate-composition volcanic fields. From W. J. Fritz, and S. Harrison, 1985, Early Tertiary volcaniclastic deposits of the northern Rocky Mountains. In R. M. Flores and S. S. Kaplan (Eds.), *Cenozoic Paleogeography of West-Central United States*, Rocky Mountain Section of SEPM, Denver, Table 3, p. 387.

Dominant Depositional Environment	Common Processes	Lithofacies Codes
Vent facies	Intrusions (dikes, sills), lava flows, vent breccia, pyroclastic flows, welded-ash composed of basalt, andesite, dacite, and rhyodacite	N/A
Proximal to vent facies: high gradient bedload streams, fans, braid plains, and sediment, debris, and mudflows in channels	Debris flow	Gms, Gmu
	Sediment flow, Hyperconcentrated stream flow, sheet flood	Gm, Sh, Gms
	Bar and dune migration, channel	Gt, St, Sr
	Air-fall ash, traction and suspension	Fh, Fla
	Lava flows	
	Welded-ash-flow tuff	Andesite, basalt, dacite, etc. Andesite, dacite, rhyodacite
Medial to vent facies: bedload streams	Bar and dune migration, aggradation in channels	Gm, Gt, Sh, Sp, St
	Sediment flow	Gm, Sh, Gms
	Air-fall ash	Fh
Distal to vent facies: bedload and mixed-load streams, floodplains, flood basins, lakes and swamps	Channel	Gt, St, Sh, Sp, Fla
	Floodplain and lacustrine vertical accretion	Sr, Sh, Fla, C
	Sediment flow	Sh, Fh
	Air-fall ash	Fh

and textures. From these data, hydrodynamic, biological, and chemical concepts can be applied to interpret the processes that formed the deposit. Such interpretations can incorporate experimental, theoretical, and field observations and so provide a much more precise interpretation than comparison to actual examples. Once this stage is reached, the next step is to use all other available information, including sedimentary associations and stratigraphic sequence, to establish the environments of deposition.

Such an approach de-emphasizes the use of actualistic facies models, which rely heavily on producing a set of criteria (a generalized stratigraphic column) to determine the depositional environment of a sequence or package of sedimentary rocks. Instead, it uses such criteria only as part of the data for interpretation of sedimentary characteristics. Interpretive power also lies in the fact that process–response modeling can be applied to sequences of rocks that have no direct counterpart in modern sedimentary environments. For example, the earliest continental environments on the earth were quite different from those forming today because of the absence of land plants and lesser amounts of oxygen in the atmosphere. Very different deposits formed that could not be interpreted by a direct comparison to modern depositional systems. However, by combining basic hydrodynamic concepts, comparison to modern continental depositional

TABLE 9.1
Classification of Depositional Environments

Geographic	Geomorphic	Dominant Processes
Continental	Glacial	I, UF
	Lacustrine	SS, W, SF
	Riverine	UF, Em, Wm
	Alluvial fan	SF, UF
	Desert dune	E
	Desert plain	E, UF
	Volcaniclastic	UF, Wm, E,
Transitional (marginal marine)	Beach offshore	W, T, SS
	Barrier island/lagoon	W, T, E, SS, B
	Estuarine	T, SS, W
	Coastal dunes	E
Marine	Shelf/carbonate platform	W, T, B
	Slope/rise/trench	SF, UG, SS, B
	Submarine fan	SF, SS, B
	Abyssal plain	SS, B
	Volcanic islands	B, SS, SG, W
	Atolls	

Explanation of symbols: UF, unidirectional current; SS, suspension sedimentation; W, wave-oscillating currents; SF, sediment flow, SG, sediment-gravity flow; E, eolian; I, ice processes; T, tidal-oscillating currents; B, biogenic deposition; m, minor constituent.

environments, and using chemical principles, such deposits can be interpreted relatively precisely. There are many examples of such limitations of specific actualistic models that emphasize the need to rely instead on process–response models. Such models work even with limitless varieties of depositional environments. Even though there are a relatively small number of major depositional environments possible on the surface of the earth (Table 9.1), the variety within those environments today and that found in 3.8 billion years of geologic history make it impossible to use only actualistic models.

Depositional environments on the earth can be classified in several different ways. By using a geographic classification, three basic environments exist (Table 9.1): continental, transitional, and marine. But within these broad categories, geomorphic subdivisions can be designated that lead to another 15 or so. Each of these could be further divided into specific types and subtypes, for example, riverine environments could be divided into types based on channel morphology (braided, meandering) or grain size (coarse, fine grained) or combinations of these too (coarse-grained braided system). Further subdivision can be made using specific elements within a system (a point bar in a fine-grained, meandering river system). Each of the major geomorphic divisions has a set of associated processes (Table 9.1). These processes become more specific for the subdivisions of the environments, resulting in a complex, interrelated framework that describes each depositional environment. Thus interrelationship is what allows the lithofacies to be used to work backward toward the environment through a process–response model. There are many textbooks and compilations on facies models, mostly emphasizing actualistic models, and we have listed these in the additional reading. These references cover all the geomorphic environments listed in Table 9.1.

We could select a number of excellent examples to present the techniques of process–response modeling but that would begin us on an intricate discussion of major environments of deposition that are well discussed in other textbooks. It would also fly in the face of the purpose of the book as stated in the preface, namely, to supply you with the basics needed to start the adventure and kindle the fires of inquiry. We end this discussion of the basics of the physical processes of stratigraphy and sedimentation with one final thought, one that we hope entreats you to venture forth into stratigraphy and sedimentation to tackle the myriad unknowns not discussed in this book and advance those that are.

EPILOGUE

Out yonder there was this huge world, which exists independently of us human beings and which stands before us like a great, eternal riddle. . . . The contemplation of this world beckoned like a liberation.

Albert Einstein

OUTSIDE READING

Blatt *et al.* (1980); Boggs (1986); Brenner (1980); Davis (1983); Dickinson (1974); Galloway and Hobday (1983); Miall (1978a, 1978b, 1984); Middleton (1973); Potter (1984); Reading (1986); Reineck and Singh (1980); Selley (1985); Walker (1984); Walther (1893–1894); Wilson (1975).

Appendix A

North American Stratigraphic Code

The North American Commission on Stratigraphic Nomenclature

Reprinted from
The American Association of Petroleum Geologists Bulletin
Volume 67, Number 5 (May 1983)

The American Association of Petroleum Geologists Bulletin
V. 67, No. 5 (May 1983), P. 841-875, 11 Figs., 2 Tables

North American Stratigraphic Code[1]

NORTH AMERICAN COMMISSION ON STRATIGRAPHIC NOMENCLATURE

FOREWORD

This code of recommended procedures for classifying and naming stratigraphic and related units has been prepared during a four-year period, by and for North American earth scientists, under the auspices of the North American Commission on Stratigraphic Nomenclature. It represents the thought and work of scores of persons, and thousands of hours of writing and editing. Opportunities to participate in and review the work have been provided throughout its development, as cited in the Preamble, to a degree unprecedented during preparation of earlier codes.

Publication of the International Stratigraphic Guide in 1976 made evident some insufficiencies of the American Stratigraphic Codes of 1961 and 1970. The Commission considered whether to discard our codes, patch them over, or rewrite them fully, and chose the last. We believe it desirable to sponsor a code of stratigraphic practice for use in North America, for we can adapt to new methods and points of view more rapidly than a worldwide body. A timely example was the recognized need to develop modes of establishing formal nonstratiform (igneous and high-grade metamorphic) rock units, an objective which is met in this Code, but not yet in the Guide.

The ways in which this Code differs from earlier American codes are evident from the Contents. Some categories have disappeared and others are new, but this Code has evolved from earlier codes and from the International Stratigraphic Guide. Some new units have not yet stood the test of long practice, and conceivably may not, but they are introduced toward meeting recognized and defined needs of the profession. Take this Code, use it, but do not condemn it because it contains something new or not of direct interest to you. Innovations that prove unacceptable to the profession will expire without damage to other concepts and procedures, just as did the geologic-climate units of the 1961 Code.

This Code is necessarily somewhat innovative because of: (1) the decision to write a new code, rather than to revise the old; (2) the open invitation to members of the geologic profession to offer suggestions and ideas, both in writing and orally; and (3) the progress in the earth sciences since completion of previous codes. This report strives to incorporate the strength and acceptance of established practice, with suggestions for meeting future needs perceived by our colleagues; its authors have attempted to bring together the good from the past, the lessons of the Guide, and carefully reasoned provisions for the immediate future.

Participants in preparation of this Code are listed in Appendix I, but many others helped with their suggestions and comments. Major contributions were made by the members, and especially the chairmen, of the named subcommittees and advisory groups under the guidance of the Code Committee, chaired by Steven S. Oriel, who also served as principal, but not sole, editor. Amidst the noteworthy contributions by many, those of James D. Aitken have been outstanding. The work was performed for and supported by the Commission, chaired by Malcolm P. Weiss from 1978 to 1982.

This Code is the product of a truly North American effort. Many former and current commissioners representing not only the ten organizational members of the North American Commission on Stratigraphic Nomenclature (Appendix II), but other institutions as well, generated the product. Endorsement by constituent organizations is anticipated, and scientific communication will be fostered if Canadian, United States, and Mexican scientists, editors, and administrators consult Code recommendations for guidance in scientific reports. The Commission will appreciate reports of formal adoption or endorsement of the Code, and asks that they be transmitted to the Chairman of the Commission (c/o American Association of Petroleum Geologists, Box 979, Tulsa, Oklahoma 74101, U.S.A.).

Any code necessarily represents but a stage in the evolution of scientific communication. Suggestions for future changes of, or additions to, the North American Stratigraphic Code are welcome. Suggested and adopted modifications will be announced to the profession, as in the past, by serial Notes and Reports published in the *Bulletin* of the American Association of Petroleum Geologists. Suggestions may be made to representatives of your association or agency who are current commissioners, or directly to the Commission itself. The Commission meets annually, during the national meetings of the Geological Society of America.

1982 NORTH AMERICAN COMMISSION
ON STRATIGRAPHIC NOMENCLATURE

[1]Manuscript received, December 20, 1982; accepted, January 21, 1983.
Copies are available at $1.00 per copy postpaid. Order from American Association of Petroleum Geologists, Box 979, Tulsa, Oklahoma 74101.

CONTENTS

North American Stratigraphic Code

North American Commission on Stratigraphic Nomenclature

North American Stratigraphic Code

North American Commission on Stratigraphic Nomenclature

North American Stratigraphic Code

North American Commission on Stratigraphic Nomenclature

PART I. PREAMBLE

BACKGROUND

PERSPECTIVE

Codes of Stratigraphic Nomenclature prepared by the American Commission on Stratigraphic Nomenclature (ACSN, 1961) and its predecessor (Committee on Stratigraphic Nomenclature, 1933) have been used widely as a basis for stratigraphic terminology. Their formulation was a response to needs recognized during the past century by government surveys (both national and local) and by editors of scientific journals for uniform standards and common procedures in defining and classifying formal rock bodies, their fossils, and the time spans represented by them. The most recent Code (ACSN, 1970) is a slightly revised version of that published in 1961, incorporating some minor amendments adopted by the Commission between 1962 and 1969. The Codes have served the profession admirably and have been drawn upon heavily for codes and guides prepared in other parts of the world (ISSC, 1976, p. 104-106). The principles embodied by any code, however, reflect the state of knowledge at the time of its preparation, and even the most recent code is now in need of revision.

New concepts and techniques developed during the past two decades have revolutionized the earth sciences. Moreover, increasingly evident have been the limitations of previous codes in meeting some needs of Precambrian and Quaternary geology and in classification of plutonic, high-grade metamorphic, volcanic, and intensely deformed rock assemblages. In addition, the important contributions of numerous international stratigraphic organizations associated with both the International Union of Geological Sciences (IUGS) and UNESCO, including working groups of the International Geological Correlation Program (IGCP), merit recognition and incorporation into a North American code.

For these and other reasons, revision of the American Code has been undertaken by committees appointed by the North American Commission on Stratigraphic Nomenclature (NACSN). The Commission, founded as the American Commission on Stratigraphic Nomenclature in 1946 (ACSN, 1947), was renamed the NACSN in 1978 (Weiss, 1979b) to emphasize that delegates from ten organizations in Canada, the United States, and Mexico represent the geological profession throughout North America (Appendix II).

Although many past and current members of the Commission helped prepare this revision of the Code, the participation of all interested geologists has been sought (for example, Weiss, 1979a). Open forums were held at the national meetings of both the Geological Society of America at San Diego in November, 1979, and the American Association of Petroleum Geologists at Denver in June, 1980, at which comments and suggestions were offered by more than 150 geologists. The resulting draft of this report was printed, through the courtesy of the Canadian Society of Petroleum Geologists, on October 1, 1981, and additional comments were invited from the profession for a period of one year before submittal of this report to the Commission for adoption. More than 50 responses were received with sufficient suggestions for improvement to prompt moderate revision of the printed draft (NACSN, 1981). We are particularly indebted to Hollis D. Hedberg and Amos Salvador for their exhaustive and perceptive reviews of early drafts of this Code, as well as to those who responded to the request for comments. Participants in the preparation and revisions of this report, and conferees, are listed in Appendix I.

Some of the expenses incurred in the course of this work were defrayed by National Science Foundation Grant EAR 7919845, for which we express appreciation. Institutions represented by the participants have been especially generous in their support.

SCOPE

The North American Stratigraphic Code seeks to describe explicit practices for classifying and naming all formally defined geologic units. *Stratigraphic procedures* and principles, although developed initially to bring order to strata and the events recorded therein, are applicable to all earth materials, not solely to strata. They promote systematic and rigorous study of the composition, geometry, sequence, history, and genesis of rocks and unconsolidated materials. They provide the framework within which time and space relations among rock bodies that constitute the Earth are ordered systematically. Stratigraphic procedures are used not only to reconstruct the history of the Earth and of extra-terrestrial bodies, but also to define the distribution and geometry of some commodities needed by society. *Stratigraphic classification* systematically arranges and partitions bodies of rock or unconsolidated materials of the Earth's crust into units based on their inherent properties or attributes.

A *stratigraphic code* or guide is a formulation of current views on stratigraphic principles and procedures designed to promote standardized classification and formal nomenclature of rock materials. It provides the basis for formalization of the language used to denote rock units and their spatial and temporal relations. To be effective, a code must be widely accepted and used; geologic organizations and journals may adopt its recommendations for nomenclatural procedure. Because any code embodies only current concepts and principles, it should have the flexibility to provide for both changes and additions to improve its relevance to new scientific problems.

Any system of nomenclature must be sufficiently explicit to enable users to distinguish objects that are embraced in a class from those that are not. This stratigraphic code makes no attempt to systematize structural, petrographic, paleontologic, or physiographic terms. Terms from these other fields that are used as part of formal stratigraphic names should be sufficiently general as to be unaffected by revisions of precise petrographic or other classifications.

The objective of a system of classification is to promote unambiguous communication in a manner not so restrictive as to inhibit scientific progress. To minimize ambiguity, a code must promote recognition of the distinction between observable features (reproducible data) and inferences or interpretations. Moreover, it should be sufficiently adaptable and flexible to promote the further development of science.

Stratigraphic classification promotes understanding of the *geometry* and *sequence* of rock bodies. The development of stratigraphy as a science required formulation of the Law of Superposition to explain sequential stratal relations. Although superposition is not applicable to many igneous, metamorphic, and tectonic rock assemblages, other criteria (such as cross-cutting relations and isotopic dating) can be used to determine sequential arrangements among rock bodies.

The term *stratigraphic unit* may be defined in several ways. Etymological emphasis requires that it be a stratum or assemblage of adjacent strata distinguished by any or several of the many properties that rocks may possess (ISSC, 1976, p. 13). The scope of stratigraphic classification and procedures, however, suggests a broader definition: a naturally occurring body of rock or rock material distinguished from adjoining rock on the basis of some stated property or properties. Commonly used properties include composition, texture, included fossils, magnetic signature, radioactivity, seismic velocity, and age. Sufficient care is required in defining the boundaries of a unit to enable others to distinguish the material body from those adjoining it. Units based on one property commonly do not coincide with those based on another and, therefore, distinctive terms are needed to identify the property used in defining each unit.

The adjective *stratigraphic* is used in two ways in the remainder of this report. In discussions of lithic (used here as synonymous with "lithologic") units, a conscious attempt is made to restrict the term to lithostratigraphic or layered rocks and sequences that obey the Law of Superposition. For nonstratiform rocks (of plutonic or tectonic origin, for example), the term *lithodemic* (see Article 27) is used. The adjective *stratigraphic* is also used in a broader sense to refer to those procedures derived from stratigraphy which are now applied to all classes of earth materials.

An assumption made in the material that follows is that the reader has some degree of familiarity with basic principles of stratigraphy as outlined, for example, by Dunbar and Rodgers (1957), Weller (1960), Shaw (1964), Matthews (1974), or the International Stratigraphic Guide (ISSC, 1976).

RELATION OF CODES TO INTERNATIONAL GUIDE

Publication of the International Stratigraphic Guide by the International Subcommission on Stratigraphic Classification (ISSC, 1976), which is being endorsed and adopted throughout the world, played a part in prompting examination of the American Stratigraphic Code and the decision to revise it.

The International Guide embodies principles and procedures that had been adopted by several national and regional stratigraphic committees and commissions. More than two decades of effort by H. D. Hedberg and other members of the Subcommission (ISSC, 1976, p. VI, 1, 3) developed the consensus required for preparation of the Guide. Although the Guide attempts to cover all kinds of rocks and the diverse ways of investigating them, it is necessarily incomplete. Mechanisms are needed to stimulate individual innovations toward promulgating new concepts, principles, and practices which subsequently may be found worthy of inclusion in later editions of the Guide. The flexibility of national and regional committees or commissions enables them to perform this function more readily than an international subcommission, even while they adopt the Guide as the international standard of stratigraphic classification.

A guiding principle in preparing this Code has been to make it as consistent as possible with the International Guide, which was endorsed by the ACSN in 1976, and at the same time to foster further innovations to meet the expanding and changing needs of earth scientists on the North American continent.

OVERVIEW

CATEGORIES RECOGNIZED

An attempt is made in this Code to strike a balance between serving the needs of those in evolving specialties and resisting the proliferation of categories of units. Consequently, more formal categories are recognized here than in previous codes or in the International Guide (ISSC, 1976). On the other hand, no special provision is made for formalizing certain kinds of units (deep oceanic, for example) which may be accommodated by available categories.

Four principal categories of units have previously been used widely in traditional stratigraphic work; these have been termed lithostratigraphic, biostratigraphic, chronostratigraphic, and geochronologic and are distinguished as follows:

1. A *lithostratigraphic unit* is a stratum or body of strata, generally but not invariably layered, generally but not invariably tabular, which conforms to the Law of Superposition and is distinguished and delimited on the basis of lithic characteristics and stratigraphic position. Example: Navajo Sandstone.
2. A *biostratigraphic unit* is a body of rock defined and characterized by its fossil content. Example: *Discoaster multiradiatus* Interval Zone.
3. A *chronostratigraphic unit* is a body of rock established to serve as the material reference for all rocks formed during the same span of time. Example: Devonian System. Each boundary of a chronostratigraphic unit is synchronous. Chronostratigraphy provides a means of organizing strata into units based on their age relations. A chronostratigraphic body also serves as the basis for defining the specific interval of geologic time, or geochronologic unit, represented by the referent.
4. A *geochronologic unit* is a division of time distinguished on the basis of the rock record preserved in a chronostratigraphic unit. Example: Devonian Period.

The first two categories are comparable in that they consist of material units defined on the basis of content. The third category differs from the first two in that it serves primarily as the standard for recognizing and isolating materials of a specific age. The fourth, in contrast, is not a material, but rather a conceptual, unit; it is a division of time. Although a geochronologic unit is not a stratigraphic body, it is so intimately tied to chronostratigraphy that the two are discussed properly together.

Properties and procedures that may be used in distinguishing geologic units are both diverse and numerous (ISSC, 1976, p. 1, 96; Harland, 1977, p. 230), but all may be assigned to the following principal classes of categories used in stratigraphic classification (Table 1), which are discussed below:

I. Material categories based on content, inherent attributes, or physical limits,
II. Categories distinguished by geologic age:
  A. Material categories used to define temporal spans, and
  B. Temporal categories.

Table 1. Categories of Units Defined*

MATERIAL CATEGORIES BASED ON CONTENT OR PHYSICAL LIMITS
Lithostratigraphic (22)
Lithodemic (31)**
Magnetopolarity (44)
Biostratigraphic (48)
Pedostratigraphic (55)
Allostratigraphic (58)
CATEGORIES EXPRESSING OR RELATED TO GEOLOGIC AGE
Material Categories Used to Define Temporal Spans
Chronostratigraphic (66)
Polarity-Chronostratigraphic (83)
Temporal (Non-Material) Categories
Geochronologic (80)
Polarity-Chronologic (88)
Diachronic (91)
Geochronometric (96)

*Numbers in parentheses are the numbers of the Articles where units are defined.

**Italicized categories are those introduced or developed since publication of the previous code (ACSN, 1970).

Material Categories Based on Content or Physical Limits

The basic building blocks for most geologic work are rock bodies defined on the basis of composition and related lithic characteristics, or on their physical, chemical, or biologic content or properties. Emphasis is placed on the relative objectivity and reproducibility of data used in defining units within each category.

Foremost properties of rocks are composition, texture, fabric, structure, and color, which together are designated *lithic characteristics*. These serve as the basis for distinguishing and defining the most fundamental of all formal units. Such units based primarily on composition are divided into two categories (Henderson and others, 1980): lithostratigraphic (Article 22) and lithodemic (defined here in Article 31). A lithostratigraphic unit obeys the Law of Superposition, whereas a lithodemic unit does not. A *lithodemic unit* is a defined body of predominantly intrusive, highly metamorphosed, or intensely deformed rock that, because it is intrusive or has lost primary structure through metamorphism or tectonism, generally does not conform to the Law of Superposition.

North American Commission on Stratigraphic Nomenclature

Recognition during the past several decades that remanent magnetism in rocks records the Earth's past magnetic characteristics (Cox, Doell, and Dalrymple, 1963) provides a powerful new tool encompassed by magnetostratigraphy (McDougall, 1977; McElhinny, 1978). *Magnetostratigraphy* (Article 43) is the study of remanent magnetism in rocks; it is the record of the Earth's magnetic polarity (or field reversals), dipole-field-pole position (including apparent polar wander), the non-dipole component (secular variation), and field intensity. Polarity is of particular utility and is used to define a *magnetopolarity unit* (Article 44) as a body of rock identified by its remanent magnetic polarity (ACSN, 1976; ISSC, 1979). Empirical demonstration of uniform polarity does not necessarily have direct temporal connotations because the remanent magnetism need not be related to rock deposition or crystallization. Nevertheless, polarity is a physical attribute that may characterize a body of rock.

Biologic remains contained in, or forming, strata are uniquely important in stratigraphic practice. First, they provide the means of defining and recognizing material units based on fossil content (biostratigraphic units). Second, the irreversibility of organic evolution makes it possible to partition enclosing strata temporally. Third, biologic remains provide important data for the reconstruction of ancient environments of deposition.

Composition also is important in distinguishing pedostratigraphic units. A *pedostratigraphic unit* is a body of rock that consists of one or more pedologic horizons developed in one or more lithic units now buried by a formally defined lithostratigraphic or allostratigraphic unit or units. A pedostratigraphic unit is the part of a buried soil characterized by one or more clearly defined soil horizons containing pedogenically formed minerals and organic compounds. Pedostratigraphic terminology is discussed below and in Article 55.

Many upper Cenozoic, especially Quaternary, deposits are distinguished and delineated on the basis of content, for which lithostratigraphic classification is appropriate. However, others are delineated on the basis of criteria other than content. To facilitate the reconstruction of geologic history, some compositionally similar deposits in vertical sequence merit distinction as separate stratigraphic units because they are the products of different processes; others merit distinction because they are of demonstrably different ages. Lithostratigraphic classification of these units is impractical and a new approach, allostratigraphic classification, is introduced here and may prove applicable to older deposits as well. An *allostratigraphic unit* is a mappable stratiform body of sedimentary rock defined and identified on the basis of bounding discontinuities (Article 58 and related Remarks).

Geologic-Climate units, defined in the previous Code (ACSN, 1970, p. 31), are abandoned here because they proved to be of dubious utility. Inferences regarding climate are subjective and too tenuous a basis for the definition of formal geologic units. Such inferences commonly are based on deposits assigned more appropriately to lithostratigraphic or allostratigraphic units and may be expressed in terms of diachronic units (defined below).

Categories Expressing or Related to Geologic Age

Time is a single, irreversible continuum. Nevertheless, various categories of units are used to define intervals of geologic time, just as terms having different bases, such as Paleolithic, Renaissance, and Elizabethan, are used to designate specific periods of human history. Different temporal categories are established to express intervals of time distinguished in different ways.

Major objectives of stratigraphic classification are to provide a basis for systematic ordering of the time and space relations of rock bodies and to establish a time framework for the discussion of geologic history. For such purposes, units of geologic time traditionally have been named to represent the span of time during which a well-described sequence of rock, or a chronostratigraphic unit, was deposited ("time units based on material referents," Fig. 1). This procedure continues, to the exclusion of other possible approaches, to be standard practice in studies of Phanerozoic rocks. Despite admonitions in previous American codes and the International Stratigraphic Guide (ISSC, 1976, p. 81) that similar procedures should be applied to the Precambrian, no comparable chronostratigraphic units, or geochronologic units derived therefrom, proposed for the Precambrian have yet been accepted worldwide. Instead, the IUGS Subcommission on Precambrian Stratigraphy (Sims, 1979) and its Working Groups (Harrison and Peterman, 1980) recommend division of Precambrian time into *geochronometric units* having no material referents.

A distinction is made throughout this report between *isochronous* and *synchronous*, as urged by Cumming, Fuller, and Porter (1959, p. 730), although the terms have been used synonymously by many. *Isochronous* means of equal duration; *synchronous* means simultaneous, or occurring at the same time. Although two rock bodies of very different ages may be formed during equal durations of time, the term *isochronous* is not applied to them in the earth sciences. Rather, isochronous bodies are those bounded by synchronous surfaces and formed during the same span of time. *Isochron*, in contrast, is used for a line connecting points of equal age on a graph representing physical or chemical phenomena; the line represents the same or equal time. The adjective *diachronous* is applied either to a rock unit with one or two bounding surfaces which are not synchronous, or to a boundary which is not synchronous (which "transgresses time").

Two classes of time units based on material referents, or stratotypes, are recognized (Fig. 1). The first is that of the traditional and conceptually isochronous units, and includes *geochronologic units*, which are based on *chronostratigraphic units*, and *polarity-geochronologic units*. These isochronous units have worldwide applicability and may be used even in areas lacking a material record of the named span of time. The second class of time units, newly defined in this Code, consists of *diachronic units* (Article 91), which are based on rock bodies known to be diachronous. In contrast to isochronous units, a diachronic term is used only where a material referent is present; a diachronic unit is coextensive with the material body or bodies on which it is based.

A *chronostratigraphic unit*, as defined above and in Article 66, is a body of rock established to serve as the material reference for all rocks formed during the same span of time; its boundaries are synchronous. It is the referent for a *geochronologic unit*, as defined above and in Article 80. Internationally accepted and traditional chronostratigraphic units were based initially on the time spans of lithostratigraphic units, biostratigraphic units, or other features of the rock record that have specific durations. In sum, they form the Standard Global Chronostratigraphic Scale (ISSC, 1976, p. 76-81; Harland, 1978), consisting of established systems and series.

A *polarity-chronostratigraphic unit* is a body of rock that contains a primary magnetopolarity record imposed when the rock was deposited or crystallized (Article 83). It serves as a material standard or referent for a part of geologic time during which the Earth's magnetic field had a characteristic polarity or sequence of polarities; that is, for a *polarity-chronologic unit* (Article 88).

A *diachronic unit* comprises the unequal spans of time represented by one or more specific diachronous rock bodies (Article 91). Such bodies may be lithostratigraphic, biostratigraphic, pedostratigraphic, allostratigraphic, or an assemblage of such units. A diachronic unit is applicable only where its material referent is present.

A *geochronometric* (or chronometric) *unit* is an isochronous direct division of geologic time expressed in years (Article 96). It has no material referent.

North American Stratigraphic Code

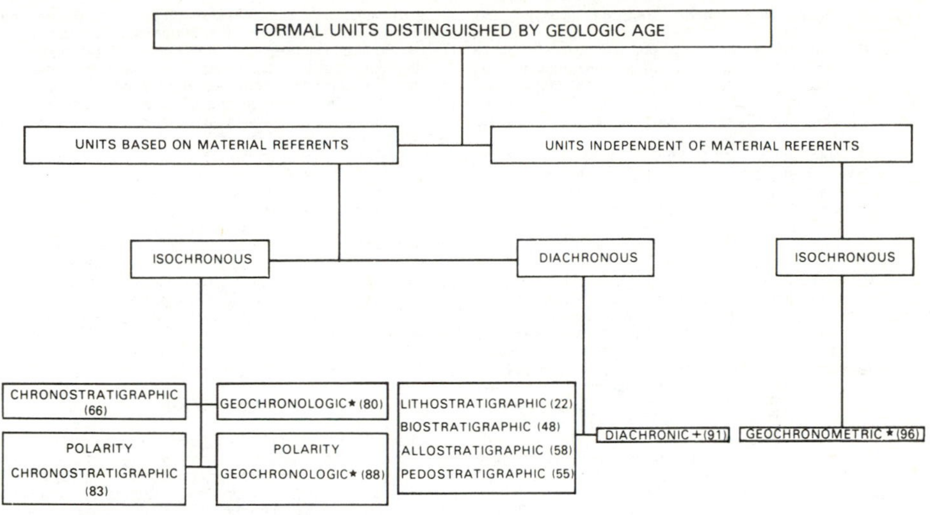

FIG. 1.—Relation of geologic time units to the kinds of rock-unit referents on which most are based.

Pedostratigraphic Terms

The definition and nomenclature for pedostratigraphic[2] units in this Code differ from those for soil-stratigraphic units in the previous Code (ACSN, 1970, Article 18), by being more specific with regard to content, boundaries, and the basis for determining stratigraphic position.

The term "soil" has different meanings to the geologist, the soil scientist, the engineer, and the layman, and commonly has no stratigraphic significance. The term *paleosol* is currently used in North America for any soil that formed on a landscape of the past; it may be a buried soil, a relict soil, or an exhumed soil (Ruhe, 1965; Valentine and Dalrymple, 1976).

A *pedologic soil* is composed of one or more soil horizons.[3] A *soil horizon* is a layer within a pedologic soil that (1) is approximately parallel to the soil surface, (2) has distinctive physical, chemical, biological, and morphological properties that differ from those of adjacent, genetically related, soil horizons, and (3) is distinguished from other soil horizons by objective compositional properties that can be observed or measured in the field. The physical boundaries of buried pedologic horizons are objective traceable boundaries with stratigraphic significance. A buried pedologic soil provides the material basis for definition of a stratigraphic unit in pedostratigraphic classification (Article 55), but a buried pedologic soil may be somewhat more inclusive than a pedostratigraphic unit. A pedologic soil may contain both an 0-horizon and the entire C-horizon (Fig. 6), whereas the former is excluded and the latter need not be included in a pedostratigraphic unit.

The definition and nomenclature for pedostratigraphic units in this Code differ from those of soil stratigraphic units proposed by the International Union for Quaternary Research and International Society of Soil Science (Parsons, 1981). The pedostratigraphic unit, geosol, also differs from the proposed INQUA-ISSS soil-stratigraphic unit, pedoderm, in several ways, the most important of which are: (1) a geosol may be in any part of the geologic column, whereas a pedoderm is a surficial soil; (2) a geosol is a buried soil, whereas a pedoderm may be a buried, relict, or exhumed soil; (3) the boundaries and stratigraphic position of a geosol are defined and delineated by criteria that differ from those for a pedoderm; and (4) a geosol may be either all or only a part of a buried soil, whereas a pedoderm is the entire soil.

The term *geosol*, as defined by Morrison (1967, p. 3), is a laterally traceable, mappable, geologic weathering profile that has a consistent stratigraphic position. The term is adopted and redefined here as the fundamental and only unit in formal pedostratigraphic classification (Article 56).

FORMAL AND INFORMAL UNITS

Although the emphasis in this Code is necessarily on formal categories of geologic units, informal nomenclature is highly useful in stratigraphic work.

Formally named units are those that are named in accordance with an established scheme of classification; the fact of formality is conveyed by capitalization of the initial letter of the *rank* or *unit* term (for example, Morrison Formation). Informal units, whose unit terms are ordinary nouns, are not protected by the stability provided by proper formalization and recommended classification procedures. Informal terms are devised for both economic and scientific reasons. Formalization is appropriate for those units requiring stability of nomenclature, particularly those likely to be extended far beyond the locality in which they were first recognized. Informal terms are appropriate for casually mentioned, innovative, and most economic units, those

[2]From Greek, *pedon*, ground or soil.

[3]As used in a geological sense, a *horizon* is a surface or line. In pedology, however, it is a body of material, and such usage is continued here.

defined by unconventional criteria, and those that may be too thin to map at usual scales.

Casually mentioned geologic units not defined in accordance with this Code are informal. For many of these, there may be insufficient need or information, or perhaps an inappropriate basis, for formal designations. Informal designations as beds or lithozones (the pebbly beds, the shaly zone, third coal) are appropriate for many such units.

Most economic units, such as aquifers, oil sands, coal beds, quarry layers, and ore-bearing "reefs," are informal, even though they may be named. Some such units, however, are so significant scientifically and economically that they merit formal recognition as beds, members, or formations.

Innovative approaches in regional stratigraphic studies have resulted in the recognition and definition of units best left as informal, at least for the time being. Units bounded by major regional unconformities on the North American craton were designated "sequences" (example: Sauk sequence) by Sloss (1963). Major unconformity-bounded units also were designated "synthems" by Chang (1975), who recommended that they be treated formally. Marker-defined units that are continuous from one lithofacies to another were designated "formats" by Forgotson (1957). The term "chronosome" was proposed by Schultz (1982) for rocks of diverse facies corresponding to geographic variations in sedimentation during an interval of deposition identified on the basis of bounding stratigraphic markers. Successions of faunal zones containing evolutionally related forms, but bounded by non-evolutionary biotic discontinuities, were termed "biomeres" (Palmer, 1965). The foregoing are only a few selected examples to demonstrate how informality provides a continuing avenue for innovation.

The terms *magnafacies* and *parvafacies*, coined by Caster (1934) to emphasize the distinction between lithostratigraphic and chronostratigraphic units in sequences displaying marked facies variation, have remained informal despite their impact on clarifying the concepts involved.

Tephrochronologic studies provide examples of informal units too thin to map at conventional scales but yet invaluable for dating important geologic events. Although some such units are named for physiographic features and places where first recognized (e.g., Guaje pumice bed, where it is not mapped as the Guaje Member of the Bandelier Tuff), others bear the same name as the volcanic vent (e.g., Huckleberry Ridge ash bed of Izett and Wilcox, 1981).

Informal geologic units are designated by ordinary nouns, adjectives or geographic terms and lithic or unit-terms that are not capitalized (chalky formation or beds, St. Francis coal).

No geologic unit should be established and defined, whether formally or informally, unless its recognition serves a clear purpose.

CORRELATION

Correlation is a procedure for demonstrating correspondence between geographically separated parts of a geologic unit. The term is a general one having diverse meanings in different disciplines. Demonstration of temporal correspondence is one of the most important objectives of stratigraphy. The term "correlation" frequently is misused to express the idea that a unit has been identified or recognized.

Correlation is used in this Code as the demonstration of correspondence between two geologic units in both some defined property and relative stratigraphic position. Because correspondence may be based on various properties, three kinds of correlation are best distinguished by more specific terms. *Lithocorrelation* links units of similar lithology and stratigraphic position (or sequential or geometric relation, for lithodemic units). *Biocorrelation* expresses similarity of fossil content and biostratigraphic position. *Chronocorrelation* expresses correspondence in age and in chronostratigraphic position.

Other terms that have been used for the similarity of content and stratal succession are homotaxy and chronotaxy. *Homotaxy* is the similarity in separate regions of the serial arrangement or succession of strata of comparable compositions or of included fossils. The term is derived from *homotaxis*, proposed by Huxley (1862, p. xlvi) to emphasize that similarity in succession does not prove age equivalence of comparable units. The term *chronotaxy* has been applied to similar stratigraphic sequences composed of units which are of equivalent age (Henbest, 1952, p. 310).

Criteria used for ascertaining temporal and other types of correspondence are diverse (ISSC, 1976, p. 86-93) and new criteria will emerge in the future. Evolving statistical tests, as well as isotopic and paleomagnetic techniques, complement the traditional paleontologic and lithologic procedures. Boundaries defined by one set of criteria need not correspond to those defined by others.

PART II. ARTICLES

INTRODUCTION

Article 1.—**Purpose.** This Code describes explicit stratigraphic procedures for classifying and naming geologic units accorded formal status. Such procedures, if widely adopted, assure consistent and uniform usage in classification and terminology and therefore promote unambiguous communication.

Article 2.—**Categories.** Categories of formal stratigraphic units, though diverse, are of three classes (Table 1). The first class is of rock-material categories based on inherent attributes or content and stratigraphic position, and includes lithostratigraphic, lithodemic, magnetopolarity, biostratigraphic, pedostratigraphic, and allostratigraphic units. The second class is of material categories used as standards for defining spans of geologic time, and includes chronostratigraphic and polarity-chronostratigraphic units. The third class is of non-material temporal categories, and includes geochronologic, polarity-chronologic, geochronometric, and diachronic units.

GENERAL PROCEDURES

DEFINITION OF FORMAL UNITS

Article 3.—**Requirements for Formally Named Geologic Units.** Naming, establishing, revising, redefining, and abandoning formal geologic units require publication in a recognized scientific medium of a comprehensive statement which includes: (i) intent to designate or modify a formal unit; (ii) designation of category and rank of unit; (iii) selection and derivation of name; (iv) specification of stratotype (where applicable); (v) description of unit; (vi) definition of boundaries; (vii) historical background; (viii) dimensions, shape, and other regional aspects; (ix) geologic age; (x) correlations; and possibly (xi) genesis (where applicable). These requirements apply to subsurface and offshore, as well as exposed, units.

Article 4.—**Publication.**[4] "Publication in a recognized scientific medium" in conformance with this Code means that a work, when first issued, must (1) be reproduced in ink on paper or by some method that assures numerous identical copies and wide distribution; (2) be issued for the purpose of scientific, public, permanent record; and (3) be readily obtainable by purchase or free distribution.

Remarks. (a) **Inadequate publication.**—The following do not constitute publication within the meaning of the Code: (1) distribution of microfilms, microcards, or matter reproduced by similar methods; (2)

[4]This article is modified slightly from a statement by the International Commission of Zoological Nomenclature (1964, p. 7-9).

North American Stratigraphic Code

Table 2. Categories and Ranks of Units Defined in This Code*

A. Material Units

LITHOSTRATIGRAPHIC	LITHODEMIC	MAGNETOPOLARITY	BIOSTRATIGRAPHIC	PEDOSTRATIGRAPHIC	ALLOSTRATIGRAPHIC
Supergroup	Supersuite (Complex)				
Group	Suite (Complex)	Polarity Superzone			Allogroup
Formation	*Lithodeme*	*Polarity zone*	*Biozone* (Interval, Assemblage or Abundance)	*Geosol*	*Alloformation*
Member (or Lens, or Tongue)		Polarity Subzone	Subbiozone		Allomember
Bed(s) or Flow(s)					

B. Temporal and Related Chronostratigraphic Units

CHRONO-STRATIGRAPHIC	GEOCHRONOLOGIC GEOCHRONOMETRIC	POLARITY CHRONO-STRATIGRAPHIC	POLARITY CHRONOLOGIC	DIACHRONIC
Eonothem	Eon	Polarity Superchronozone	Polarity Superchron	
Erathem (Supersystem)	Era (Superperiod)			
System (Subsystem)	*Period* (Subperiod)	*Polarity Chronozone*	*Polarity Chron*	*Episode*
Series	Epoch			Phase (Diachron)
Stage (Substage)	Age (Subage)	Polarity Subchronozone	Polarity Subchron	Span (Diachron)
Chronozone	Chron			Cline

*Fundamental units are italicized.

distribution to colleagues or students of a note, even if printed, in explanation of an accompanying illustration; (3) distribution of proof sheets; (4) open-file release; (5) theses, dissertations, and dissertation abstracts; (6) mention at a scientific or other meeting; (7) mention in an abstract, map explanation, or figure caption; (8) labeling of a rock specimen in a collection; (9) mere deposit of a document in a library; (10) anonymous publication; or (11) mention in the popular press or in a legal document.

(b). **Guidebooks.**—A guidebook with distribution limited to participants of a field excursion does not meet the test of availability. Some organizations publish and distribute widely large editions of serial guidebooks that include refereed regional papers; although these do meet the tests of scientific purpose and availability, and therefore constitute valid publication, other media are preferable.

Article 5.—**Intent and Utility.** To be valid, a new unit must serve a clear purpose and be duly proposed and duly described, and the intent to establish it must be specified. Casual mention of a unit, such as "the granite exposed near the Middleville schoolhouse," does not establish a new formal unit, nor does mere use in a table, columnar section, or map.

Remark. (a) **Demonstration of purpose served.**—The initial definition or revision of a named geologic unit constitutes, in essence, a proposal. As such, it lacks status until use by others demonstrates that a clear purpose has been served. A unit becomes established through repeated demonstration of its utility. The decision not to use a newly proposed or a newly revised term requires a full discussion of its unsuitability.

Article 6.—**Category and Rank.** The category and rank of a new or revised unit must be specified.

Remark. (a) **Need for specification.**—Many stratigraphic controversies have arisen from confusion or misinterpretation of the category of a unit (for example, lithostratigraphic vs. chronostratigraphic). Specification and unambiguous description of the category is of paramount importance. Selection and designation of an appropriate rank from the distinctive terminology developed for each category help serve this function (Table 2).

Article 7.—**Name.** The name of a formal geologic unit is compound. For most categories, the name of a unit should consist of a geographic name combined with an appropriate rank (Wasatch Formation) or descriptive term (Viola Limestone). Biostratigraphic units are designated by appropriate biologic forms (*Exus albus* Assemblage Biozone). Worldwide chronostratigraphic units bear long established and generally accepted names of diverse origins (Triassic System). The first letters of all words used in the names of formal geologic units are capitalized (except for the trivial species and subspecies terms in the name of a biostratigraphic unit).

Remarks. (a) **Appropriate geographic terms.**—Geographic names derived from permanent natural or artificial features at or near which the unit is present are preferable to those derived from impermanent features such as farms, schools, stores, churches, crossroads, and small communities. Appropriate names may be selected from those shown on topographic, state, provincial, county, forest service, hydrographic, or comparable maps, particularly those showing names approved by a national board for geographic names. The generic part of a geographic name, e.g., river, lake, village, should be omitted from new terms, unless required to distinguish between two otherwise identical names (e.g., Redstone Formation and Redstone River Formation). Two names should not be derived from the same geographic feature. A unit should not be named for the source of its components; for example, a deposit inferred to have been derived from the Keewatin glaciation center should not be designated the "Keewatin Till."

(b) **Duplication of names.**—Responsibility for avoiding duplication, either in use of the same name for different units (homonymy) or in use of different names for the same unit (synonomy), rests with the proposer. Although the same geographic term has been applied to different categories of units (example: the lithostratigraphic Word Formation and the chronostratigraphic Wordian Stage) now entrenched in the literature, the practice is undesirable. The extensive geologic nomenclature of North America, including not only names but also nomenclatural history of formal units, is recorded in compendia maintained by the Committee on Stratigraphic Nomenclature of the Geological Survey of Canada, Ottawa, Ontario; by the Geologic Names Committee of the United States Geological Survey, Reston, Virginia; by the Instituto de Geología, Ciudad Universitaria, México, D.F.; and by many state and provincial geological surveys. These organizations respond to inquiries regarding the availability of names, and some are prepared to reserve names for units likely to be defined in the next year or two.

(c) **Priority and preservation of established names.**—Stability of nomenclature is maintained by use of the rule of priority and by preservation of well-established names. Names should not be modified without explaining the need. Priority in publication is to be respected, but priority alone does not justify displacing a well-established name by one neither well-known nor commonly used; nor should an inadequately established name be preserved merely on the basis of priority. Redefinitions in precise terms are preferable to abandonment of the names of well-established units which may have been defined imprecisely but nonetheless in conformance with older and less stringent standards.

(d) **Differences of spelling and changes in name.**—The geographic component of a well-established stratigraphic name is not changed due to differences in spelling or changes in the name of a geographic feature. The name Bennett Shale, for example, used for more than half a century, need not be altered because the town is named Bennet. Nor should the Mauch Chunk Formation be changed because the town has been renamed Jim Thorpe. Disappearance of an impermanent geographic feature, such as a town, does not affect the name of an established geologic unit.

(e) **Names in different countries and different languages.**—For geologic units that cross local and international boundaries, a single name for each is preferable to several. Spelling of a geographic name commonly conforms to the usage of the country and linguistic group involved. Although geographic names are not translated (Cuchillo is not translated to Knife), lithologic or rank terms are (Edwards Limestone, Caliza Edwards; Formacion La Casita, La Casita Formation).

Article 8.—**Stratotypes.** The designation of a unit or boundary stratotype (type section or type locality) is essential in the definition of most formal geologic units. Many kinds of units are best defined by reference to an accessible and specific sequence of rock that may be examined and studied by others. A stratotype is the standard (original or subsequently designated) for a named geologic unit or boundary and constitutes the basis for definition or recognition of that unit or boundary; therefore, it must be illustrative and representative of the concept of the unit or boundary being defined.

Remarks. (a) **Unit stratotypes.**—A unit stratotype is the type section for a stratiform deposit or the type area for a nonstratiform body that serves as the standard for definition and recognition of a geologic unit. The upper and lower limits of a unit stratotype are designated points in a specific sequence or locality and serve as the standards for definition and recognition of a stratigraphic unit's boundaries.

(b) **Boundary stratotype.**—A boundary stratotype is the type locality for the boundary reference point for a stratigraphic unit. Both boundary stratotypes for any unit need not be in the same section or region. Each boundary stratotype serves as the standard for definition and recognition of the base of a stratigraphic unit. The top of a unit may be defined by the boundary stratotype of the next higher stratigraphic unit.

(c) **Type locality.**—A type locality is the specified geographic locality where the stratotype of a formal unit or unit boundary was originally defined and named. A type area is the geographic territory encompassing the type locality. Before the concept of a stratotype was developed, only type localities and areas were designated for many geologic units which are now long- and well-established. Stratotypes, though now mandatory in defining most stratiform units, are impractical in definitions of many large nonstratiform rock bodies whose diverse major components may be best displayed at several reference localities.

(d) **Composite-stratotype.**—A composite-stratotype consists of several reference sections (which may include a type section) required to demonstrate the range or totality of a stratigraphic unit.

(e) **Reference sections.**—Reference sections may serve as invaluable standards in definitions or revisions of formal geologic units. For those well-established stratigraphic units for which a type section never was specified, a principal reference section (lectostratotype of ISSC, 1976, p. 26) may be designated. A principal reference section (neostratotype of ISSC, 1976, p. 26) also may be designated for those units or boundaries whose stratotypes have been destroyed, covered, or otherwise made inaccessible. Supplementary reference sections often are designated to illustrate the diversity or heterogeneity of a defined unit or some critical feature not evident or exposed in the stratotype. Once a unit or boundary stratotype section is designated, it is never abandoned or changed; however, if a stratotype proves inadequate, it may be supplemented by a principal reference section or by several reference sections that may constitute a composite-stratotype.

(f) **Stratotype descriptions.**—Stratotypes should be described both geographically and geologically. Sufficient geographic detail must be included to enable others to find the stratotype in the field, and may consist of maps and/or aerial photographs showing location and access, as well as appropriate coordinates or bearings. Geologic information should include thickness, descriptive criteria appropriate to the recognition of the unit and its boundaries, and discussion of the relation of the unit to other geologic units of the area. A carefully measured and described section provides the best foundation for definition of stratiform units. Graphic profiles, columnar sections, structure-sections, and photographs are useful supplements to a description; a geologic map of the area including the type locality is essential.

Article 9.—**Unit Description.** A unit proposed for formal status should be described and defined so clearly that any subsequent investigator can recognize that unit unequivocally. Distinguishing features that characterize a unit may include any or several of the following: composition, texture, primary structures, structural attitudes, biologic remains, readily apparent mineral composition (e.g., calcite vs. dolomite), geochemistry, geophysical properties (including magnetic signatures), geomorphic expression, unconformable or cross-cutting relations, and age. Although all distinguishing features pertinent to the unit category should be described sufficiently to characterize the unit, those not pertinent to the category (such as age and inferred genesis for lithostratigraphic units, or lithology for biostratigraphic units) should not be made part of the definition.

Article 10.—**Boundaries.** The criteria specified for the recognition of boundaries between adjoining geologic units are of paramount importance because they provide the basis for scientific reproducibility of results. Care is required in describing the criteria, which must be appropriate to the category of unit involved.

Remarks. (a) **Boundaries between intergradational units.**—Contacts between rocks of markedly contrasting composition are appropriate boundaries of lithic units, but some rocks grade into, or intertongue with, others of different lithology. Consequently, some boundaries are necessarily arbitrary as, for example, the top of the uppermost limestone in a sequence of interbedded limestone and shale. Such arbitrary boundaries commonly are diachronous.

(b) **Overlaps and gaps.**—The problem of overlaps and gaps between long-established adjacent chronostratigraphic units is being addressed by international IUGS and IGCP working groups appointed to deal with various parts of the geologic column. The procedure recommended by the Geological Society of London (George and others, 1969; Holland and others, 1978), of defining only the basal boundaries of chronostratigraphic units, has been widely adopted (e.g., McLaren, 1977) to resolve the problem. Such boundaries are defined by a carefully selected and agreed-upon boundary-stratotype (marker-point type section or "golden spike") which becomes the standard for the base of a chronostratigraphic unit. The concept of the mutual-boundary stratotype (ISSC, 1976, p. 84-86), based on the assumption of continuous deposition in selected sequences, also has been used to define chronostratigraphic units.

North American Stratigraphic Code

Although international chronostratigraphic units of series and higher rank are being redefined by IUGS and IGCP working groups, there may be a continuing need for some provincial series. Adoption of the basal boundary-stratotype concept is urged.

Article 11.—**Historical Background.** A proposal for a new name must include a nomenclatorial history of rocks assigned to the proposed unit, describing how they were treated previously and by whom (references), as well as such matters as priorities, possible synonymy, and other pertinent considerations. Consideration of the historical background of an older unit commonly provides the basis for justifying definition of a new unit.

Article 12.—**Dimensions and Regional Relations.** A perspective on the magnitude of a unit should be provided by such information as may be available on the geographic extent of a unit; observed ranges in thickness, composition, and geomorphic expression; relations to other kinds and ranks of stratigraphic units; correlations with other nearby sequences; and the bases for recognizing and extending the unit beyond the type locality. If the unit is not known anywhere but in an area of limited extent, informal designation is recommended.

Article 13.—**Age.** For most formal material geologic units, other than chronostratigraphic and polarity-chronostratigraphic, inferences regarding geologic age play no proper role in their definition. Nevertheless, the age, as well as the basis for its assignment, are important features of the unit and should be stated. For many lithodemic units, the age of the protolith should be distinguished from that of the metamorphism or deformation. If the basis for assigning an age is tenuous, a doubt should be expressed.

Remarks. (a) **Dating.**—The geochronologic ordering of the rock record, whether in terms of radioactive-decay rates or other processes, is generally called "dating." However, the use of the noun "date" to mean "isotopic age" is not recommended. Similarly, the term "absolute age" should be suppressed in favor of "isotopic age" for an age determined on the basis of isotopic ratios. The more inclusive term "numerical age" is recommended for all ages determined from isotopic ratios, fission tracks, and other quantifiable age-related phenomena.

(b) **Calibration**—The dating of chronostratigraphic boundaries in terms of numerical ages is a special form of dating for which the word "calibration" should be used. The geochronologic time-scale now in use has been developed mainly through such calibration of chronostratigraphic sequences.

(c) **Convention and abbreviations.**—The age of a stratigraphic unit or the time of a geologic event, as commonly determined by numerical dating or by reference to a calibrated time-scale, may be expressed in years before the present. The unit of time is the modern year as presently recognized worldwide. Recommended (but not mandatory) abbreviations for such ages are SI (International System of Units) multipliers coupled with "a" for annum: ka, Ma, and Ga[5] for kilo-annum (10^3 years), Mega-annum (10^6 years), and Giga-annum (10^9 years), respectively. Use of these terms after the age value follows the convention established in the field of C-14 dating. The "present" refers to 1950 AD, and such qualifiers as "ago" or "before the present" are omitted after the value because measurement of the duration from the present to the past is implicit in the designation. In contrast, the duration of a remote interval of geologic time, as a number of years, should not be expressed by the same symbols. Abbreviations for numbers of years, without reference to the present, are informal (e.g., y or yr for years; my, m.y., or m.yr. for millions of years; and so forth, as preference dictates). For example, boundaries of the Late Cretaceous Epoch currently are calibrated at 63 Ma and 96 Ma, but the interval of time represented by this epoch is 33 m.y.

(d) **Expression of "age" of lithodemic units.**—The adjectives "early," "middle," and "late" should be used with the appropriate geochronologic term to designate the age of lithodemic units. For example, a granite dated isotopically at 510 Ma should be referred to using the geochronologic term "Late Cambrian granite" rather than either the chronostratigraphic term "Upper Cambrian granite" or the more cumbersome designation "granite of Late Cambrian age."

Article 14.—**Correlation.** Information regarding spatial and temporal counterparts of a newly defined unit beyond the type area provides readers with an enlarged perspective. Discussions of criteria used in correlating a unit with those in other areas should make clear the distinction between data and inferences.

Article 15.—**Genesis.** Objective data are used to define and classify geologic units and to express their spatial and temporal relations. Although many of the categories defined in this Code (e.g., lithostratigraphic group, plutonic suite) have genetic connotations, inferences regarding geologic history or specific environments of formation may play no proper role in the definition of a unit. However, observations, as well as inferences, that bear on genesis are of great interest to readers and should be discussed.

Article 16.—**Subsurface and Subsea Units.** The foregoing procedures for establishing formal geologic units apply also to subsurface and offshore or subsea units. Complete lithologic and paleontologic descriptions or logs of the samples or cores are required in written or graphic form, or both. Boundaries and divisions, if any, of the unit should be indicated clearly with their depths from an established datum.

Remarks. (a) **Naming subsurface units.**—A subsurface unit may be named for the borehole (Eagle Mills Formation), oil field (Smackover Limestone), or mine which is intended to serve as the stratotype, or for a nearby geographic feature. The hole or mine should be located precisely, both with map and exact geographic coordinates, and identified fully (operator or company, farm or lease block, dates drilled or mined, surface elevation and total depth, etc).

(b) **Additional recommendations.**—Inclusion of appropriate borehole geophysical logs is urged. Moreover, rock and fossil samples and cores and all pertinent accompanying materials should be stored, and available for examination, at appropriate federal, state, provincial, university, or museum depositories. For offshore or subsea units (Clipperton Formation of Tracey and others, 1971, p. 22; Argo Salt of McIver, 1972, p. 57), the names of the project and vessel, depth of sea floor, and pertinent regional sampling and geophysical data should be added.

(c) **Seismostratigraphic units.**—High-resolution seismic methods now can delineate stratal geometry and continuity at a level of confidence not previously attainable. Accordingly, seismic surveys have come to be the principal adjunct of the drill in subsurface exploration. On the other hand, the method identifies rock types only broadly and by inference. Thus, formalization of units known only from seismic profiles is inappropriate. Once the stratigraphy is calibrated by drilling, the seismic method may provide objective well-to-well correlations.

REVISION AND ABANDONMENT OF FORMAL UNITS

Article 17.—**Requirements for Major Changes.** Formally defined and named geologic units may be redefined, revised, or abandoned, but revision and abandonment require as much justification as establishment of a new unit.

Remark. (a) **Distinction between redefinition and revision.**—Redefinition of a unit involves changing the view or emphasis on the content of the unit without changing the boundaries or rank, and differs only slightly from redescription. Neither redefinition nor redescription is considered revision. A redescription corrects an inadequate or inaccurate description, whereas a redefinition may change a descriptive (for example, lithologic) designation. Revision involves either minor changes in the definition of one or both boundaries or in the rank of a unit (normally, elevation to a higher rank). Correction of a misidentification of a unit outside its type area is neither redefinition nor revision.

[5] Note that the initial letters of Mega- and Giga- are capitalized, but that of kilo- is not, by SI convention.

North American Commission on Stratigraphic Nomenclature

Article 18.—**Redefinition.** A correction or change in the descriptive term applied to a stratigraphic or lithodemic unit is a redefinition which does not require a new geographic term.

Remarks. (a) **Change in lithic designation.**—Priority should not prevent more exact lithic designation if the original designation is not everywhere applicable; for example, the Niobrara Chalk changes gradually westward to a unit in which shale is prominent, for which the designation "Niobrara Shale" or "Formation" is more appropriate. Many carbonate formations originally designated "limestone" or "dolomite" are found to be geographically inconsistent as to prevailing rock type. The appropriate lithic term or "formation" is again preferable for such units.

(b) **Original lithic designation inappropriate.**—Restudy of some long-established lithostratigraphic units has shown that the original lithic designation was incorrect according to modern criteria; for example, some "shales" have the chemical and mineralogical composition of limestone, and some rocks described as felsic lavas now are understood to be welded tuffs. Such new knowledge is recognized by changing the lithic designation of the unit, while retaining the original geographic term. Similarly, changes in the classification of igneous rocks have resulted in recognition that rocks originally described as quartz monzonite now are more appropriately termed granite. Such lithic designations may be modernized when the new classification is widely adopted. If heterogeneous bodies of plutonic rock have been misleadingly identified with a single compositional term, such as "gabbro," the adoption of a neutral term, such as "intrusion" or "pluton," may be advisable.

Article 19.—**Revision.** Revision involves either minor changes in the definition of one or both boundaries of a unit, or in the unit's rank.

Remarks. (a) **Boundary change.**—Revision is justifiable if a minor change in boundary or content will make a unit more natural and useful. If revision modifies only a minor part of the content of a previously established unit, the original name may be retained.

(b) **Change in rank.**—Change in rank of a stratigraphic or temporal unit requires neither redefinition of its boundaries nor alteration of the geographic part of its name. A member may become a formation or vice versa, a formation may become a group or vice versa, and a lithodeme may become a suite or vice versa.

(c) **Examples of changes from area to area.**—The Conasauga Shale is recognized as a formation in Georgia and as a group in eastern Tennessee; the Osgood Formation, Laurel Limestone, and Waldron Shale in Indiana are classed as members of the Wayne Formation in a part of Tennessee; the Virgelle Sandstone is a formation in western Montana and a member of the Eagle Sandstone in central Montana; the Skull Creek Shale and the Newcastle Sandstone in North Dakota are members of the Ashville Formation in Manitoba.

(d) **Example of change in single area.**—The rank of a unit may be changed without changing its content. For example, the Madison Limestone of early work in Montana later became the Madison Group, containing several formations.

(e) **Retention of type section.**—When the rank of a geologic unit is changed, the original type section or type locality is retained for the newly ranked unit (see Article 22c).

(f) **Different geographic name for a unit and its parts.**—In changing the rank of a unit, the same name may not be applied both to the unit as a whole and to a part of it. For example, the Astoria Group should not contain an Astoria Sandstone, nor the Washington Formation, a Washington Sandstone Member.

(g) **Undesirable restriction.**—When a unit is divided into two or more of the same rank as the original, the original name should not be used for any of the divisions. Retention of the old name for one of the units precludes use of the name in a term of higher rank. Furthermore, in order to understand an author's meaning, a later reader would have to know about the modification and its date, and whether the author is following the original or the modified usage. For these reasons, the normal practice is to raise the rank of an established unit when units of the same rank are recognized and mapped within it.

Article 20.—**Abandonment.** An improperly defined or obsolete stratigraphic, lithodemic, or temporal unit may be formally abandoned, provided that (a) sufficient justification is presented to demonstrate a concern for nomenclatural stability, and (b) recommendations are made for the classification and nomenclature to be used in its place.

Remarks. (a) **Reasons for abandonment.**—A formally defined unit may be abandoned by the demonstration of synonymy or homonymy, of assignment to an improper category (for example, definition of a lithostratigraphic unit in a chronostratigraphic sense), or of other direct violations of a stratigraphic code or procedures prevailing at the time of the original definition. Disuse, or the lack of need or useful purpose for a unit, may be a basis for abandonment; so, too, may widespread misuse in diverse ways which compound confusion. A unit also may be abandoned if it proves impracticable, neither recognizable nor mappable elsewhere.

(b) **Abandoned names.**—A name for a lithostratigraphic or lithodemic unit, once applied and then abandoned, is available for some other unit only if the name was introduced casually, or if it has been published only once in the last several decades and is not in current usage, and if its reintroduction will cause no confusion. An explanation of the history of the name and of the new usage should be a part of the designation.

(c) **Obsolete names.**—Authors may refer to national and provincial records of stratigraphic names to determine whether a name is obsolete (see Article 7b).

(d) **Reference to abandoned names.**—When it is useful to refer to an obsolete or abandoned formal name, its status is made clear by some such term as "abandoned" or "obsolete," and by using a phrase such as "La Plata Sandstone of Cross (1898)". (The same phrase also is used to convey that a named unit has not yet been adopted for usage by the organization involved.)

(e) **Reinstatement.**—A name abandoned for reasons that seem valid at the time, but which subsequently are found to be erroneous, may be reinstated. Example: the Washakie Formation, defined in 1869, was abandoned in 1918 and reinstated in 1973.

CODE AMENDMENT

Article 21.—**Procedure for Amendment.** Additions to, or changes of, this Code may be proposed in writing to the Commission by any geoscientist at any time. If accepted for consideration by a majority vote of the Commission, they may be adopted by a two-thirds vote of the Commission at an annual meeting not less than a year after publication of the proposal.

FORMAL UNITS DISTINGUISHED BY CONTENT, PROPERTIES, OR PHYSICAL LIMITS

LITHOSTRATIGRAPHIC UNITS

Nature and Boundaries

Article 22.—**Nature of Lithostratigraphic Units.** A lithostratigraphic unit is a defined body of sedimentary, extrusive igneous, metasedimentary, or metavolcanic strata which is distinguished and delimited on the basis of lithic characteristics and stratigraphic position. A lithostratigraphic unit generally conforms to the Law of Superposition and commonly is stratified and tabular in form.

Remarks. (a) **Basic units.**—Lithostratigraphic units are the basic units of general geologic work and serve as the foundation for delineating strata, local and regional structure, economic resources, and geologic history in regions of stratified rocks. They are recognized and defined by observable rock characteristics; boundaries may be placed at clearly distinguished contacts or drawn arbitrarily within a zone of gradation. Lithification or cementation is not a necessary property; clay, gravel, till, and other unconsolidated deposits may constitute valid lithostratigraphic units.

(b) **Type section and locality.**—The definition of a lithostratigraphic unit should be based, if possible, on a stratotype consisting of readily accessible rocks in place, e.g., in outcrops, excavations, and mines, or of rocks accessible only to remote sampling devices, such as those in drill holes and underwater. Even where remote methods are used, definitions must be based on lithic criteria and not on the geophysical characteristics of the rocks, nor the implied age of their contained fossils. Definitions

North American Stratigraphic Code

must be based on descriptions of actual rock material. Regional validity must be demonstrated for all such units. In regions where the stratigraphy has been established through studies of surface exposures, the naming of new units in the subsurface is justified only where the subsurface section differs materially from the surface section, or where there is doubt as to the equivalence of a subsurface and a surface unit. The establishment of subsurface reference sections for units originally defined in outcrop is encouraged.

(c) **Type section never changed.**—The definition and name of a lithostratigraphic unit are established at a type section (or locality) that, once specified, must not be changed. If the type section is poorly designated or delimited, it may be redefined subsequently. If the originally specified stratotype is incomplete, poorly exposed, structurally complicated, or unrepresentative of the unit, a principal reference section or several reference sections may be designated to supplement, but not to supplant, the type section (Article 8e).

(d) **Independence from inferred geologic history.**—Inferred geologic history, depositional environment, and biological sequence have no place in the definition of a lithostratigraphic unit, which must be based on composition and other lithic characteristics; nevertheless, considerations of well-documented geologic history properly may influence the choice of vertical and lateral boundaries of a new unit. Fossils may be valuable during mapping in distinguishing between two lithologically similar, noncontiguous lithostratigraphic units. The fossil content of a lithostratigraphic unit is a legitimate lithic characteristic; for example, oyster-rich sandstone, coquina, coral reef, or graptolitic shale. Moreover, otherwise similar units, such as the Formación Mendez and Formación Velasco mudstones, may be distinguished on the basis of coarseness of contained fossils (foraminifera).

(e) **Independence from time concepts.**—The boundaries of most lithostratigraphic units may transgress time horizons, but some may be approximately synchronous. Inferred time-spans, however measured, play no part in differentiating or determining the boundaries of any lithostratigraphic unit. Either relatively short or relatively long intervals of time may be represented by a single unit. The accumulation of material assigned to a particular unit may have begun or ended earlier in some localities than in others; also, removal of rock by erosion, either within the time-span of deposition of the unit or later, may reduce the time-span represented by the unit locally. The body in some places may be entirely younger than in other places. On the other hand, the establishment of formal units that straddle known, identifiable, regional disconformities is to be avoided, if at all possible. Although concepts of time or age play no part in defining lithostratigraphic units nor in determining their boundaries, evidence of age may aid recognition of similar lithostratigraphic units at localities far removed from the type sections or areas.

(f) **Surface form.**—Erosional morphology or secondary surface form may be a factor in the recognition of a lithostratigraphic unit, but properly should play a minor part at most in the definition of such units. Because the surface expression of lithostratigraphic units is an important aid in mapping, it is commonly advisable, where other factors do not countervail, to define lithostratigraphic boundaries so as to coincide with lithic changes that are expressed in topography.

(g) **Economically exploited units.**—Aquifers, oil sands, coal beds, and quarry layers are, in general, informal units even though named. Some such units, however, may be recognized formally as beds, members, or formations because they are important in the elucidation of regional stratigraphy.

(h) **Instrumentally defined units.**—In subsurface investigations, certain bodies of rock and their boundaries are widely recognized on borehole geophysical logs showing their electrical resistivity, radioactivity, density, or other physical properties. Such bodies and their boundaries may or may not correspond to formal lithostratigraphic units and their boundaries. Where other considerations do not countervail, the boundaries of subsurface units should be defined so as to correspond to useful geophysical markers; nevertheless, units defined exclusively on the basis of remotely sensed physical properties, although commonly useful in stratigraphic analysis, stand completely apart from the hierarchy of formal lithostratigraphic units and are considered informal.

(i) **Zone.**—As applied to the designation of lithostratigraphic units, the term "zone" is informal. Examples are "producing zone," "mineralized zone," "metamorphic zone," and "heavy-mineral zone." A zone may include all or parts of a bed, a member, a formation, or even a group.

(j) **Cyclothems.**—Cyclic or rhythmic sequences of sedimentary rocks, whose repetitive divisions have been named cyclothems, have been recognized in sedimentary basins around the world. Some cyclothems have been identified by geographic names, but such names are considered informal. A clear distinction must be maintained between the division of a stratigraphic column into cyclothems and its division into groups, formations, and members. Where a cyclothem is identified by a geographic name, the word *cyclothem* should be part of the name, and the geographic term should not be the same as that of any formal unit embraced by the cyclothem.

(k) **Soils and paleosols.**—Soils and paleosols are layers composed of the in-situ products of weathering of older rocks which may be of diverse composition and age. Soils and paleosols differ in several respects from lithostratigraphic units, and should not be treated as such (see "Pedostratigraphic Units," Articles 55 et seq).

(l) **Depositional facies.**—Depositional facies are informal units, whether objective (conglomeratic, black shale, graptolitic) or genetic and environmental (platform, turbiditic, fluvial), even when a geographic term has been applied, e.g., Lantz Mills facies. Descriptive designations convey more information than geographic terms and are preferable.

Article 23.—**Boundaries.** Boundaries of lithostratigraphic units are placed at positions of lithic change. Boundaries are placed at distinct contacts or may be fixed arbitrarily within zones of gradation (Fig. 2a). Both vertical and lateral boundaries are based on the lithic criteria that provide the greatest unity and utility.

Remarks. (a) **Boundary in a vertically gradational sequence.**—A named lithostratigraphic unit is preferably bounded by a single lower and a single upper surface so that the name does not recur in a normal stratigraphic succession (see Remark b). Where a rock unit passes vertically into another by intergrading or interfingering of two or more kinds of rock, unless the gradational strata are sufficiently thick to warrant designation of a third, independent unit, the boundary is necessarily arbitrary and should be selected on the basis of practicality (Fig. 2b). For example, where a shale unit overlies a unit of interbedded limestone and shale, the boundary commonly is placed at the top of the highest readily traceable limestone bed. Where a sandstone unit grades upward into shale, the boundary may be so gradational as to be difficult to place even arbitrarily; ideally it should be drawn at the level where the rock is composed of one-half of each component. Because of creep in outcrops and caving in boreholes, it is generally best to define such arbitrary boundaries by the highest occurrence of a particular rock type, rather than the lowest.

(b) **Boundaries in lateral lithologic change.**—Where a unit changes laterally through abrupt gradation into, or intertongues with, a markedly different kind of rock, a new unit should be proposed for the different rock type. An arbitrary lateral boundary may be placed between the two equivalent units. Where the area of lateral intergradation or intertonguing is sufficiently extensive, a transitional interval of interbedded rocks may constitute a third independent unit (Fig. 2c). Where tongues (Article 25b) of formations are mapped separately or otherwise set apart without being formally named, the unmodified formation name should not be repeated in a normal stratigraphic sequence, although the modified name may be repeated in such phrases as "lower tongue of Mancos Shale" and "upper tongue of Mancos Shale." To show the order of superposition on maps and cross sections, the unnamed tongues may be distinguished informally (Fig. 2d) by number, letter, or other means. Such relationships may also be dealt with informally through the recognition of depositional facies (Article 22-1).

(c) **Key beds used for boundaries.**—Key beds (Article 26b) may be used as boundaries for a formal lithostratigraphic unit where the internal lithic characteristics of the unit remain relatively constant. Even though bounding key beds may be traceable beyond the area of the diagnostic overall rock type, geographic extension of the lithostratigraphic unit bounded thereby is not necessarily justified. Where the rock between key beds becomes drastically different from that of the type locality, a new name should be applied (Fig. 2e), even though the key beds are continuous (Article 26b). Stratigraphic and sedimentologic studies of stratigraphic units (usually informal) bounded by key beds may be very informative and useful, especially in subsurface work where the key beds may be recognized by their geophysical signatures. Such units, however, may be a kind of chronostratigraphic, rather than lithostratigraphic, unit (Article 75, 75c), although others are diachronous because one, or both, of the key beds are also diachronous.

(d) **Unconformities as boundaries.**—Unconformities, where recognizable objectively on lithic criteria, are ideal boundaries for lithostratigraphic units. However, a sequence of similar rocks may include an

North American Commission on Stratigraphic Nomenclature

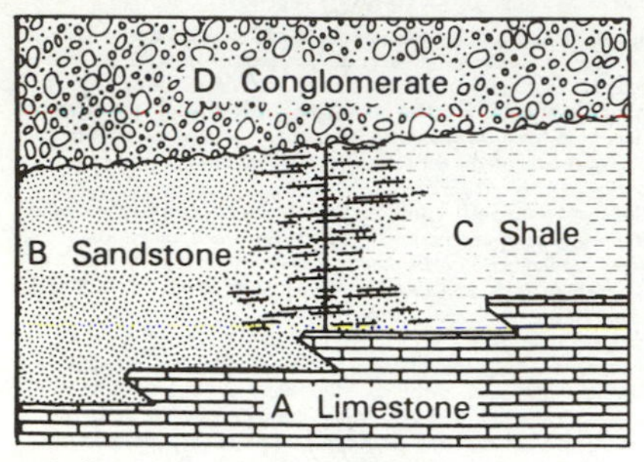

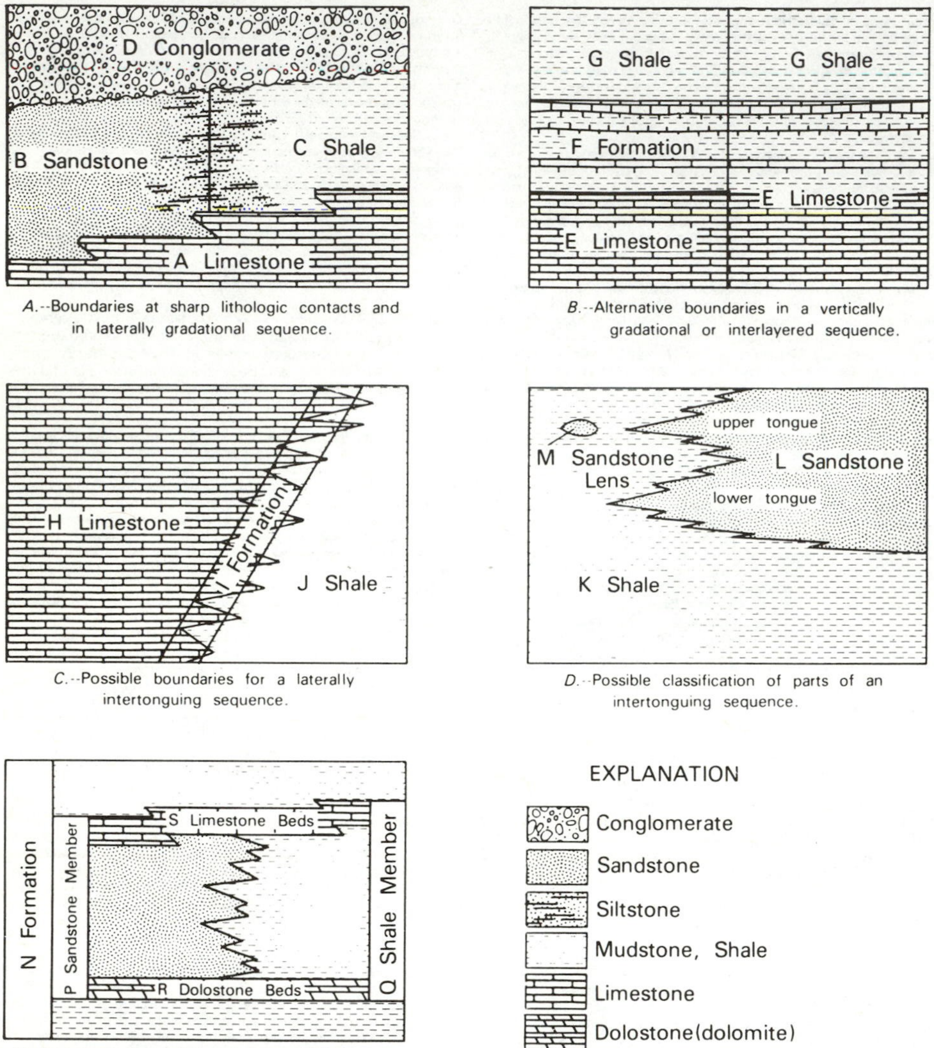

A.--Boundaries at sharp lithologic contacts and in laterally gradational sequence.

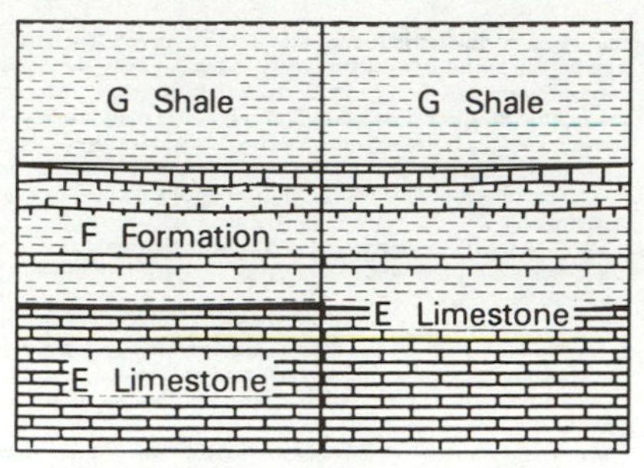

B.--Alternative boundaries in a vertically gradational or interlayered sequence.

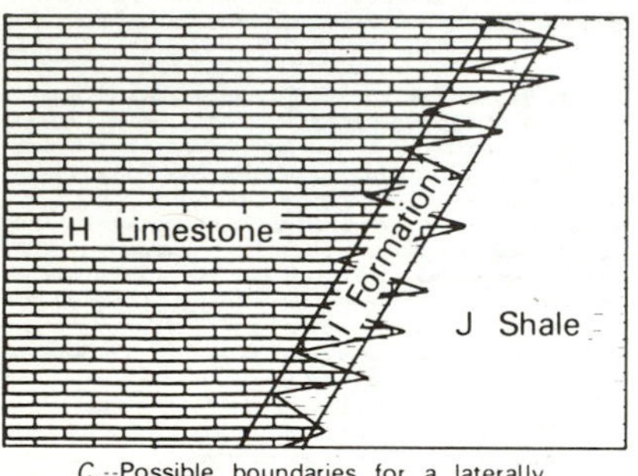

C.--Possible boundaries for a laterally intertonguing sequence.

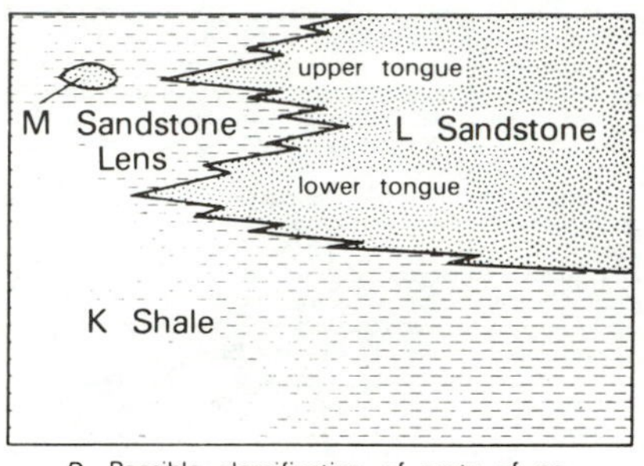

D.--Possible classification of parts of an intertonguing sequence.

E.--Key beds, here designated the R Dolostone Beds and the S Limestone Beds, are used as boundaries to distinguish the Q Shale Member from the other parts of the N Formation. A lateral change in composition between the key beds requires that another name, P Sandstone Member, be applied. The key beds are part of each member.

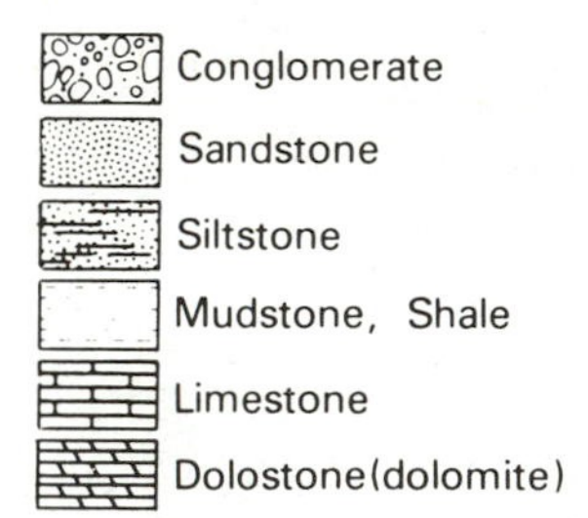

FIG. 2.—Diagrammatic examples of lithostratigraphic boundaries and classification.

obscure unconformity so that separation into two units may be desirable but impracticable. If no lithic distinction adequate to define a widely recognizable boundary can be made, only one unit should be recognized, even though it may include rock that accumulated in different epochs, periods, or eras.

(e) **Correspondence with genetic units.**—The boundaries of lithostratigraphic units should be chosen on the basis of lithic changes and, where feasible, to correspond with the boundaries of genetic units, so that subsequent studies of genesis will not have to deal with units that straddle formal boundaries.

Ranks of Lithostratigraphic Units

Article 24.—**Formation.** The formation is the fundamental unit in lithostratigraphic classification. A formation is a body of rock identified by lithic characteristics and stratigraphic position; it is prevailingly but not necessarily tabular and is mappable at the Earth's surface or traceable in the subsurface.

Remarks. (a) **Fundamental unit.**—Formations are the basic lithostratigraphic units used in describing and interpreting the geology of a region. The limits of a formation normally are those surfaces of lithic change that give it the greatest practicable unity of constitution. A formation may represent a long or short time interval, may be composed of materials from one or several sources, and may include breaks in deposition (see Article 23d).

(b) **Content.**—A formation should possess some degree of internal lithic homogeneity or distinctive lithic features. It may contain between its upper and lower limits (i) rock of one lithic type, (ii) repetitions of two or more lithic types, or (iii) extreme lithic heterogeneity which in itself may constitute a form of unity when compared to the adjacent rock units.

(c) **Lithic characteristics.**—Distinctive lithic characteristics include chemical and mineralogical composition, texture, and such supplementary features as color, primary sedimentary or volcanic structures, fossils (viewed as rock-forming particles), or other organic content (coal, oil-shale). A unit distinguishable only by the taxonomy of its fossils is not a lithostratigraphic but a biostratigraphic unit (Article 48). Rock type may be distinctively represented by electrical, radioactive, seismic, or other properties (Article 22h), but these properties by themselves do not describe adequately the lithic character of the unit.

(d) **Mappability and thickness.**—The proposal of a new formation must be based on tested mappability. Well-established formations commonly are divisible into several widely recognizable lithostratigraphic units; where formal recognition of these smaller units serves a useful purpose, they may be established as members and beds, for which the requirement of mappability is not mandatory. A unit formally recognized as a formation in one area may be treated elsewhere as a group, or as a member of another formation, without change of name. Example: the Niobrara is mapped at different places as a member of the Mancos Shale, of the Cody Shale, or of the Colorado Shale, and also as the Niobrara Formation, as the Niobrara Limestone, and as the Niobrara Shale.

Thickness is not a determining parameter in dividing a rock succession into formations; the thickness of a formation may range from a feather edge at its depositional or erosional limit to thousands of meters elsewhere. No formation is considered valid that cannot be delineated at the scale of geologic mapping practiced in the region when the formation is proposed. Although representation of a formation on maps and cross sections by a labeled line may be justified, proliferation of such exceptionally thin units is undesirable. The methods of subsurface mapping permit delineation of units much thinner than those usually practicable for surface studies; before such thin units are formalized, consideration should be given to the effect on subsequent surface and subsurface studies.

(e) **Organic reefs and carbonate mounds.**—Organic reefs and carbonate mounds ("buildups") may be distinguished formally, if desirable, as formations distinct from their surrounding, thinner, temporal equivalents. For the requirements of formalization, see Article 30f.

(f) **Interbedded volcanic and sedimentary rock.**—Sedimentary rock and volcanic rock that are interbedded may be assembled into a formation under one name which should indicate the predominant or distinguishing lithology, such as Mindego Basalt.

(g) **Volcanic rock.**—Mappable distinguishable sequences of stratified volcanic rock should be treated as formations or lithostratigraphic units of higher or lower rank. A small intrusive component of a dominantly stratiform volcanic assemblage may be treated informally.

(h) **Metamorphic rock.**—Formations composed of low-grade metamorphic rock (defined for this purpose as rock in which primary structures are clearly recognizable) are, like sedimentary formations, distinguished mainly by lithic characteristics. The mineral facies may differ from place to place, but these variations do not require definition of a new formation. High-grade metamorphic rocks whose relation to established formations is uncertain are treated as lithodemic units (see Articles 31 et seq).

Article 25.—**Member.** A member is the formal lithostratigraphic unit next in rank below a formation and is always a part of some formation. It is recognized as a named entity within a formation because it possesses characteristics distinguishing it from adjacent parts of the formation. A formation need not be divided into members unless a useful purpose is served by doing so. Some formations may be divided completely into members; others may have only certain parts designated as members; still others may have no members. A member may extend laterally from one formation to another.

Remarks. (a) **Mapping of members.**—A member is established when it is advantageous to recognize a particular part of a heterogeneous formation. A member, whether formally or informally designated, need not be mappable at the scale required for formations. Even if all members of a formation are locally mappable, it does not follow that they should be raised to formational rank, because proliferation of formation names may obscure rather than clarify relations with other areas.

(b) **Lens and tongue.**—A geographically restricted member that terminates on all sides within a formation may be called a lens (lentil). A wedging member that extends outward beyond a formation or wedges ("pinches") out within another formation may be called a tongue.

(c) **Organic reefs and carbonate mounds.**—Organic reefs and carbonate mounds may be distinguished formally, if desirable, as members within a formation. For the requirements of formalization, see Article 30f.

(d) **Division of members.**—A formally or informally recognized division of a member is called a bed or beds, except for volcanic flow rocks, for which the smallest formal unit is a flow. Members may contain beds or flows, but may never contain other members.

(e) **Laterally equivalent members.**—Although members normally are in vertical sequence, laterally equivalent parts of a formation that differ recognizably may also be considered members.

Article 26.—**Bed(s).** A bed, or beds, is the smallest formal lithostratigraphic unit of sedimentary rocks.

Remarks. (a) **Limitations.**—The designation of a bed or a unit of beds as a formally named lithostratigraphic unit generally should be limited to certain distinctive beds whose recognition is particularly useful. Coal beds, oil sands, and other beds of economic importance commonly are named, but such units and their names usually are not a part of formal stratigraphic nomenclature (Articles 22g and 30g).

(b) **Key or marker beds.**—A key or marker bed is a thin bed of distinctive rock that is widely distributed. Such beds may be named, but usually are considered informal units. Individual key beds may be traced beyond the lateral limits of a particular formal unit (Article 23c).

Article 27.—**Flow.** A flow is the smallest formal lithostratigraphic unit of volcanic flow rocks. A flow is a discrete, extrusive, volcanic body distinguishable by texture, composition, order of superposition, paleomagnetism, or other objective criteria. It is part of a member and thus is equivalent in rank to a bed or beds of sedimentary-rock classification. Many flows are informal units. The designation and naming of flows as formal rock-stratigraphic units should be limited to those that are distinctive and widespread.

Article 28.—**Group.** A group is the lithostratigraphic unit next higher in rank to formation; a group may consist entirely of named formations, or alternatively, need not be composed entirely of named formations.

Remarks. (a) **Use and content.**—Groups are defined to express the natural relationships of associated formations. They are useful in small-scale mapping and regional stratigraphic analysis. In some reconnaissance work, the term "group" has been applied to lithostratigraphic units that appear to be divisible into formations, but have not yet been so divided. In such cases, formations may be erected subsequently for one or all of the practical divisions of the group.

(b) **Change in component formations.**—The formations making up a group need not necessarily be everywhere the same. The Rundle Group, for example, is widespread in western Canada and undergoes several changes in formational content. In southwestern Alberta, it comprises the Livingstone, Mount Head, and Etherington Formations in the Front Ranges, whereas in the foothills and subsurface of the adjacent plains, it comprises the Pekisko, Shunda, Turner Valley, and Mount Head Formations. However, a formation or its parts may not be assigned to two vertically adjacent groups.

(c) **Change in rank.**—The wedge-out of a component formation or formations may justify the reduction of a group to formation rank, retaining the same name. When a group is extended laterally beyond where it is divided into formations, it becomes in effect a formation, even if it is still called a group. When a previously established formation is divided into two or more component units that are given formal formation rank, the old formation, with its old geographic name, should be raised to group status. Raising the rank of the unit is preferable to restricting the old name to a part of its former content, because a change in rank leaves the sense of a well-established unit unchanged (Articles 19b, 19g).

Article 29.—**Supergroup.** A supergroup is a formal assemblage of related or superposed groups, or of groups and formations. Such units have proved useful in regional and provincial syntheses. Supergroups should be named only where their recognition serves a clear purpose.

Remark. (a) **Misuse of "series" for group or supergroup.**—Although "series" is a useful general term, it is applied formally only to a chronostratigraphic unit and should not be used for a lithostratigraphic unit. The term "series" should no longer be employed for an assemblage of formations or an assemblage of formations and groups, as it has been, especially in studies of the Precambrian. These assemblages are groups or supergroups.

Lithostratigraphic Nomenclature

Article 30.—**Compound Character.** The formal name of a lithostratigraphic unit is compound. It consists of a geographic name combined with a descriptive lithic term or with the appropriate rank term, or both. Initial letters of all words used in forming the names of formal rock-stratigraphic units are capitalized.

Remarks. (a) **Omission of part of a name.**—Where frequent repetition would be cumbersome, the geographic name, the lithic term, or the rank term may be used alone, once the full name has been introduced; as "the Burlington," "the limestone," or "the formation," for the Burlington Limestone.

(b) **Use of simple lithic terms.**—The lithic part of the name should indicate the predominant or diagnostic lithology, even if subordinate lithologies are included. Where a lithic term is used in the name of a lithostratigraphic unit, the simplest generally acceptable term is recommended (for example, limestone, sandstone, shale, tuff, quartzite). Compound terms (for example, clay shale) and terms that are not in common usage (for example, calcirudite, orthoquartzite) should be avoided. Combined terms, such as "sand and clay," should not be used for the lithic part of the names of lithostratigraphic units, nor should an adjective be used between the geographic and the lithic terms, as "Chattanooga Black Shale" and "Biwabik Iron-Bearing Formation."

(c) **Group names.**—A group name combines a geographic name with the term "group," and no lithic designation is included; for example, San Rafael Group.

(d) **Formation names.**—A formation name consists of a geographic name followed by a lithic designation or by the word "formation." Examples: Dakota Sandstone, Mitchell Mesa Rhyolite, Monmouth Formation, Halton Till.

(e) **Member names.**—All member names include a geographic term and the word "member;" some have an intervening lithic designation, if useful; for example, Wedington Sandstone Member of the Fayetteville Shale. Members designated solely by lithic character (for example, siliceous shale member), by position (upper, lower), or by letter or number, are informal.

(f) **Names of reefs.**—Organic reefs identified as formations or members are formal units only where the name combines a geographic name with the appropriate rank term, e.g., Leduc Formation (a name applied to the several reefs enveloped by the Ireton Formation), Rainbow Reef Member.

(g) **Bed and flow names.**—The names of beds or flows combine a geographic term, a lithic term, and the term "bed" or "flow;" for example, Knee Hills Tuff Bed, Ardmore Bentonite Beds, Negus Variolitic Flows.

(h) **Informal units.**—When geographic names are applied to such informal units as oil sands, coal beds, mineralized zones, and informal members (see Articles 22g and 26a), the unit term should not be capitalized. A name is not necessarily formal because it is capitalized, nor does failure to capitalize a name render it informal. Geographic names should be combined with the terms "formation" or "group" only in formal nomenclature.

(i) **Informal usage of identical geographic names.**—The application of identical geographic names to several minor units in one vertical sequence is considered informal nomenclature (lower Mount Savage coal, Mount Savage fireclay, upper Mount Savage coal, Mount Savage rider coal, and Mount Savage sandstone). The application of identical geographic names to the several lithologic units constituting a cyclothem likewise is considered informal.

(j) **Metamorphic rock.**—Metamorphic rock recognized as a normal stratified sequence, commonly low-grade metavolcanic or metasedimentary rocks, should be assigned to named groups, formations, and members, such as the Deception Rhyolite, a formation of the Ash Creek Group, or the Bonner Quartzite, a formation of the Missoula Group. High-grade metamorphic and metasomatic rocks are treated as lithodemes and suites (see Articles 31, 33, 35).

(k) **Misuse of well-known name.**—A name that suggests some well-known locality, region, or political division should not be applied to a unit typically developed in another less well-known locality of the same name. For example, it would be inadvisable to use the name "Chicago Formation" for a unit in California.

LITHODEMIC UNITS

Nature and Boundaries

Article 31.—**Nature of Lithodemic Units.** A lithodemic[6] unit is a defined body of predominantly intrusive, highly deformed, and/or highly metamorphosed rock, distinguished and delimited on the basis of rock characteristics. In contrast to lithostratigraphic units, a lithodemic unit generally does not conform to the Law of Superposition. Its contacts with other rock units may be sedimentary, extrusive, intrusive, tectonic, or metamorphic (Fig. 3).

Remarks. (a) **Recognition and definition.**—Lithodemic units are defined and recognized by observable rock characteristics. They are the practical units of general geological work in terranes in which rocks generally lack primary stratification; in such terranes they serve as the foundation for studying, describing, and delineating lithology, local and regional structure, economic resources, and geologic history.

(b) **Type and reference localities.**—The definition of a lithodemic unit should be based on as full a knowledge as possible of its lateral and vertical variations and its contact relationships. For purposes of nomenclatural stability, a type locality and, wherever appropriate, reference localities should be designated.

(c) **Independence from inferred geologic history.**—Concepts based on inferred geologic history properly play no part in the definition of a lithodemic unit. Nevertheless, where two rock masses are lithically similar but display objective structural relations that preclude the possibility of their being even broadly of the same age, they should be assigned to different lithodemic units.

(d) **Use of "zone."**—As applied to the designation of lithodemic units, the term "zone" is informal. Examples are: "mineralized zone," "contact zone," and "pegmatitic zone."

6From the Greek *demas, -os:* "living body, frame".

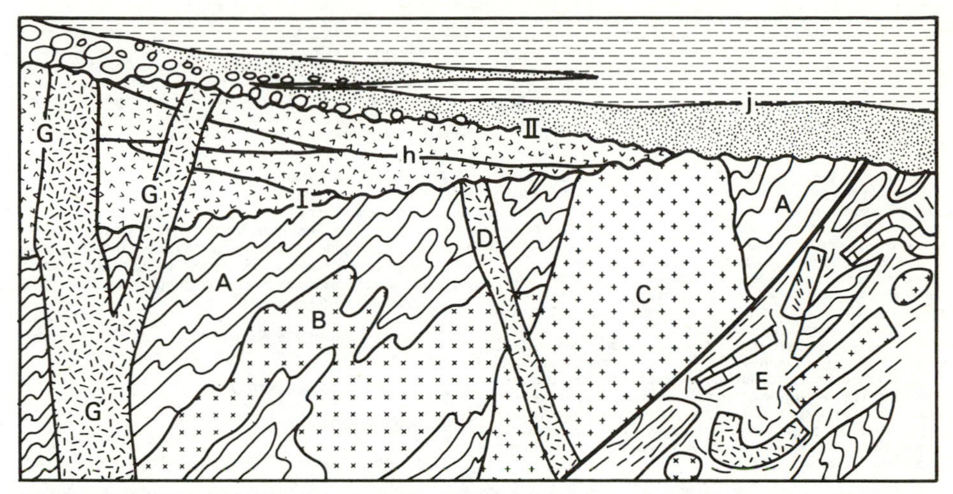

FIG. 3.—Lithodemic (upper case) and lithostratigraphic (lower case) units. A *lithodeme* of *gneiss* (A) contains an *intrusion* of diorite (B) that was deformed with the gneiss. A and B may be treated jointly as a *complex*. A younger *granite* (C) is cut by a dike of *syenite* (D), that is cut in turn by unconformity I. All the foregoing are in fault contact with a *structural complex* (E). A *volcanic complex* (G) is built upon unconformity I, and its feeder dikes cut the unconformity. Laterally equivalent volcanic strata in orderly, mappable succession (h) are treated as lithostratigraphic units. A *gabbro* feeder (G′), to the volcanic complex, where surrounded by gneiss is readily distinguished as a separate lithodeme and named as a *gabbro* or an *intrusion*. All the foregoing are overlain, at unconformity II, by sedimentary rocks (j) divided into formations and members.

Article 32.—**Boundaries.** Boundaries of lithodemic units are placed at positions of lithic change. They may be placed at clearly distinguished contacts or within zones of gradation. Boundaries, both vertical and lateral, are based on the lithic criteria that provide the greatest unity and practical utility. Contacts with other lithodemic and lithostratigraphic units may be depositional, intrusive, metamorphic, or tectonic.

Remark. (a) **Boundaries within gradational zones.**—Where a lithodemic unit changes through gradation into, or intertongues with, a rock-mass with markedly different characteristics, it is usually desirable to propose a new unit. It may be necessary to draw an arbitrary boundary within the zone of gradation. Where the area of intergradation or intertonguing is sufficiently extensive, the rocks of mixed character may constitute a third unit.

Ranks of Lithodemic Units

Article 33.—**Lithodeme.** The lithodeme is the fundamental unit in lithodemic classification. A lithodeme is a body of intrusive, pervasively deformed, or highly metamorphosed rock, generally non-tabular and lacking primary depositional structures, and characterized by lithic homogeneity. It is mappable at the Earth's surface and traceable in the subsurface. For cartographic and hierarchical purposes, it is comparable to a formation (see Table 2).

Remarks. (a) **Content.**—A lithodeme should possess distinctive lithic features and some degree of internal lithic homogeneity. It may consist of (i) rock of one type, (ii) a mixture of rocks of two or more types, or (iii) extreme heterogeneity of composition, which may constitute in itself a form of unity when compared to adjoining rock-masses (see also "complex," Article 37).

(b) **Lithic characteristics.**—Distinctive lithic characteristics may include mineralogy, textural features such as grain size, and structural features such as schistose or gneissic structure. A unit distinguishable from its neighbors only by means of chemical analysis is informal.

(c) **Mappability.**—Practicability of surface or subsurface mapping is an essential characteristic of a lithodeme (see Article 24d).

Article 34.—**Division of Lithodemes.** Units below the rank of lithodeme are informal.

Article 35.—**Suite.** A *suite* (metamorphic suite, intrusive suite, plutonic suite) is the lithodemic unit next higher in rank to lithodeme. It comprises two or more associated lithodemes of the same class (e.g., plutonic, metamorphic). For cartographic and hierarchical purposes, suite is comparable to group (see Table 2).

Remarks. (a) **Purpose.**—Suites are recognized for the purpose of expressing the natural relations of associated lithodemes having significant lithic features in common, and of depicting geology at compilation scales too small to allow delineation of individual lithodemes. Ideally, a suite consists entirely of named lithodemes, but may contain both named and unnamed units.

(b) **Change in component units.**—The named and unnamed units constituting a suite may change from place to place, so long as the original sense of natural relations and of common lithic features is not violated.

(c) **Change in rank.**—Traced laterally, a suite may lose all of its formally named divisions but remain a recognizable, mappable entity. Under such circumstances, it may be treated as a lithodeme but retain the same name. Conversely, when a previously established lithodeme is divided into two or more mappable divisions, it may be desirable to raise its rank to suite, retaining the original geographic component of the name. To avoid confusion, the original name should not be retained for one of the divisions of the original unit (see Article 19g).

Article 36.—**Supersuite.** A supersuite is the unit next higher in rank to a suite. It comprises two or more suites or complexes having a degree of natural relationship to one another, either in the vertical or the lateral sense. For cartographic and hierarchical purposes, supersuite is similar in rank to supergroup.

Article 37.—**Complex.** An assemblage or mixture of rocks of *two or more genetic classes*, i.e., igneous, sedimentary, or metamorphic, with or without highly complicated structure, may be named a *complex*. The term "complex" takes the place of the lithic or rank term (for example, Boil Mountain Complex, Franciscan Complex) and, although unranked, commonly is comparable to suite or supersuite and is named in the same manner (Articles 41, 42).

Remarks (a) **Use of "complex."**—Identification of an assemblage of diverse rocks as a complex is useful where the mapping of each separate lithic component is impractical at ordinary mapping scales. "Complex" is unranked but commonly comparable to suite or supersuite; therefore, the term may be retained if subsequent, detailed mapping distinguishes some or all of the component lithodemes or lithostratigraphic units.

(b) **Volcanic complex.**—Sites of persistent volcanic activity commonly are characterized by a diverse assemblage of extrusive volcanic rocks, related intrusions, and their weathering products. Such an assemblage may be designated a *volcanic complex*.

(c) **Structural complex.**—In some terranes, tectonic processes (e.g., shearing, faulting) have produced heterogeneous mixtures or disrupted bodies of rock in which some individual components are too small to be mapped. *Where there is no doubt that the mixing or disruption is due to tectonic processes,* such a mixture may be designated as a structural complex, whether it consists of two or more classes of rock, or a single class only. A simpler solution for some mapping purposes is to indicate intense deformation by an overprinted pattern.

(d) **Misuse of "complex"**.—Where the rock assemblage to be united under a single, formal name consists of diverse types of a *single class* of rock, as in many terranes that expose a variety of either intrusive igneous or high-grade metamorphic rocks, the term "intrusive suite," "plutonic suite," or "metamorphic suite" should be used, rather than the unmodified term "complex." Exceptions to this rule are the terms *structural complex* and *volcanic complex* (see Remarks c and b, above).

Article 38.—**Misuse of "Series" for Suite, Complex, or Supersuite.** The term "series" has been employed for an assemblage of lithodemes or an assemblage of lithodemes and suites, especially in studies of the Precambrian. This practice now is regarded as improper; these assemblages are suites, complexes, or supersuites. The term "series" also has been applied to a sequence of rocks resulting from a succession of eruptions or intrusions. In these cases a different term should be used; "group" should replace "series" for volcanic and low-grade metamorphic rocks, and "intrusive suite" or "plutonic suite" should replace "series" for intrusive rocks of group rank.

Lithodemic Nomenclature

Article 39.—**General Provisions.** The formal name of a lithodemic unit is compound. It consists of a geographic name combined with a descriptive or appropriate rank term. The principles for the selection of the geographic term, concerning suitability, availability, priority, etc, follow those established in Article 7, where the rules for capitalization are also specified.

Article 40.—**Lithodeme Names.** The name of a lithodeme combines a geographic term with a lithic or descriptive term, e.g., Killarney Granite, Adamant Pluton, Manhattan Schist, Skaergaard Intrusion, Duluth Gabbro. The term *formation* should not be used.

Remarks. (a) **Lithic term.**—The lithic term should be a common and familiar term, such as schist, gneiss, gabbro. Specialized terms and terms not widely used, such as websterite and jacupirangite, and compound terms, such as graphitic schist and augen gneiss, should be avoided.

(b) **Intrusive and plutonic rocks.**—Because many bodies of intrusive rock range in composition from place to place and are difficult to characterize with a single lithic term, and because many bodies of plutonic rock are considered not to be intrusions, latitude is allowed in the choice of a lithic or descriptive term. Thus, the descriptive term should preferably be compositional (e.g., gabbro, granodiorite), but may, if necessary, denote form (e.g., dike, sill), or be neutral (e.g., intrusion, pluton[7]). In any event, specialized compositional terms not widely used are to be avoided, as are form terms that are not widely used, such as bysmalith and chonolith. Terms implying genesis should be avoided as much as possible, because interpretations of genesis may change.

Article 41.—**Suite Names.** The name of a suite combines a geographic term, the term "suite," and an adjective denoting the fundamental character of the suite; for example, Idaho Springs Metamorphic Suite, Tuolumne Intrusive Suite, Cassiar Plutonic Suite. The geographic name of a suite may not be the same as that of a component lithodeme (see Article 19f). Intrusive assemblages, however, may share the same geographic name if an intrusive lithodeme is representative of the suite.

Article 42.—**Supersuite Names.** The name of a supersuite combines a geographic term with the term "supersuite."

MAGNETOSTRATIGRAPHIC UNITS

Nature and Boundaries

Article 43.—**Nature of Magnetostratigraphic Units.** A magnetostratigraphic unit is a body of rock unified by specified remanent-magnetic properties and is distinct from underlying and overlying magnetostratigraphic units having different magnetic properties.

Remarks. (a) **Definition.**—Magnetostratigraphy is defined here as all aspects of stratigraphy based on remanent magnetism (paleomagnetic signatures). Four basic paleomagnetic phenomena can be determined or inferred from remanent magnetism: polarity, dipole-field-pole position (including apparent polar wander), the non-dipole component (secular variation), and field intensity.

(b) **Contemporaneity of rock and remanent magnetism.**—Many paleomagnetic signatures reflect earth magnetism at the time the rock formed. Nevertheless, some rocks have been subjected subsequently to physical and/or chemical processes which altered the magnetic properties. For example, a body of rock may be heated above the blocking temperature or Curie point for one or more minerals, or a ferromagnetic mineral may be produced by low-temperature alteration long after the enclosing rock formed, thus acquiring a component of remanent magnetism reflecting the field at the time of alteration, rather than the time of original rock deposition or crystallization.

(c) **Designations and scope.**—The prefix *magneto* is used with an appropriate term to designate the aspect of remanent magnetism used to define a unit. The terms "magnetointensity" or "magnetosecular-variation" are possible examples. This Code considers only polarity reversals, which now are recognized widely as a stratigraphic tool. However, apparent-polar-wander paths offer increasing promise for correlations within Precambrian rocks.

Article 44.—**Definition of Magnetopolarity Unit.** A magnetopolarity unit is a body of rock unified by its remanent magnetic polarity and distinguished from adjacent rock that has different polarity.

Remarks. (a) **Nature.**—Magnetopolarity is the record in rocks of the polarity history of the Earth's magnetic-dipole field. Frequent past reversals of the polarity of the Earth's magnetic field provide a basis for magnetopolarity stratigraphy.

(b) **Stratotype.**—A stratotype for a magnetopolarity unit should be designated and the boundaries defined in terms of recognized lithostratigraphic and/or biostratigraphic units in the stratotype. The formal definition of a magnetopolarity unit should meet the applicable specific requirements of Articles 3 to 16.

(c) **Independence from inferred history.**—Definition of a magnetopolarity unit does not require knowledge of the time at which the unit acquired its remanent magnetism; its magnetism may be primary or secondary. Nevertheless, the unit's present polarity is a property that may be

[7]Pluton—a mappable body of plutonic rock.

ascertained and confirmed by others.

(d) **Relation to lithostratigraphic and biostratigraphic units.**—Magnetopolarity units resemble lithostratigraphic and biostratigraphic units in that they are defined on the basis of an objective recognizable property, but differ fundamentally in that most magnetopolarity unit boundaries are thought not to be time transgressive. Their boundaries may coincide with those of lithostratigraphic or biostratigraphic units, or be parallel to but displaced from those of such units, or be crossed by them.

(e) **Relation of magnetopolarity units to chronostratigraphic units.**—Although transitions between polarity reversals are of global extent, a magnetopolarity unit does not contain within itself evidence that the polarity is primary, or criteria that permit its unequivocal recognition in chronocorrelative strata of other areas. Other criteria, such as paleontologic or numerical age, are required for both correlation and dating. Although polarity reversals are useful in recognizing chronostratigraphic units, magnetopolarity alone is insufficient for their definition.

Article 45.—**Boundaries.** The upper and lower limits of a magnetopolarity unit are defined by boundaries marking a change of polarity. Such boundaries may represent either a depositional discontinuity or a magnetic-field transition. The boundaries are either polarity-reversal horizons or polarity transition-zones, respectively.

Remark. (a) **Polarity-reversal horizons and transition-zones.**—A polarity-reversal horizon is either a single, clearly definable surface or a thin body of strata constituting a transitional interval across which a change in magnetic polarity is recorded. Polarity-reversal horizons describe transitional intervals of 1 m or less; where the change in polarity takes place over a stratigraphic interval greater than 1 m, the term "polarity transition-zone" should be used. Polarity-reversal horizons and polarity transition-zones provide the boundaries for polarity zones, although they may also be contained within a polarity zone where they mark an internal change subsidiary in rank to those at its boundaries.

Ranks of Magnetopolarity Units

Article 46.—**Fundamental Unit.** A polarity zone is the fundamental unit of magnetopolarity classification. A polarity zone is a unit of rock characterized by the polarity of its magnetic signature. Magnetopolarity zone, rather than polarity zone, should be used where there is risk of confusion with other kinds of polarity.

Remarks. (a) **Content.**—A polarity zone should possess some degree of internal homogeneity. It may contain rocks of (1) entirely or predominantly one polarity, or (2) mixed polarity.

(b) **Thickness and duration.**—The thickness of rock of a polarity zone or the amount of time represented should play no part in the definition of the zone. The polarity signature is the essential property for definition.

(c) **Ranks.**—When continued work at the stratotype for a polarity zone, or new work in correlative rocks elsewhere, reveals smaller polarity units, these may be recognized formally as polarity subzones. If it should prove necessary or desirable to group polarity zones, these should be termed polarity superzones. The rank of a polarity unit may be changed when deemed appropriate.

Magnetopolarity Nomenclature

Article 47.—**Compound Name.** The formal name of a magnetopolarity zone should consist of a geographic name and the term *Polarity Zone*. The term may be modified by *Normal*, *Reversed*, or *Mixed* (example: Deer Park Reversed Polarity Zone). In naming or revising magnetopolarity units, appropriate parts of Articles 7 and 19 apply. The use of informal designations, e.g., numbers or letters, is not precluded.

BIOSTRATIGRAPHIC UNITS

Nature and Boundaries

Article 48.—**Nature of Biostratigraphic Units.** A biostratigraphic unit is a body of rock defined or characterized by its fossil content. The basic unit in biostratigraphic classification is the biozone, of which there are several kinds.

Remarks. (a) **Enclosing strata.**—Fossils that define or characterize a biostratigraphic unit commonly are contemporaneous with the body of rock that contains them. Some biostratigraphic units, however, may be represented only by their fossils, preserved in normal stratigraphic succession (e.g., on hardgrounds, in lag deposits, in certain types of remanié accumulations), which alone represent the rock of the biostratigraphic unit. In addition, some strata contain fossils derived from older or younger rocks or from essentially coeval materials of different facies; such fossils should not be used to define a biostratigraphic unit.

(b) **Independence from lithostratigraphic units.**—Biostratigraphic units are based on criteria which differ fundamentally from those for lithostratigraphic units. Their boundaries may or may not coincide with the boundaries of lithostratigraphic units, but they bear no inherent relation to them.

(c) **Independence from chronostratigraphic units.**—The boundaries of most biostratigraphic units, unlike the boundaries of chronostratigraphic units, are both characteristically and conceptually diachronous. An exception is an abundance biozone boundary that reflects a mass-mortality event. The vertical and lateral limits of the rock body that constitutes the biostratigraphic unit represent the limits in distribution of the defining biotic elements. The lateral limits never represent, and the vertical limits rarely represent, regionally synchronous events. Nevertheless, biostratigraphic units are effective for interpreting chronostratigraphic relations.

Article 49.—**Kinds of Biostratigraphic Units.** Three principal kinds of biostratigraphic units are recognized: *interval, assemblage,* and *abundance* biozones.

Remark: (a) **Boundary definitions.**—Boundaries of interval zones are defined by lowest and/or highest occurrences of single taxa; boundaries of some kinds of assemblage zones (Oppel or concurrent range zones) are defined by lowest and/or highest occurrences of more than one taxon; and boundaries of abundance zones are defined by marked changes in relative abundances of preserved taxa.

Article 50.—**Definition of Interval Zone.** An interval zone (or subzone) is the body of strata between two specified, documented lowest and/or highest occurrences of single taxa.

Remarks. (a) **Interval zone types.**—Three basic types of interval zones are recognized (Fig. 4). These include the range zones and interval zones of the International Stratigraphic Guide (ISSC, 1976, p. 53, 60) and are:

1. The interval between the documented lowest and highest occurrences of a single taxon (Fig. 4A). This is the *taxon range zone* of ISSC (1976, p. 53).

2. The interval included between the documented lowest occurrence of one taxon and the documented highest occurrence of another taxon (Fig. 4B). When such occurrences result in stratigraphic overlap of the taxa (Fig. 4B-1), the interval zone is the *concurrent range zone* of ISSC (1976, p. 55), that involves only two taxa. When such occurrences do not result in stratigraphic overlap (Fig. 4B-2), but are used to partition the range of a third taxon, the interval is the *partial range zone* of George and others (1969).

3. The interval between documented successive lowest occurrences or successive highest occurrences of two taxa (Fig. 4C). When the interval is between successive documented lowest occurrences within an evolutionary lineage (Fig. 4C-1), it is the *lineage zone* of ISSC (1976, p. 58). When the interval is between successive lowest occurrences of unrelated taxa or between successive highest occurrences of either related or unrelated taxa (Fig. 4C-2), it is a kind of *interval zone* of ISSC (1976, p. 60).

(b) **Unfossiliferous intervals.**—Unfossiliferous intervals between or within biozones are the *barren interzones* and *intrazones* of ISSC (1976, p. 49).

Article 51.—**Definition of Assemblage Zone.** An assemblage zone is a biozone characterized by the association of three or more taxa. It may be based on all kinds of fossils present, or restricted to only certain kinds of fossils.

North American Commission on Stratigraphic Nomenclature

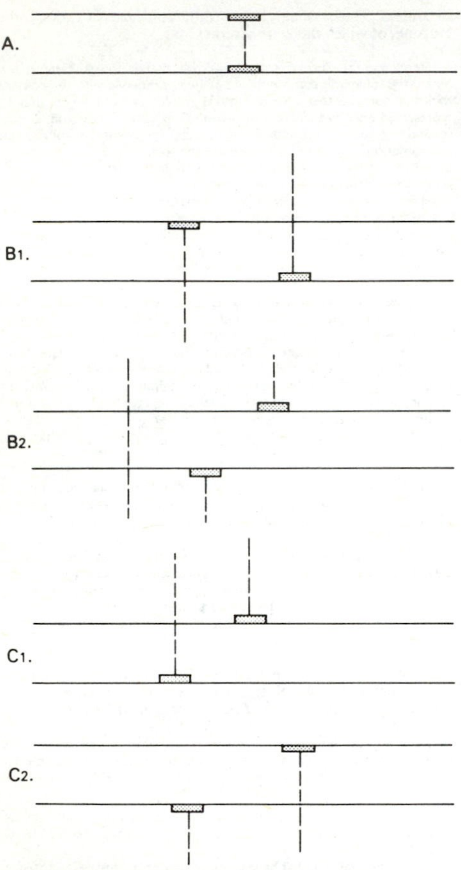

FIG. 4.—Examples of biostratigraphic interval zones.
Vertical broken lines indicate ranges of taxa; bars indicate lowest or highest documented occurrences.

Remarks. (a) **Assemblage zone contents.**—An assemblage zone may consist of a geographically or stratigraphically restricted assemblage, or may incorporate two or more contemporaneous assemblages with shared characterizing taxa (*composite assemblage zones* of Kauffman, 1969) (Fig. 5c).

(b) **Assemblage zone types.**—In practice, two assemblage zone concepts are used:

1. The *assemblage zone* (or cenozone) of ISSC (1976, p. 50), which is characterized by taxa without regard to their range limits (Fig. 5a). Recognition of this type of assemblage zone can be aided by using techniques of multivariate analysis. Careful designation of the characterizing taxa is especially important.

2. The *Oppel zone*, or the *concurrent range zone* of ISSC (1976, p. 55, 57), a type of zone characterized by more than two taxa and having boundaries based on two or more documented first and/or last occurrences of the included characterizing taxa (Fig. 5b).

Article 52.—**Definition of Abundance Zone.** An abundance zone is a biozone characterized by quantitatively distinctive maxima of relative abundance of one or more taxa. This is the *acme zone* of ISSC (1976, p. 59).

Remark. (a) **Ecologic controls.**—The distribution of biotic assemblages used to characterize some assemblage and abundance biozones may reflect strong local ecological control. Biozones based on such assemblages are included within the concept of ecozones (Vella, 1964), and are informal.

Ranks of Biostratigraphic Units

Article 53.—**Fundamental Unit.** The fundamental unit of biostratigraphic classification is a biozone.

Remarks. (a) **Scope.**—A single body of rock may be divided into various kinds and scales of biozones or subzones, as discussed in the International Stratigraphic Guide (ISSC, 1976, p. 62). Such usage is recommended if it will promote clarity, but only the unmodified term *biozone* is accorded formal status.

(b) **Divisions.**—A biozone may be completely or partly divided into formally designated sub-biozones (subzones), if such divisions serve a useful purpose.

Biostratigraphic Nomenclature

Article 54.—**Establishing Formal Units.** Formal establishment of a biozone or subzone must meet the requirements of Article 3 and requires a unique name, a description of its content and its boundaries, reference to a stratigraphic sequence in which the zone is characteristically developed, and a discussion of its spatial extent.

Remarks. (a) **Name.**—The name, which is compound and designates the kind of biozone, may be based on:

1. One or two characteristic and common taxa that are restricted to the biozone, reach peak relative abundance within the biozone, or have their total stratigraphic overlap within the biozone. These names most commonly are those of genera or subgenera, binomial designations of species, or trinomial designations of subspecies. If names of the nominate taxa change, names of the zones should be changed accordingly. Generic or subgeneric names may be abbreviated. Trivial species or subspecies names should not be used alone because they may not be unique.

2. Combinations of letters derived from taxa which characterize the biozone. However, alpha-numeric code designations (e.g., N1, N2, N3...) are informal and not recommended because they do not lend themselves readily to subsequent insertions, combinations, or eliminations. Biozonal systems based *only* on simple progressions of letters or numbers (e.g., A, B, C, or 1, 2, 3) are also not recommended.

(b) **Revision.**—Biozones and subzones are established empirically and may be modified on the basis of new evidence. Positions of established biozone or subzone boundaries may be stratigraphically refined, new characterizing taxa may be recognized, or original characterizing taxa may be superseded. If the concept of a particular biozone or subzone is substantially modified, a new unique designation is required to avoid ambiguity in subsequent citations.

(c) **Specifying kind of zone.**—Initial designation of a formally proposed biozone or subzone as an abundance zone, or as one of the types of interval zones, or assemblage zones (Articles 49-52), is strongly recommended. Once the type of biozone is clearly identified, the designation may be dropped in the remainder of a text (e.g., *Exus albus* taxon range zone to *Exus albus* biozone).

(d) **Defining taxa.**—Initial description or subsequent emendation of a biozone or subzone requires designation of the defining and characteristic taxa, and/or the documented first and last occurrences which mark the biozone or subzone boundaries.

(e) **Stratotypes.**—The geographic and stratigraphic position and boundaries of a formally proposed biozone or subzone should be defined precisely or characterized in one or more designated reference sections. Designation of a stratotype for each new biostratigraphic unit and of reference sections for emended biostratigraphic units is required.

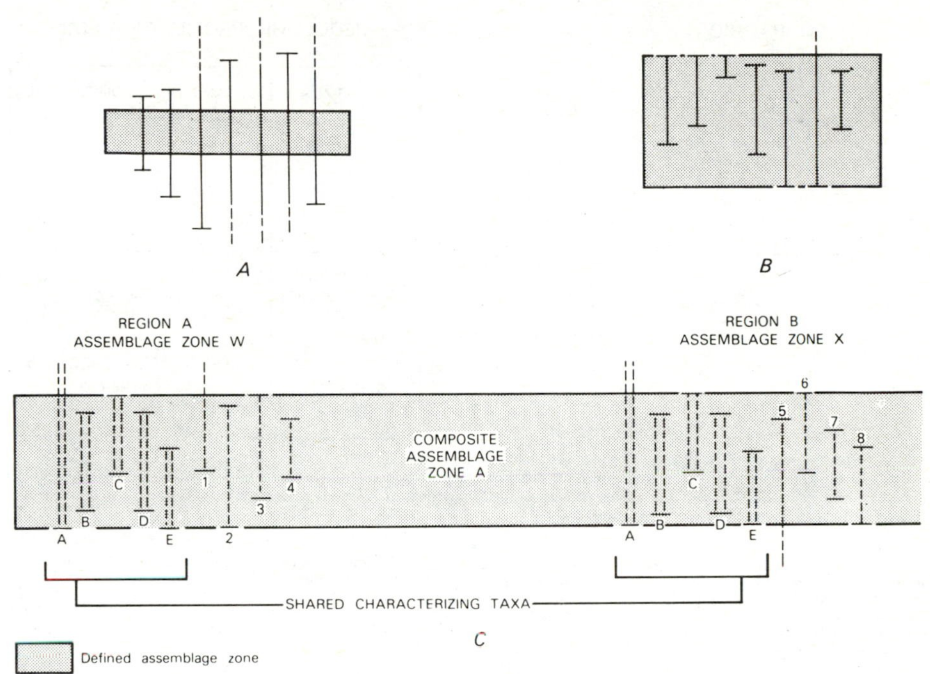

FIG. 5.—Examples of assemblage zone concepts.

PEDOSTRATIGRAPHIC UNITS

Nature and Boundaries

Article 55.—**Nature of Pedostratigraphic Units.** A pedostratigraphic unit is a body of rock that consists of one or more pedologic horizons developed in one or more lithostratigraphic, allostratigraphic, or lithodemic units (Fig. 6) and is overlain by one or more formally defined lithostratigraphic or allostratigraphic units.

Remarks. (a) **Definition.**—A pedostratigraphic[8] unit is a buried, traceable, three-dimensional body of rock that consists of one or more differentiated pedologic horizons.

(b) **Recognition.**—The distinguishing property of a pedostratigraphic unit is the presence of one or more distinct, differentiated, pedologic horizons. Pedologic horizons are products of soil development (pedogenesis) which occurred subsequent to formation of the lithostratigraphic, allostratigraphic, or lithodemic unit or units on which the buried soil was formed; these units are the parent materials in which pedogenesis occurred. Pedologic horizons are recognized in the field by diagnostic features such as color, soil structure, organic-matter accumulation, texture, clay coatings, stains, or concretions. Micromorphology, particle size, clay mineralogy, and other properties determined in the laboratory also may be used to identify and distinguish pedostratigraphic units.

(c) **Boundaries and stratigraphic position.**—The upper boundary of a pedostratigraphic unit is the top of the uppermost pedologic horizon formed by pedogenesis in a buried soil profile. The lower boundary of pedostratigraphic unit is the lowest *definite* physical boundary of a ped logic horizon within a buried soil profile. The stratigraphic position of pedostratigraphic unit is determined by its relation to overlying ar underlying stratigraphic units (see Remark d).

(d) **Traceability.**—Practicability of subsurface tracing of the upp boundary of a buried soil is essential in establishing a pedostratigrapl unit because (1) few buried soils are exposed continuously for great d tances, (2) the physical and chemical properties of a specific pedostra graphic unit may vary greatly, both vertically and laterally, from place place, and (3) pedostratigraphic units of different stratigraphic signi cance in the same region generally do not have unique identifying physi and chemical characteristics. Consequently, extension of a pedostra graphic unit is accomplished by lateral tracing of the contact betwee buried soil and an overlying, formally defined lithostratigraphic or all tratigraphic unit, or between a soil and two or more demonstrably cor ative stratigraphic units.

(e) **Distinction from pedologic soils.**—Pedologic soils may incl organic deposits (e.g., litter zones, peat deposits, or swamp deposits) t overlie or grade laterally into differentiated buried soils. The orga deposits are not products of pedogenesis, and O horizons are not inclu in a pedostratigraphic unit (Fig. 6); they may be classified as biostr graphic or lithostratigraphic units. Pedologic soils also include the en C horizon of a soil. The C horizon in pedology is not rigidly defined; merely the part of a soil profile that underlies the B horizon. The bas the C horizon in many soil profiles is gradational or unidentifiable; c monly it is placed arbitrarily. The need for clearly defined and easily ognized physical boundaries for a stratigraphic unit requires that lower boundary of a pedostratigraphic unit be defined as the lowest *c nite* physical boundary of a pedologic horizon in a buried soil profile, part or all of the C horizon may be excluded from a pedostratigra unit.

[8]Terminology related to pedostratigraphic classification is summarized on page 850.

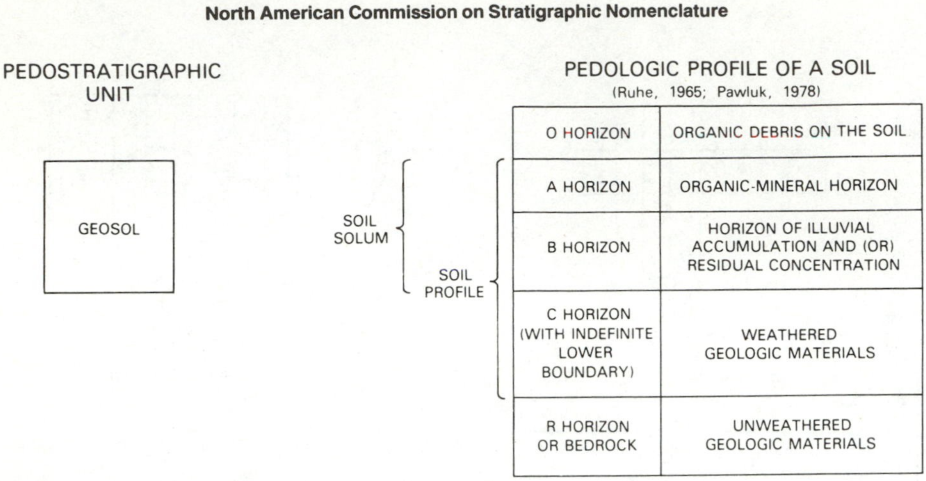

PEDOSTRATIGRAPHIC UNIT

GEOSOL

PEDOLOGIC PROFILE OF A SOIL
(Ruhe, 1965; Pawluk, 1978)

SOIL SOLUM (A and B horizons)

SOIL PROFILE (A, B and C horizons)

O HORIZON	ORGANIC DEBRIS ON THE SOIL
A HORIZON	ORGANIC-MINERAL HORIZON
B HORIZON	HORIZON OF ILLUVIAL ACCUMULATION AND (OR) RESIDUAL CONCENTRATION
C HORIZON (WITH INDEFINITE LOWER BOUNDARY)	WEATHERED GEOLOGIC MATERIALS
R HORIZON OR BEDROCK	UNWEATHERED GEOLOGIC MATERIALS

FIG. 6.—Relationship between pedostratigraphic units and pedologic profiles.
The base of a geosol is the lowest clearly defined physical boundary of a pedologic horizon in a buried soil profile. In this example it is the lower boundary of the B horizon because the base of the C horizon is not a clearly defined physical boundary. In other profiles the base may be the lower boundary of a C horizon.

(f) **Relation to saprolite and other weathered materials.**—A material derived by in situ weathering of lithostratigraphic, allostratigraphic, and(or) lithodemic units (e.g., saprolite, bauxite, residuum) may be the parent material in which pedologic horizons form, but is not a pedologic soil. A pedostratigraphic unit may be based on the pedologic horizons of a buried soil developed in the product of in-situ weathering, such as saprolite. The parents of such a pedostratigraphic unit are both the saprolite and, indirectly, the rock from which it formed.

(g) **Distinction from other stratigraphic units.**—A pedostratigraphic unit differs from other stratigraphic units in that (1) it is a product of surface alteration of one or more older material units by specific processes (pedogenesis), (2) its lithology and other properties differ markedly from those of the parent material(s), and (3) a single pedostratigraphic unit may be formed in situ in parent material units of diverse compositions and ages.

(h) **Independence from time concepts.**—The boundaries of a pedostratigraphic unit are time-transgressive. Concepts of time spans, however measured, play no part in defining the boundaries of a pedostratigraphic unit. Nonetheless, evidence of age, whether based on fossils, numerical ages, or geometrical or other relationships, may play an important role in distinguishing and identifying non-contiguous pedostratigraphic units at localities away from the type areas. The name of a pedostratigraphic unit should be chosen from a geographic feature in the type area, and not from a time span.

Pedostratigraphic Nomenclature and Unit

Article 56.—**Fundamental Unit.** The fundamental and only unit in pedostratigraphic classification is a geosol.

Article 57.—**Nomenclature.**—The formal name of a pedostratigraphic unit consists of a geographic name combined with the term "geosol." Capitalization of the initial letter in each word serves to identify formal usage. The geographic name should be selected in accordance with recommendations in Article 7 and should not duplicate the name of another formal geologic unit. Names based on subjacent and superjacent rock units, for example the super-Wilcox-sub-Claiborne soil, are informal, as are those with time connotations (post-Wilcox-pre-Claiborne soil).

Remarks. (a) **Composite geosols.**—Where the horizons of two or more merged or "welded" buried soils can be distinguished, formal names of pedostratigraphic units based on the horizon boundaries can be retained. Where the horizon boundaries of the respective merged or "welded" soils cannot be distinguished, formal pedostratigraphic classification is abandoned and a combined name such as Hallettville-Jamesville geosol may be used informally.

(b) **Characterization.**—The physical and chemical properties of a pedostratigraphic unit commonly vary vertically and laterally throughout the geographic extent of the unit. A pedostratigraphic unit is characterized by the *range* of physical and chemical properties of the unit in the type area, rather than by "typical" properties exhibited in a type section. Consequently, a pedostratigraphic unit is characterized on the basis of a composite stratotype (Article 8d).

(c) **Procedures for establishing formal pedostratigraphic units.**—A formal pedostratigraphic unit may be established in accordance with the applicable requirements of Article 3, and additionally by describing major soil horizons in each soil facies.

ALLOSTRATIGRAPHIC UNITS

Nature and Boundaries

Article 58.—**Nature of Allostratigraphic Units.** An allostratigraphic[9] unit is a mappable stratiform body of sedimentary rock that is defined and identified on the basis of its bounding discontinuities.

Remarks. (a) **Purpose.**—Formal allostratigraphic units may be defined to distinguish between different (1) superposed discontinuity-bounded deposits of similar lithology (Figs. 7, 9), (2) contiguous discontinuity-bounded deposits of similar lithology (Fig. 8), or (3) geographically separated discontinuity-bounded units of similar lithology (Fig. 9), or to distinguish as single units discontinuity-bounded deposits characterized by lithic heterogeneity (Fig. 8).

(b) **Internal characteristics.**—Internal characteristics (physical, chemical, and paleontological) may vary laterally and vertically throughout the unit.

(c) **Boundaries.**—Boundaries of allostratigraphic units are laterally traceable discontinuities (Figs. 7, 8, and 9).

(d) **Mappability.**—A formal allostratigraphic unit must be mappable

[9]From the Greek *allo*: "other, different."

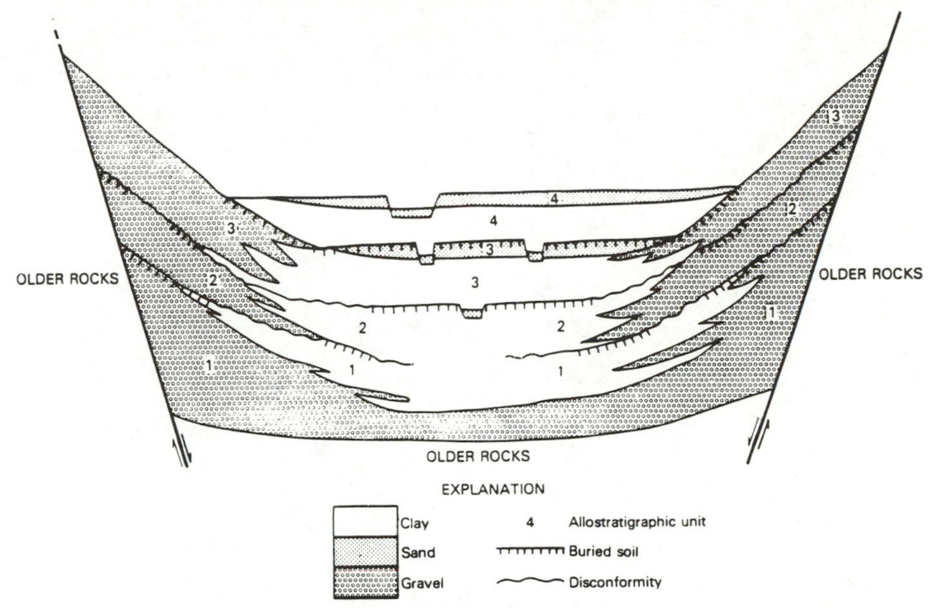

FIG. 7.—Example of allostratigraphic classification of alluvial and lacustrine deposits in a graben.

The alluvial and lacustrine deposits may be included in a single formation, or may be separated laterally into formations distinguished on the basis of contrasting texture (gravel, clay). Textural changes are abrupt and sharp, both vertically and laterally. The gravel deposits and clay deposits, respectively, are lithologically similar and thus cannot be distinguished as members of a formation. Four allostratigraphic units, each including two or three textural facies, may be defined on the basis of laterally traceable discontinuities (buried soils and disconformities).

at the scale practiced in the region where the unit is defined.

(e) **Type locality and extent.**—A type locality and type area must be designated; a composite stratotype or a type section and several reference sections are desirable. An allostratigraphic unit may be laterally contiguous with a formally defined lithostratigraphic unit; a vertical cut-off between such units is placed where the units meet.

(f) **Relation to genesis.**—Genetic interpretation is an inappropriate basis for defining an allostratigraphic unit. However, genetic interpretation may influence the choice of its boundaries.

(g) **Relation to geomorphic surfaces.**—A geomorphic surface may be used as a boundary of an allostratigraphic unit, but the unit should not be given the geographic name of the surface.

(h) **Relation to soils and paleosols.**—Soils and paleosols are composed of products of weathering and pedogenesis and differ in many respects from allostratigraphic units, which are depositional units (see "Pedostratigraphic Units," Article 55). The upper boundary of a surface or buried soil may be used as a boundary of an allostratigraphic unit.

(i) **Relation to inferred geologic history.**—Inferred geologic history is not used to define an allostratigraphic unit. However, well-documented geologic history may influence the choice of the unit's boundaries.

(j) **Relation to time concepts.**—Inferred time spans, however measured, are not used to define an allostratigraphic unit. However, age relationships may influence the choice of the unit's boundaries.

(k) **Extension of allostratigraphic units.**—An allostratigraphic unit is extended from its type area by tracing the boundary discontinuities or by tracing or matching the deposits between the discontinuities.

Ranks of Allostratigraphic Units

Article 59.—**Hierarchy.** The hierarchy of allostratigraphic units, in order of decreasing rank, is allogroup, alloformation, and allomember.

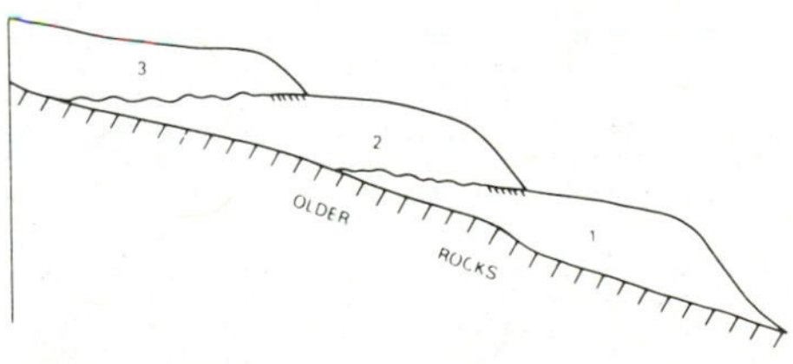

FIG. 8.—Example of allostratigraphic classification of contiguous deposits of similar lithology.

Allostratigraphic units 1, 2, and 3 are physical records of three glaciations. They are lithologically similar, reflecting derivation from the same bedrock, and constitute a single lithostratigraphic unit.

Remarks. (a) **Alloformation.**—The alloformation is the fundamental unit in allostratigraphic classification. An alloformation may be completely or only partly divided into allomembers, if some useful purpose is served, or it may have no allomembers.

(b) **Allomember.**—An allomember is the formal allostratigraphic unit next in rank below an alloformation.

(c) **Allogroup.**—An allogroup is the allostratigraphic unit next in rank above an alloformation. An allogroup is established only if a unit of that rank is essential to elucidation of geologic history. An allogroup may consist entirely of named alloformations or, alternatively, may contain one or more named alloformations which jointly do not comprise the entire allogroup.

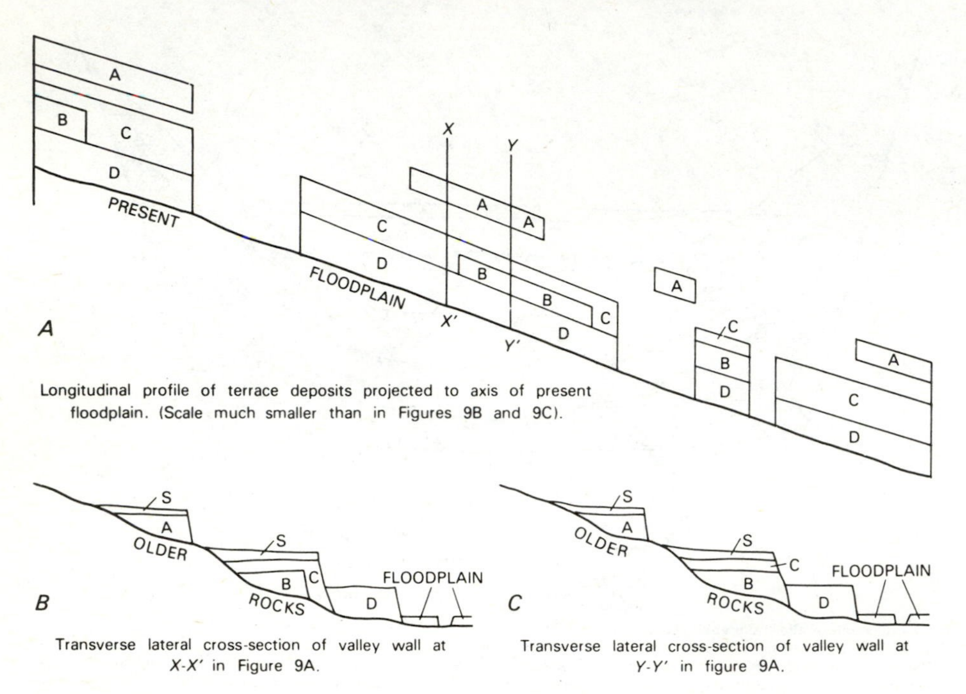

A
B
C
D
PRESENT
FLOODPLAIN
X
Y
X'
Y'
A
Longitudinal profile of terrace deposits projected to axis of present floodplain. (Scale much smaller than in Figures 9B and 9C).
S
OLDER
ROCKS
FLOODPLAIN
B
Transverse lateral cross-section of valley wall at X-X' in Figure 9A.
C
Transverse lateral cross-section of valley wall at Y-Y' in figure 9A.

FIG. 9.—Example of allostratigraphic classification of lithologically similar, discontinuous terrace deposits.

A, B, C, and D are terrace gravel units of similar lithology at different topographic positions on a valley wall. The deposits may be defined as separate formal allostratigraphic units if such units are useful and if bounding discontinuities can be traced laterally. Terrace gravels of the same age commonly are separated geographically by exposures of older rocks. Where the bounding discontinuities cannot be traced continuously, they may be extended geographically on the basis of objective correlation of internal properites of the deposits other than lithology (e.g., fossil content, included tephras), topographic position, numerical ages, or relative-age criteria (e.g., soils or other weathering phenomena). The criteria for such extension should be documented. Slope deposits and eolian deposits (S) that mantle terrace surfaces may be of diverse ages and are not included in a terrace-gravel allostratigraphic unit. A single terrace surface may be underlain by more than one allostratigraphic unit (units B and C in sections b and c).

(d) **Changes in rank.**—The principles and procedures for elevation and reduction in rank of formal allostratigraphic units are the same as those in Articles 19b, 19g, and 28.

Allostratigraphic Nomenclature

Article 60.—**Nomenclature.** The principles and procedures for naming allostratigraphic units are the same as those for naming of lithostratigraphic units (see Articles 7, 30).

Remark. (a) **Revision.**—Allostratigraphic units may be revised or otherwise modified in accordance with the recommendations in Articles 17 to 20.

FORMAL UNITS DISTINGUISHED BY AGE

GEOLOGIC-TIME UNITS

Nature and Types

Article 61.—**Types.** Geologic-time units are conceptual, rather than material, in nature. Two types are recognized: those based on material standards or referents (specific rock sequences or bodies), and those independent of material referents (Fig. 1).

Units Based on Material Referents

Article 62.—**Types Based on Referents.** Two types of formal geologic-time units based on material referents are recognized: they are isochronous and diachronous units.

Article 63.—**Isochronous Categories.** Isochronous time units and the material bodies from which they are derived are twofold: geochronologic units (Article 80), which are based on corresponding material chronostratigraphic units (Article 66), and polarity-geochronologic units (Article 88), based on corresponding material polarity-chronostratigraphic units (Article 83).

Remark. (a) **Extent.**—Isochronous units are applicable worldwide; they may be referred to even in areas lacking a material record of the named span of time. The duration of the time may be represented by a unit-stratotype referent. The beginning and end of the time are represented by point-boundary-stratotypes either in a single stratigraphic sequence or in separate stratotype sections (Articles 8b, 10b).

North American Stratigraphic Code

Article 64.—**Diachronous Categories.** Diachronic units (Article 91) are time units corresponding to diachronous material allostratigraphic units (Article 58), pedostratigraphic units (Article 55), and most lithostratigraphic (Article 22) and biostratigraphic (Article 48) units.

Remarks. (a) **Diachroneity.**—Some lithostratigraphic and biostratigraphic units are clearly diachronous, whereas others have boundaries which are not demonstrably diachronous within the resolving power of available dating methods. The latter commonly are treated as isochronous and are used for purposes of chronocorrelation (see biochronozone, Article 75). However, the assumption of isochroneity must be tested continually.

(b) **Extent.**—Diachronic units are coextensive with the diachronous material stratigraphic units on which they are based and are not used beyond the extent of their material referents.

Units Independent of Material Referents

Article 65.—**Numerical Divisions of Time.** Isochronous geologic-time units based on numerical divisions of time in years are geochronometric units (Article 96) and have no material referents.

CHRONOSTRATIGRAPHIC UNITS

Nature and Boundaries

Article 66.—**Definition.** A chronostratigraphic unit is a body of rock established to serve as the material reference for all rocks formed during the same span of time. Each of its boundaries is synchronous. The body also serves as the basis for defining the specific interval of time, or geochronologic unit (Article 80), represented by the referent.

Remarks. (a) **Purposes.**—Chronostratigraphic classification provides a means of establishing the temporally sequential order of rock bodies. Principal purposes are to provide a framework for (1) temporal correlation of the rocks in one area with those in another, (2) placing the rocks of the Earth's crust in a systematic sequence and indicating their relative position and age with respect to earth history as a whole, and (3) constructing an internationally recognized Standard Global Chronostratigraphic Scale.

(b) **Nature.**—A chronostratigraphic unit is a material unit and consists of a body of strata formed during a specific time span. Such a unit represents all rocks, and only those rocks, formed during that time span.

(c) **Content.**—A chronostratigraphic unit may be based upon the time span of a biostratigraphic unit, a lithic unit, a magnetopolarity unit, or any other feature of the rock record that has a time range. Or it may be any arbitrary but specified sequence of rocks, provided it has properties allowing chronocorrelation with rock sequences elsewhere.

Article 67.—**Boundaries.** Boundaries of chronostratigraphic units should be defined in a designated stratotype on the basis of observable paleontological or physical features of the rocks.

Remark. (a) **Emphasis on lower boundaries of chronostratigraphic units.**—Designation of point boundaries for both base and top of chronostratigraphic units is not recommended, because subsequent information on relations between successive units may identify overlaps or gaps. One means of minimizing or eliminating problems of duplication or gaps in chronostratigraphic successions is to define formally as a point-boundary stratotype only the base of the unit. Thus, a chronostratigraphic unit with its base defined at one locality, will have its top defined by the base of an overlying unit at the same, but more commonly another, locality (Article 8b).

Article 68.—**Correlation.** Demonstration of time equivalence is required for geographic extension of a chronostratigraphic unit from its type section or area. Boundaries of chronostratigraphic units can be extended only within the limits of resolution of available means of chronocorrelation, which currently include paleontology, numerical dating, remanent magnetism, thermoluminescence, relative-age criteria (examples are superposition and cross-cutting relations), and such indirect and inferential physical criteria as climatic changes, degree of weathering, and relations to unconformities. Ideally, the boundaries of chronostratigraphic units are independent of lithology, fossil content, or other material bases of stratigraphic division, but, in practice, the correlation or geographic extension of these boundaries relies at least in part on such features. Boundaries of chronostratigraphic units commonly are intersected by boundaries of most other kinds of material units.

Ranks of Chronostratigraphic Units

Article 69.—**Hierarchy.** The hierarchy of chronostratigraphic units, in order of decreasing rank, is eonothem, erathem, system, series, and stage. Of these, system is the primary unit of worldwide major rank; its primacy derives from the history of development of stratigraphic classification. All systems and units of higher rank are divided completely into units of the next lower rank. Chronozones are non-hierarchical and commonly lower-rank chronostratigraphic units. Stages and chronozones in sum do not necessarily equal the units of next higher rank and need not be contiguous. The rank and magnitude of chronostratigraphic units are related to the time interval represented by the units, rather than to the thickness or areal extent of the rocks on which the units are based.

Article 70.—**Eonothem.** The unit highest in rank is eonothem. The Phanerozoic Eonothem encompasses the Paleozoic, Mesozoic, and Cenozoic Erathems. Although older rocks have been assigned heretofore to the Precambrian Eonothem, they also have been assigned recently to other (Archean and Proterozoic) eonothems by the IUGS Precambrian Subcommission. The span of time corresponding to an eonothem is an *eon*.

Article 71.—**Erathem.** An erathem is the formal chronostratigraphic unit of rank next lower to eonothem and consists of several adjacent systems. The span of time corresponding to an erathem is an *era*.

Remark. (a) **Names.**—Names given to traditional Phanerozoic erathems were based upon major stages in the development of life on Earth: Paleozoic (old), Mesozoic (intermediate), and Cenozoic (recent) life. Although somewhat comparable terms have been applied to Precambrian units, the names and ranks of Precambrian divisions are not yet universally agreed upon and are under consideration by the IUGS Subcommission on Precambrian Stratigraphy.

Article 72.—**System.** The unit of rank next lower to erathem is the system. Rocks encompassed by a system represent a time-span and an episode of Earth history sufficiently great to serve as a worldwide chronostratigraphic reference unit. The temporal equivalent of a system is a *period*.

Remark. (a) **Subsystem and supersystem.**—Some systems initially established in Europe later were divided or grouped elsewhere into units ranked as systems. *Subsystems* (Mississippian Subsystem of the Carboniferous System) and *supersystems* (Karoo Supersystem) are more appropriate.

Article 73.—**Series.** Series is a conventional chronostratigraphic unit that ranks below a system and always is a division of a system. A series commonly constitutes a major unit of chronostratigraphic correlation within a province, between provinces, or between continents. Although many European series are being adopted increasingly for dividing systems on other continents, provincial series of regional scope continue to be useful. The temporal equivalent of a series is an *epoch*.

Article 74.—**Stage.** A stage is a chronostratigraphic unit of smaller scope and rank than a series. It is most commonly of greatest use in intra-continental classification and correlation, although it has the potential for worldwide recognition. The geochronologic equivalent of stage is *age*.

North American Commission on Stratigraphic Nomenclature

Remark. (a) **Substage.**—Stages may be, but need not be, divided completely into substages.

Article 75.—**Chronozone.** A chronozone is a non-hierarchical, but commonly small, formal chronostratigraphic unit, and its boundaries may be independent of those of ranked units. Although a chronozone is an isochronous unit, it may be based on a biostratigraphic unit (example: *Cardioceras cordatum* Biochronozone), a lithostratigraphic unit (Woodbend Lithochronozone), or a magnetopolarity unit (Gilbert Reversed-Polarity Chronozone). Modifiers (litho-, bio-, polarity) used in formal names of the units need not be repeated in general discussions where the meaning is evident from the context, e.g., *Exus albus* Chronozone.

Remarks. (a) **Boundaries of chronozones.**—The base and top of a *chronozone* correspond in the unit's stratotype to the observed, defining, physical and paleontological features, but they are extended to other areas by any means available for recognition of synchroneity. The temporal equivalent of a chronozone is a chron.

(b) **Scope.**—The scope of the non-hierarchical chronozone may range markedly, depending upon the purpose for which it is defined either formally or informally. The informal "biochronozone of the ammonites," for example, represents a duration of time which is enormous and exceeds that of a system. In contrast, a biochronozone defined by a species of limited range, such as the *Exus albus* Chronozone, may represent a duration equal to or briefer than that of a stage.

(c) **Practical utility.**—Chronozones, especially thin and informal biochronozones and lithochronozones bounded by key beds or other "markers," are the units used most commonly in industry investigations of selected parts of the stratigraphy of economically favorable basins. Such units are useful to define geographic distributions of lithofacies or biofacies, which provide a basis for genetic interpretations and the selection of targets to drill.

Chronostratigraphic Nomenclature

Article 76.—**Requirements.** Requirements for establishing a formal chronostratigraphic unit include: (i) statement of intention to designate such a unit; (ii) selection of name; (iii) statement of kind and rank of unit; (iv) statement of general concept of unit including historical background, synonymy, previous treatment, and reasons for proposed establishment; (v) description of characterizing physical and/or biological features; (vi) designation and description of boundary type sections, stratotypes, or other kinds of units on which it is based; (vii) correlation and age relations; and (viii) publication in a recognized scientific medium as specified in Article 4.

Article 77.—**Nomenclature.** A formal chronostratigraphic unit is given a compound name, and the initial letter of all words, except for trivial taxonomic terms, is capitalized. Except for chronozones (Article 75), names proposed for new chronostratigraphic units should not duplicate those for other stratigraphic units. For example, naming a new chronostratigraphic unit simply by adding "-an" or "-ian" to the name of a lithostratigraphic unit is improper.

Remarks. (a) **Systems and units of higher rank.**—Names that are generally accepted for systems and units of higher rank have diverse origins, and they also have different kinds of endings (Paleozoic, Cambrian, Cretaceous, Jurassic, Quaternary).

(b) **Series and units of lower rank.**—Series and units of lower rank are commonly known either by geographic names (Virgilian Series, Ochoan Series) or by names of their encompassing units modified by the capitalized adjectives Upper, Middle, and Lower (Lower Ordovician). Names of chronozones are derived from the unit on which they are based (Article 75). For series and stage, a geographic name is preferable because it may be related to a type area. For geographic names, the adjectival endings -an or -ian are recommended (Cincinnatian Series), but it is permissible to use the geographic name without any special ending, if more euphonious. Many series and stage names already in use have been based on lithic units (groups, formations, and members) and bear the names of these units (Wolfcampian Series, Claibornian Stage). Nevertheless, a stage preferably should have a geographic name not previously used in stratigraphic nomenclature. Use of internationally accepted (mainly European) stage names is preferable to the proliferation of others.

Article 78.—**Stratotypes.** An ideal stratotype for a chronostratigraphic unit is a completely exposed unbroken and continuous sequence of fossiliferous stratified rocks extending from a well-defined lower boundary to the base of the next higher unit. Unfortunately, few available sequences are sufficiently complete to define stages and units of higher rank, which therefore are best defined by boundary-stratotypes (Article 8b).

Boundary-stratotypes for major chronostratigraphic units ideally should be based on complete sequences of either fossiliferous monofacial marine strata or rocks with other criteria for chronocorrelation to permit widespread tracing of synchronous horizons. Extension of synchronous surfaces should be based on as many indicators of age as possible.

Article 79.—**Revision of units.** Revision of a chronostratigraphic unit without changing its name is allowable but requires as much justification as the establishment of a new unit (Articles 17, 19, and 76). Revision or redefinition of a unit of system or higher rank requires international agreement. If the definition of a chronostratigraphic unit is inadequate, it may be clarified by establishment of boundary stratotypes in a principal reference section.

GEOCHRONOLOGIC UNITS

Nature and Boundaries

Article 80.—**Definition and Basis.** Geochronologic units are divisions of time traditionally distinguished on the basis of the rock record as expressed by chronostratigraphic units. A geochronologic unit is not a stratigraphic unit (i.e., it is not a material unit), but it corresponds to the time span of an established chronostratigraphic unit (Articles 65 and 66), and its beginning and ending corresponds to the base and top of the referent.

Ranks and Nomenclature of Geochronologic Units

Article 81.—**Hierarchy.** The hierarchy of geochronologic units in order of decreasing rank is *eon, era, period, epoch,* and *age.* Chron is a non-hierarchical, but commonly brief, geochronologic unit. Ages in sum do not necessarily equal epochs and need not form a continuum. An eon is the time represented by the rocks constituting an eonothem; era by an erathem; period by a system; epoch by a series; age by a stage; and chron by a chronozone.

Article 82.—**Nomenclature.** Names for periods and units of lower rank are identical with those of the corresponding chronostratigraphic units; the names of some eras and eons are independently formed. Rules of capitalization for chronostratigraphic units (Article 77) apply to geochronologic units. The adjectives Early, Middle, and Late are used for the geochronologic epochs equivalent to the corresponding chronostratigraphic Lower, Middle, and Upper series, where these are formally established.

POLARITY-CHRONOSTRATIGRAPHIC UNITS

Nature and Boundaries

Article 83.—**Definition.** A polarity-chronostratigraphic unit is a body of rock that contains the primary magnetic-polarity record imposed when the rock was deposited, or crystallized, during a specific interval of geologic time.

Remarks. (a) **Nature.**—Polarity-chronostratigraphic units depend fundamentally for definition on actual sections or sequences, or measurements on individual rock units, and without these standards they are meaningless. They are based on material units, the polarity zones of magnetopolarity classification. Each polarity-chronostratigraphic unit is the record of the time during which the rock formed and the Earth's magnetic field had a designated polarity. Care should be taken to define polarity-chronologic units in terms of polarity-chronostratigraphic units, and not vice versa.

(b) **Principal purposes.**—Two principal purposes are served by polarity-chronostratigraphic classification: (1) correlation of rocks at one place with those of the same age and polarity at other places; and (2) delineation of the polarity history of the Earth's magnetic field.

(c) **Recognition.**—A polarity-chronostratigraphic unit may be extended geographically from its type locality only with the support of physical and/or paleontologic criteria used to confirm its age.

Article 84.—**Boundaries.** The boundaries of a polarity chronozone are placed at polarity-reversal horizons or polarity transition-zones (see Article 45).

Ranks and Nomenclature of Polarity-Chronostratigraphic Units

Article 85.—**Fundamental Unit.** The polarity chronozone consists of rocks of a specified primary polarity and is the fundamental unit of worldwide polarity-chronostratigraphic classification.

Remarks. (a) **Meaning of term.**—A polarity chronozone is the worldwide body of rock strata that is collectively defined as a polarity-chronostratigraphic unit.

(b) **Scope.**—Individual polarity zones are the basic building blocks of polarity chronozones. Recognition and definition of polarity chronozones may thus involve step-by-step assembly of carefully dated or correlated individual polarity zones, especially in work with rocks older than the oldest ocean-floor magnetic anomalies. This procedure is the method by which the Brunhes, Matuyama, Gauss, and Gilbert Chronozones were recognized (Cox, Doell, and Dalrymple, 1963) and defined originally (Cox, Doell, and Dalrymple, 1964).

(c) **Ranks.**—Divisions of polarity chronozones are designated polarity subchronozones. Assemblages of polarity chronozones may be termed polarity superchronozones.

Article 86.—**Establishing Formal Units.** Requirements for establishing a polarity-chronostratigraphic unit include those specified in Articles 3 and 4, and also (1) definition of boundaries of the unit, with specific references to designated sections and data; (2) distinguishing polarity characteristics, lithologic descriptions, and included fossils; and (3) correlation and age relations.

Article 87.—**Name.** A formal polarity-chronostratigraphic unit is given a compound name beginning with that for a named geographic feature; the second component indicates the normal, reversed, or mixed polarity of the unit, and the third component is *chronozone*. The initial letter of each term is capitalized. If the same geographic name is used for both a magnetopolarity zone and a polarity-chronostratigraphic unit, the latter should be distinguished by an -an or -ian ending. Example: Tetonian Reversed-Polarity Chronozone.

Remarks: (a) **Preservation of established name.**—A particularly well-established name should not be displaced, either on the basis of priority, as described in Article 7c, or because it was not taken from a geographic feature. Continued use of Brunhes, Matuyama, Gauss, and Gilbert, for example, is endorsed so long as they remain valid units.

(b) **Expression of doubt.**—Doubt in the assignment of polarity zones to polarity-chronostratigraphic units should be made explicit if criteria of time equivalence are inconclusive.

POLARITY-CHRONOLOGIC UNITS

Nature and Boundaries

Article 88.—**Definition.** Polarity-chronologic units are divisions of geologic time distinguished on the basis of the record of magnetopolarity as embodied in polarity-chronostratigraphic units. No special kind of magnetic time is implied; the designations used are meant to convey the parts of geologic time during which the Earth's magnetic field had a characteristic polarity or sequence of polarities. These units correspond to the time spans represented by polarity chronozones, e.g., Gauss Normal Polarity Chronozone. They are not material units.

Ranks and Nomenclature of Polarity-Chronologic Units

Article 89.—**Fundamental Unit.** The polarity chron is the fundamental unit of geologic time designating the time span of a polarity chronozone.

Remark. (a) **Hierarchy.**—Polarity-chronologic units of decreasing hierarchical ranks are polarity superchron, polarity chron, and polarity subchron.

Article 90.—**Nomenclature.** Names for polarity chronologic units are identical with those of corresponding polarity-chronostratigraphic units, except that the term chron (or superchron, etc) is substituted for chronozone (or superchronozone, etc).

DIACHRONIC UNITS

Nature and Boundaries

Article 91.—**Definition.** A diachronic unit comprises the unequal spans of time represented either by a specific lithostratigraphic, allostratigraphic, biostratigraphic, or pedostratigraphic unit, or by an assemblage of such units.

Remarks. (a) **Purposes.**—Diachronic classification provides (1) a means of comparing the spans of time represented by stratigraphic units with diachronous boundaries at different localities, (2) a basis for broadly establishing in time the beginning and ending of deposition of diachronous stratigraphic units at different sites, (3) a basis for inferring the rate of change in areal extent of depositional processes, (4) a means of determining and comparing rates and durations of deposition at different localities, and (5) a means of comparing temporal and spatial relations of diachronous stratigraphic units (Watson and Wright, 1980).

(b) **Scope.**—The scope of a diachronic unit is related to (1) the relative magnitude of the transgressive division of time represented by the stratigraphic unit or units on which it is based and (2) the areal extent of those units. A diachronic unit is not extended beyond the geographic limits of the stratigraphic unit or units on which it is based.

(c) **Basis.**—The basis for a diachronic unit is the diachronous referent.

(d) **Duration.**—A diachronic unit may be of equal duration at different places despite differences in the times at which it began and ended at those places.

Article 92.—**Boundaries.** The boundaries of a diachronic unit are the times recorded by the beginning and end of deposition of the material referent at the point under consideration (Figs. 10, 11).

Remark. (a) **Temporal relations.**—One or both of the boundaries of a diachronic unit are demonstrably time-transgressive. The varying time significance of the boundaries is defined by a series of boundary reference sections (Article 8b, 8e). The duration and age of a diachronic unit differ from place to place (Figs. 10, 11).

Ranks and Nomenclature of Diachronic Units

Article 93.—**Ranks.** A diachron is the fundamental and non-hierarchical diachronic unit. If a hierarchy of diachronic units is needed, the terms episode, phase, span, and cline, in order of decreasing rank, are recommended. The rank of a hierarchical

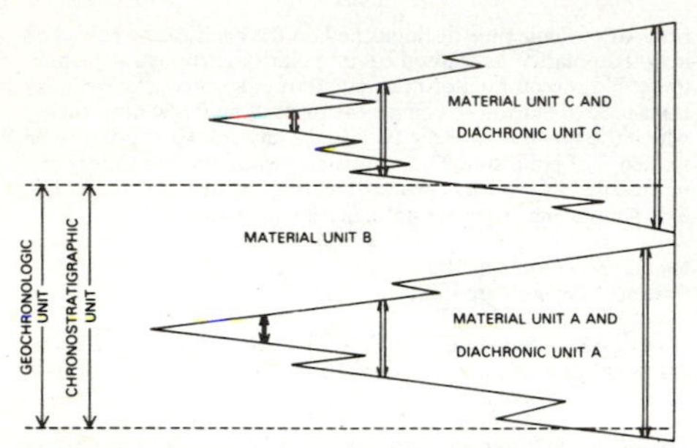

FIG. 10.—Comparison of geochronologic, chronostratigraphic, and diachronic units.

unit is determined by the scope of the unit (Article 91 b), and not by the time span represented by the unit at a particular place.

Remarks. (a) **Diachron.**—Diachrons may differ greatly in magnitude because they are the spans of time represented by individual or grouped lithostratigraphic, allostratigraphic, biostratigraphic, and(or) pedostratigraphic units.

(b) **Hierarchical ordering permissible.**—A hierarchy of diachronic units may be defined if the resolution of spatial and temporal relations of diachronous stratigraphic units is sufficiently precise to make the hierarchy useful (Watson and Wright, 1980). Although all hierarchical units of rank lower than episode are part of a unit next higher in rank, not all parts of an episode, phase, or span need be represented by a unit of lower rank.

(c) **Episode.**—An episode is the unit of highest rank and greatest scope in hierarchical classification. If the "Wisconsinan Age" were to be redefined as a diachronic unit, it would have the rank of episode.

Article 94.—**Name.** The name for a diachronic unit should be compound, consisting of a geographic name followed by the term diachron or a hierarchical rank term. Both parts of the compound name are capitalized to indicate formal status. If the diachronic unit is defined by a single stratigraphic unit, the geographic name of the unit may be applied to the diachronic unit. Otherwise, the geographic name of the diachronic unit should not duplicate that of another formal stratigraphic unit. Genetic terms (e.g., alluvial, marine) or climatic terms (e.g., glacial, interglacial) are not included in the names of diachronic units.

Remarks. (a) **Formal designation of units.**—Diachronic units should be formally defined and named only if such definition is useful.

(b) **Inter-regional extension of geographic names.**—The geographic name of a diachronic unit may be extended from one region to another if the stratigraphic units on which the diachronic unit is based extend across the regions. If different diachronic units in contiguous regions eventually prove to be based on laterally continuous stratigraphic units, one name should be applied to the unit in both regions. If two names have been applied, one name should be abandoned and the other formally extended. Rules of priority (Article 7d) apply. Priority in publication is to be respected, but priority alone does not justify displacing a well-established name by one not well-known or commonly used.

(c) **Change from geochronologic to diachronic classification.**—Lithostratigraphic units have served as the material basis for widely accepted chronostratigraphic and geochronologic classifications of Quaternary nonmarine deposits, such as the classifications of Frye et al (1968), Willman and Frye (1970), and Dreimanis and Karrow (1972). In practice, time-parallel horizons have been extended from the stratotypes on the basis of markedly time-transgressive lithostratigraphic and pedostratigraphic unit boundaries. The time ("geochronologic") units, defined on the basis of the stratotype sections but extended on the basis of diachronous stratigraphic boundaries, are diachronic units. Geographic names established for such "geochronologic" units may be used in diachronic classification if (1) the chronostratigraphic and geochronologic classifications are formally abandoned and diachronic classifications are proposed to replace the former "geochronologic" classifications, and (2) the units are redefined as formal diachronic units. Preservation of well-established names in these specific circumstances retains the intent and purpose of the names and the units, retains the practical significance of the units, enhances communication, and avoids proliferation of nomenclature.

Article 95.—**Establishing Formal Units.** Requirements for establishing a formal diachronic unit, in addition to those in Article 3, include (1) specification of the nature, stratigraphic relations, and geographic or areal relations of the stratigraphic unit or units that serve as a basis for definition of the unit, and (2) specific designation and description of multiple reference sections that illustrate the temporal and spatial relations of the defining stratigraphic unit or units and the boundaries of the unit or units.

Remark. (a) **Revision or abandonment.**—Revision or abandonment of the stratigraphic unit or units that serve as the material basis for defini-

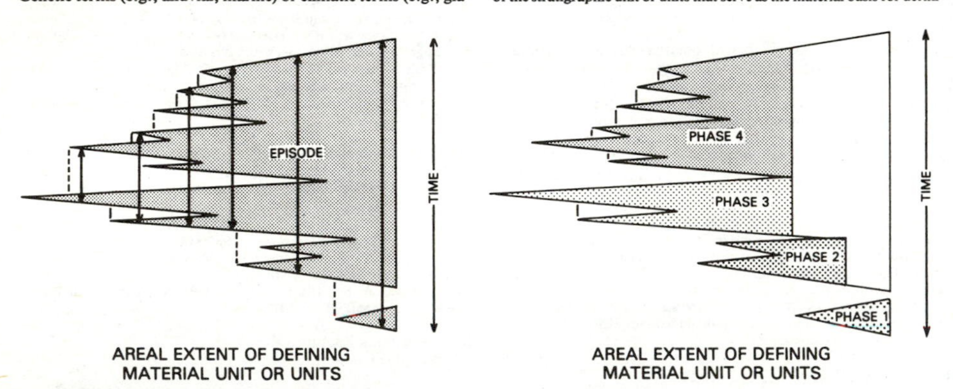

FIG. 11.—Schematic relation of phases to an episode.

Parts of a phase similarly may be divided into spans, and spans into clines. Formal definition of spans and clines is unnecessary in most diachronic unit hierarchies.

tion of a diachronic unit may require revision or abandonment of the diachronic unit. Procedure for revision must follow the requirements for establishing a new diachronic unit.

GEOCHRONOMETRIC UNITS

Nature and Boundaries

Article 96.—**Definition.** Geochronometric units are units established through the direct division of geologic time, expressed in years. Like geochronologic units (Article 80), geochronometric units are abstractions, i.e., they are not material units. Unlike geochronologic units, geochronometric units are not based on the time span of designated chronostratigraphic units (stratotypes), but are simply time divisions of convenient magnitude for the purpose for which they are established, such as the development of a time scale for the Precambrian. Their boundaries are arbitrarily chosen or agreed-upon ages in years.

Ranks and Nomenclature of Geochronometric Units

Article 97.—**Nomenclature.** Geochronologic rank terms (eon, era, period, epoch, age, and chron) may be used for geochronometric units when such terms are formalized. For example, Archean Eon and Proterozoic Eon, as recognized by the IUGS Subcommission on Precambrian Stratigraphy, are formal geochronometric units in the sense of Article 96, distinguished on the basis of an arbitrarily chosen boundary at 2.5 Ga. Geochronometric units are not defined by, but may have, corresponding chronostratigraphic units (eonothem, erathem, system, series, stage, and chronozone).

PART III: ADDENDA

REFERENCES[10]

American Commission on Stratigraphic Nomenclature, 1947, Note 1—Organization and objectives of the Stratigraphic Commission: American Association of Petroleum Geologists Bulletin, v. 31, no. 3, p. 513-518.

——— ,1961, Code of Stratigraphic Nomenclature: American Association of Petroleum Geologists Bulletin, v. 45, no. 5, p. 645-665.

——— ,1970, Code of Stratigraphic Nomenclature (2d ed.): American Association of Petroleum Geologists, Tulsa, Okla., 45 p.

———,1976, Note 44—Application for addition to code concerning magnetostratigraphic units: American Association of Petroleum Geologists Bulletin, v. 60, no. 2, p. 273-277.

Caster, K. E., 1934, The stratigraphy and paleontology of northwestern Pennsylvania, Part 1, Stratigraphy: Bulletins of American Paleontology, v. 21, 185 p.

Chang, K. H., 1975, Unconformity-bounded stratigraphic units: Geological Society of America Bulletin, v. 86, no. 11, p. 1544-1552.

Committee on Stratigraphic Nomenclature, 1933, Classification and nomenclature of rock units: Geological Society of America Bulletin, v. 44, no. 2, p. 423-459, and American Association of Petroleum Geologists Bulletin, v. 17, no. 7, p. 843-868.

Cox, A. V., R. R. Doell, and G. B. Dalrymple, 1963, Geomagnetic polarity epochs and Pleistocene geochronometry: Nature, v. 198, p. 1049-1051.

——— ,1964, Reversals of the Earth's magnetic field: Science, v. 144, no. 3626, p. 1537-1543.

Cross, C. W., 1898, Geology of the Telluride area: U.S. Geological Survey 18th Annual Report, pt. 3, p. 759.

Cumming, A. D., J. G. C. M. Fuller, and J. W. Porter, 1959, Separation of strata: Paleozoic limestones of the Williston basin: American Journal of Science, v. 257, no. 10, p. 722-733.

[10]Readers are reminded of the extensive and noteworthy bibliography of contributions to stratigraphic principles, classification, and terminology cited by the International Stratigraphic Guide (ISSC, 1976, p. 111-187).

Dreimanis, Aleksis, and P. F. Karrow, 1972, Glacial history of the Great Lakes-St. Lawrence region, the classification of the Wisconsin(an) Stage, and its correlatives: International Geologic Congress, 24th Session, Montreal, 1972, Section 12, Quaternary Geology, p. 5-15.

Dunbar, C. O., and John Rodgers, 1957, Principles of stratigraphy: Wiley, New York, 356 p.

Forgotson, J. M., Jr., 1957, Nature, usage and definition of marker-defined vertically segregated rock units: American Association of Petroleum Geologists Bulletin, v. 41, no. 9, p. 2108-2113.

Frye, J. C., H. B. Willman, Meyer Rubin, and R. F. Black, 1968, Definition of Wisconsinan Stage: U.S. Geological Survey Bulletin 1274-E, 22 p.

George, T. N., and others, 1969, Recommendations on stratigraphical usage: Geological Society of London, Proceedings no. 1656, p. 139-166.

Harland, W. B., 1977, Essay review [of] International Stratigraphic Guide, 1976: Geology Magazine, v. 114, no. 3, p. 229-235.

——— ,1978, Geochronologic scales, *in* G. V. Cohee et al, eds., Contributions to the Geologic Time Scale: American Association of Petroleum Geologists, Studies in Geology, no. 6, p. 9-32.

Harrison, J. E., and Z. E. Peterman, 1980, North American Commission on Stratigraphic Nomenclature Note 52—A preliminary proposal for a chronometric time scale for the Precambrian of the United States and Mexico: Geological Society of America Bulletin, v. 91, no. 6, p. 377-380.

Henbest, L. G., 1952, Significance of evolutionary explosions for diastrophic division of Earth history: Journal of Paleontology, v. 26, p. 299-318.

Henderson, J. B., W. G. E. Caldwell, and J. E. Harrison, 1980, North American Commission on Stratigraphic Nomenclature, Report 8—Amendment of code concerning terminology for igneous and high-grade metamorphic rocks: Geological Society of America Bulletin, v. 91, no. 6, p. 374-376.

Holland, C. H., and others, 1978, A guide to stratigraphical procedure: Geological Society of London, Special Report 10, p. 1-18.

Huxley, T. H., 1862, The anniversary address: Geological Society of London, Quarterly Journal, v. 18, p. xl-liv.

International Commission on Zoological Nomenclature, 1964: International Code of Zoological Nomenclature adopted by the XV International Congress of Zoology: International Trust for Zoological Nomenclature, London, 176 p.

International Subcommission on Stratigraphic Classification (ISSC), 1976, International Stratigraphic Guide (H. D. Hedberg, ed.): John Wiley and Sons, New York, 200 p.

International Subcommission on Stratigraphic Classification, 1979, Magnetostratigraphy polarity units—a supplementary chapter of the ISSC International Stratigraphic Guide: Geology, v. 7, p. 578-583.

Izett, G. A., and R. E. Wilcox, 1981, Map showing the distribution of the Huckleberry Ridge, Mesa Falls, and Lava Creek volcanic ash beds (Pearlette family ash beds) of Pliocene and Pleistocene age in the western United States and southern Canada: U. S. Geological Survey Miscellaneous Geological Investigations Map I-1325.

Kauffman, E. G., 1969, Cretaceous marine cycles of the Western Interior: Mountain Geologist: Rocky Mountain Association of Geologists, v. 6, no. 4, p. 227-245.

Matthews, R. K., 1974, Dynamic stratigraphy—an introduction to sedimentation and stratigraphy: Prentice-Hall, New Jersey, 370 p.

McDougall, Ian, 1977, The present status of the geomagnetic polarity time scale: Research School of Earth Sciences, Australian National University, Publication no. 1288, 34 p.

McElhinny, M. W., 1978, The magnetic polarity time scale; prospects and possibilities in magnetostratigraphy, *in* G. V. Cohee et al, eds., Contributions to the Geologic Time Scale, American Association of Petroleum Geologists, Studies in Geology, no. 6, p. 57-65.

McIver, N. L., 1972, Cenozoic and Mesozoic stratigraphy of the Nova Scotia shelf: Canadian Journal of Earth Science, v. 9, p. 54-70.

McLaren, D. J., 1977, The Silurian-Devonian Boundary Committee. A final report, *in* A. Martinsson, ed., The Silurian-Devonian boundary: IUGS Series A, no. 5, p. 1-34.

Morrison, R. B., 1967, Principles of Quaternary soil stratigraphy, *in* R. B. Morrison and H. E. Wright, Jr., eds., Quaternary soils: Reno, Nevada, Center for Water Resources Research, Desert Research Institute, Univ. Nevada, p. 1-69.

North American Commission on Stratigraphic Nomenclature, 1981, Draft North American Stratigraphic Code: Canadian Society of

North American Commission on Stratigraphic Nomenclature

Petroleum Geologists, Calgary, 63 p.
Palmer, A. R., 1965, Biomere-a new kind of biostratigraphic unit: Journal of Paleontology, v. 39, no. 1, p. 149-153.
Parsons, R. B., 1981, Proposed soil-stratigraphic guide, *in* International Union for Quaternary Research and International Society of Soil Science: INQUA Commission 6 and ISSS Commission 5 Working Group, Pedology, Report, p. 6-12.
Pawluk, S., 1978, The pedogenic profile in the stratigraphic section, *in* W. C. Mahaney, ed., Quaternary soils: Norwich, England, GeoAbstracts, Ltd., p. 61-75.
Ruhe, R. V., 1965, Quaternary paleopedology, *in* H. E. Wright, Jr., and D. G. Frey, eds., The Quaternary of the United States: Princeton, N.J., Princeton University Press, p. 755-764.
Schultz, E. H., 1982, The chronosome and supersome--terms proposed for low-rank chronostratigraphic units: Canadian Petroleum Geology, v. 30, no. 1, p. 29-33.
Shaw, A. B., 1964, Time in stratigraphy: McGraw-Hill, New York, 365 p.
Sims, P. K., 1979, Precambrian subdivided: Geotimes, v. 24, no. 12, p. 15.
Sloss, L. L., 1963, Sequences in the cratonic interior of North America: Geological Society of America Bulletin, v. 74, no. 2, p. 94-114.
Tracey, J. I., Jr., and others, 1971, Initial reports of the Deep Sea Drilling Project, v. 8: U.S. Government Printing Office, Washington, 1037 p.
Valentine, K. W. G., and J. B. Dalrymple, 1976, Quaternary buried paleosols: A critical review: Quaternary Research, v. 6, p. 209-222.
Vella, P., 1964, Biostratigraphic units: New Zealand Journal of Geology and Geophysics, v. 7, no. 3, p. 615-625.
Watson, R. A., and H. E. Wright, Jr., 1980, The end of the Pleistocene: A general critique of chronostratigraphic classification: Boreas, v. 9, p. 153-163.
Weiss, M. P., 1979a, Comments and suggestions invited for revision of American Stratigraphic Code: Geological Society of America, News and Information, v. 1, no. 7, p. 97-99.
——— ,1979b, Stratigraphic Commission Note 50--Proposal to change name of Commission: American Association of Petroleum Geologists Bulletin, v. 63, no. 10, p. 1986.
Weller, J. M., 1960, Stratigraphic principles and practice: Harper and Brothers, New York, 725 p.
Willman, H. B., and J. C. Frye, 1970, Pleistocene stratigraphy of Illinois: Illinois State Geological Survey Bulletin 94, 204 p.

APPENDIX I: PARTICIPANTS AND CONFEREES IN CODE REVISION

Code Committee

Steven S. Oriel (U.S. Geological Survey), chairman, Hubert Gabrielse (Geological Survey of Canada), William W. Hay (Joint Oceanographic Institutions), Frank E. Kottlowski (New Mexico Bureau of Mines), John B. Patton (Indiana Geological Survey).

Lithostratigraphic Subcommittee

James D. Aitken (Geological Survey of Canada), chairman, Monti Lerand (Gulf Canada Resources, Ltd.), Mitchell W. Reynolds (U.S. Geological Survey), Robert J. Weimer (Colorado School of Mines), Malcolm P. Weiss (Northern Illinois University).

Biostratigraphic Subcommittee

Allison R. (Pete) Palmer (Geological Society of America), chairman, Ismael Ferrusquia (University of Mexico), Joseph E. Hazel (U.S. Geological Survey), Erle G. Kauffman (University of Colorado), Colin McGregor (Geological Survey of Canada), Michael A. Murphy (University of California, Riverside), Walter C. Sweet (Ohio State University).

Chronostratigraphic Subcommittee

Zell E. Peterman (U.S. Geological Survey), chairman, Zoltan de Cserna (Sociedad Geológica Mexicana), Edward H. Schultz (Suncor, Inc., Calgary), Norman F. Sohl (U.S. Geological Survey), John A. Van Couvering (American Museum of Natural History).

Plutonic-Metamorphic Advisory Group

Jack E. Harrison (U.S. Geological Survey), chairman, John B. Henderson (Geological Survey of Canada), Harold L. James (retired), Leon T. Silver (California Institute of Technology), Paul C. Bateman (U.S. Geological Survey).

Magnetostratigraphic Advisory Group

Roger W. Macqueen (University of Waterloo), chairman, G. Brent Dalrymple (U.S. Geological Survey), Walter F. Fahrig (Geological Survey of Canada), J. M. Hall (Dalhousie University).

Volcanic Advisory Group

Richard V. Fisher (University of California, Santa Barbara), chairman, Thomas A. Steven (U.S. Geological Survey), Donald A. Swanson (U.S. Geological Survey).

Tectonostratigraphic Advisory Group

Darrel S. Cowan (University of Washington), chairman, Thomas W. Donnelly (State University of New York at Binghamton), Michael W. Higgins and David L. Jones (U.S. Geological Survey), Harold Williams (Memorial University, Newfoundland).

Quaternary Advisory Group

Norman P. Lasca (University of Wisconsin-Milwaukee), chairman, Mark M. Fenton (Alberta Research Council), David S. Fullerton (U.S. Geological Survey), Robert J. Fulton (Geological Survey of Canada), W. Hilton Johnson (University of Illinois), Paul F. Karrow (University of Waterloo), Gerald M. Richmond (U.S. Geological Survey).

Conferees

W. G. E. Caldwell (University of Saskatchewan), Lucy E. Edwards (U.S. Geological Survey), Henry H. Gray (Indiana Geological Survey), Hollis D. Hedberg (Princeton University), Lewis H. King (Geological Survey of Canada), Rudolph W. Kopf (U.S. Geological Survey), Jerry A. Lineback (Robertson Research U.S.), Marjorie E. MacLachlan (U.S. Geological Survey), Amos Salvador (University of Texas, Austin), Brian R. Shaw (Samson Resources, Inc.), Ogden Tweto (U.S. Geological Survey).

APPENDIX II: 1977-1982 COMPOSITION OF THE NORTH AMERICAN COMMISSION ON STRATIGRAPHIC NOMENCLATURE

Each Commissioner is appointed, with few exceptions, to serve a 3-year term (shown by such numerals as 80-82 for 1980-1982) and a few are reappointed.

American Association of Petroleum Geologists

Timothy A. Anderson (Gulf Oil Co.) 77-83, Orlo E. Childs (Texas Tech University) 76-79, Kenneth J. Englund (U.S. Geological Survey) 74-77, Susan Longacre (Getty Oil Co.) 78-84, Donald E. Owen (Cities Service Co.) 79-82, Grant Steele (Gulf Oil Co.) 75-78.

Association of American State Geologists

Larry D. Fellows (Arizona Bureau of Geology) 81-82, Lee C. Gerhard (North Dakota Geological Survey) 79-81, Donald C. Haney (Kentucky Geological Survey) 80-83, Wallace B. Howe (Missouri Division of Geology) 74-77, Robert R. Jordan (Delaware Geological Survey) 78-84, vice-chairman, Frank E. Kottlowski (New Mexico Bureau of Mines) 76-79, Meredith E. Ostrom (Wisconsin Geological Survey) 77-80, John B. Patton (Indiana Geological Survey) 75-78.

North American Stratigraphic Code

Geological Society of America

Clarence A. Hall, Jr. (University of California, Los Angeles) 78-81, Jack E. Harrison (U.S. Geological Survey) 74-77, William W. Hay (University of Miami) 75-78, Robert S. Houston (University of Wyoming) 77-80, Michael A. Murphy (University of California, Riverside) 81-84, Allison R. Palmer (Geological Society of America) 80-83, Malcolm P. Weiss (Northern Illinois University) 76-82, chairman.

United States Geological Survey

Earl E. Brabb (Menlo Park) 78-82, David S. Fullerton (Denver) 78-84, E. Dale Jackson (Menlo Park) 76-78, Kenneth L. Pierce (Denver) 75-78, Norman F. Sohl (Washington) 74-83.

Geological Survey of Canada

James D. Aitken (Calgary) 75-78, Kenneth D. Card (Kanata) 80-83, Donald G. Cook (Calgary) 78-81, Robert J. Fulton (Ottawa) 81-84, John B. Henderson (Ottawa) 74-77, Lewis H. King (Dartmouth) 79-82, Maurice B. Lambert (Ottawa) 77-80, Christopher J. Yorath (Sydney) 76-79.

Canadian Society of Petroleum Geologists

Roland F. deCaen (Union Oil Co. of Canada) 79-82, J. Ross McWhae (Petro Canada Exploration) 77-80, Edward H. Schultz (Suncor, Inc.) 74-77, 80-83, Ulrich Wissner (Union Oil Co. of Canada) 76-79.

Geological Association of Canada

W. G. E. Caldwell (University of Saskatchewan) 76-79, R. K. Jull (University of Windsor) 78-79, Paul S. Karrow (University of Waterloo) 81-84, Alfred C. Lenz (University of Western Ontario) 79-81, David E. Pearson (British Columbia Mines and Petroleum Resources) 79-81, Paul E. Schenk (Dalhousie University) 75-78.

Asociación Mexicana de Geólogos Petróleros

Jose Carillo Bravo (Petróleos Mexicanos) 78-81, Baldomerro Carrasco V., 75-78.

Sociedad Geólogica Mexicana

Zoltan de Cserna (Universidad Nacional Autónoma de México) 76-82.

Instituto de Geología de la Universidad Nacional Autónoma de México

Ismael Ferrusquia Villafranca (Universidad Nacional Autónoma de México) 76-81, Fernando Ortega Gutiérrez (Universidad Nacional Autónoma de México) 81-84.

APPENDIX III: REPORTS AND NOTES OF THE AMERICAN COMMISSION ON STRATIGRAPHIC NOMENCLATURE

Reports (formal declarations, opinions, and recommendations)

1. Moore, Raymond C., Declaration on naming of subsurface stratigraphic units: AAPG Bulletin, v. 33, no. 7, p. 1280-1282, 1949.
2. Hedberg, Hollis D., Nature, usage, and nomenclature of time-stratigraphic and geologic-time units: AAPG Bulletin, v. 36, no. 8, p. 1627-1638, 1952.
3. Harrison, J. M., Nature, usage, and nomenclature of time-stratigraphic and geologic-time units as applied to the Precambrian: AAPG Bulletin, v. 39, no. 9, p. 1859-1861, 1955.
4. Cohee, George V., and others, Nature, usage, and nomenclature of rock-stratigraphic units: AAPG Bulletin, v. 40, no. 8, p. 2003-2014, 1956.
5. McKee, Edwin D., Nature, usage and nomenclature of biostratigraphic units: AAPG Bulletin, v. 41, no. 8, p. 1877-1889, 1957.
6. Richmond, Gerald M., Application of stratigraphic classification and nomenclature to the Quaternary: AAPG Bulletin, v. 43, no. 3, pt. I, p. 663-675, 1959.
7. Lohman, Kenneth E., Function and jurisdictional scope of the American Commission on Stratigraphic Nomenclature: AAPG Bulletin, v. 47, no. 5, p. 853-855, 1963.
8. Henderson, John B., W. G. E. Caldwell, and Jack E. Harrison, Amendment of code concerning terminology for igneous and high-grade metamorphic rocks: GSA Bulletin, pt. I, v. 91, no. 6, p. 374-376, 1980.
9. Harrison, Jack E., and Zell E. Peterman, Adoption of geochronometric units for divisions of Precambrian time: AAPG Bulletin, v. 66, no. 6, p. 801-802, 1982.

Notes (informal statements, discussions, and outlines of problems)

1. Organization and objectives of the Stratigraphic Commission: AAPG Bulletin, v. 31, no. 3, p. 513-518, 1947.
2. Nature and classes of stratigraphic units: AAPG Bulletin, v. 31, no. 3, p. 519-528, 1947.
3. Moore, Raymond C., Rules of geologic nomenclature of the Geological Survey of Canada: AAPG Bulletin, v. 32, no. 3, p. 366-367, 1948.
4. Jones, Wayne V., and Raymond C. Moore, Naming of subsurface stratigraphic units: AAPG Bulletin, v. 32, no. 3, p. 367-371, 1948.
5. Flint, Richard Foster, and Raymond C. Moore, Definition and adoption of the terms stage and age: AAPG Bulletin, v. 32, no. 3, p. 372-376, 1948.
6. Moore, Raymond C., Discussion of nature and classes of stratigraphic units: AAPG Bulletin, v. 21, no. 3, p. 376-381, 1948.
7. Records of the Stratigraphic Commission for 1947-1948: AAPG Bulletin, v. 33, no. 7, p. 1271-1273, 1949.
8. Australian Code of Stratigraphical Nomenclature: AAPG Bulletin, v. 33, no. 7, p. 1273-1276, 1949.
9. The Pliocene-Pleistocene boundary: AAPG Bulletin, v. 33, no. 7, p. 1276-1280, 1949.
10. Moore, Raymond C., Should additional categories of stratigraphic units be recognized?: AAPG Bulletin, v. 34, no. 12, p. 2360-2361, 1950.
11. Moore, Raymond C., Records of the Stratigraphic Commission for 1949-1950: AAPG Bulletin, v. 35, no. 5, p. 1074-1076, 1951.
12. Moore, Raymond C., Divisions of rocks and time: AAPG Bulletin, v. 35, no. 5, p. 1076, 1951.
13. Williams, James Steele, and Aureal T. Cross, Third Congress of Carboniferous Stratigraphy and Geology: AAPG Bulletin, v. 36, no. 1, p. 169-172, 1952.
14. Official report of round table conference on stratigraphic nomenclature at Third Congress of Carboniferous Stratigraphy and Geology, Heerlen, Netherlands, June 26-28, 1951: AAPG Bulletin, v. 36, no. 10, p. 2044-2048, 1952.
15. Records of the Stratigraphic Commission for 1951-1952: AAPG Bulletin, v. 37, no. 5, p. 1078-1080, 1953.
16. Records of the Stratigraphic Commission for 1953-1954: AAPG Bulletin, v. 39, no. 9, p. 1861-1863, 1955.
17. Suppression of homonymous and obsolete stratigraphic names: AAPG Bulletin, v. 40, no. 12, p. 2953-2954, 1956.
18. Gilluly, James, Records of the Stratigraphic Commission for 1955-1956: AAPG Bulletin, v. 41, no. 1, p. 130-133, 1957.
19. Richmond, Gerald M., and John C. Frye, Status of soils in stratigraphic nomenclature: AAPG Bulletin, v. 31, no. 4, p. 758-763, 1957.
20. Frye, John C., and Gerald M. Richmond, Problems in applying standard stratigraphic practice in nonmarine Quaternary deposits: AAPG Bulletin, v. 42, no. 8, p. 1979-1983, 1958.
21. Frye, John C., Preparation of new stratigraphic code by American Commission on Stratigraphic Nomenclature: AAPG Bulletin, v. 42, no. 8, p. 1984-1986, 1958.
22. Records of the Stratigraphic Commission for 1957-1958: AAPG Bulletin, v. 43, no. 8, p. 1967-1971, 1959.
23. Rodgers, John, and Richard B. McConnell, Need for rock-stratigraphic units larger than group: AAPG Bulletin, v. 43, no. 8, p. 1971-1975, 1959.
24. Wheeler, Harry E., Unconformity-bounded units in stratigraphy: AAPG Bulletin, v. 43, no. 8, p. 1975-1977, 1959.
25. Bell, W. Charles, and others, Geochronologic and chronostratigraphic units: AAPG Bulletin, v. 45, no. 5, p. 666-670, 1961.
26. Records of the Stratigraphic Commission for 1959-1960: AAPG Bul-

North American Commission on Stratigraphic Nomenclature

letin, v. 45, no. 5, p. 670-673, 1961.
27. Frye, John C., and H. B. Willman, Morphostratigraphic units in Pleistocene stratigraphy: AAPG Bulletin, v. 46, no. 1, p. 112-113, 1962.
28. Shaver, Robert H., Application to American Commission on Stratigraphic Nomenclature for an amendment of Article 4f of the Code of Stratigraphic Nomenclature on informal status of named aquifers, oil sands, coal beds, and quarry layers: AAPG Bulletin, v. 46, no. 10, p. 1935, 1962.
29. Patton, John B., Records of the Stratigraphic Commission for 1961-1962: AAPG Bulletin, v. 47, no. 11, p. 1987-1991, 1963.
30. Richmond, Gerald M., and John G. Fyles, Application to American Commission on Stratigraphic Nomenclature for an amendment of Article 31, Remark (b) of the Code of Stratigraphic Nomenclature on misuse of the term "stage": AAPG Bulletin, v. 48, no. 5, p. 710-711, 1964.
31. Cohee, George V., Records of the Stratigraphic Commission for 1963-1964: AAPG Bulletin, v. 49, no. 3, pt. I of II, p. 296-300, 1965.
32. International Subcommission on Stratigraphic Terminology, Hollis D. Hedberg, ed., Definition of geologic systems: AAPG Bulletin, v. 49, no. 10, p. 1694-1703, 1965.
33. Hedberg, Hollis D., Application to American Commission on Stratigraphic Nomenclature for amendments to Articles 29, 31, and 37 to provide for recognition of erathem, substage, and chronozone as time-stratigraphic terms in the Code of Stratigraphic Nomenclature: AAPG Bulletin, v. 50, no. 3, p. 560-561, 1966.
34. Harker, Peter, Records of the Stratigraphic Commission for 1964-1966: AAPG Bulletin, v. 51, no. 9, p. 1862-1869, 1967.
35. DeFord, Ronald K., John A. Wilson, and Frederick M. Swain, Application to American Commission on Stratigraphic Nomenclature for an amendment of Article 3 and Article 13, Remarks (c) and (e), of the Code of Stratigraphic Nomenclature to disallow recognition of new stratigraphic names that appear only in abstracts, guidebooks, microfilms, newspapers, or in commercial or trade journals: AAPG Bulletin, v. 51, no. 9, p. 1868-1869, 1967.
36. Cohee, George V., Ronald K. DeFord, and H. B. Willman, Amendment of Article 5, Remarks (a) and (e) of the Code of Stratigraphic Nomenclature for treatment of geologic names in a gradational or interfingering relationship of rock-stratigraphic units: AAPG Bulletin, v. 53, no. 9, p. 2005-2006, 1969.
37. Kottlowski, Frank E., Records of the Stratigraphic Commission for 1966-1968: AAPG Bulletin, v. 53, no. 10, p. 2179-2186, 1969.
38. Andrews, J., and K. Jinghwa Hsü, A recommendation to the American Commission on Stratigraphic Nomenclature concerning nomenclatural problems of submarine formations: AAPG Bulletin, v. 54, no. 9, p. 1746-1747, 1970.
39. Wilson, John Andrew, Records of the Stratigraphic Commission for 1968-1970: AAPG Bulletin, v. 55, no. 10, p. 1866-1872, 1971.
40. James, Harold L., Subdivision of Precambrian: An interim scheme to be used by U.S. Geological Survey: AAPG Bulletin, v. 56, no. 6, p. 1128-1133, 1972.
41. Oriel, Steven S., Application for amendment of Article 8 of code, concerning smallest formal rock-stratigraphic unit: AAPG Bulletin, v. 59, no. 1, p. 134-135, 1975.
42. Oriel, Steven S., Records of Stratigraphic Commission for 1970-1972: AAPG Bulletin, v. 59, no. 1, p. 135-139, 1975.
43. Oriel, Steven S., and Virgil E. Barnes, Records of Stratigraphic Commission for 1972-1974: AAPG Bulletin, v. 59, no. 10, p. 2031-2036, 1975.
44. Oriel, Steven S., Roger W. Macqueen, John A. Wilson, and G. Brent Dalrymple, Application for addition to code concerning magnetostratigraphic units: AAPG Bulletin, v. 60, no. 2, p. 273-277, 1976.
45. Sohl, Norman F., Application for amendment concerning terminology for igneous and high-grade metamorphic rocks: AAPG Bulletin, v. 61, no. 2, p. 248-251, 1977.
46. Sohl, Norman F., Application for amendment of Articles 8 and 10 of code, concerning smallest formal rock-stratigraphic unit: AAPG Bulletin, v. 61, no. 2, p. 252, 1977.
47. Macqueen, Roger W., and Steven S. Oriel, Application for amendment of Articles 27 and 34 of stratigraphic code to introduce point-boundary stratotype concept: AAPG Bulletin, v. 61, no. 7, p. 1083-1085, 1977.
48. Sohl, Norman F., Application for amendment of Code of Stratigraphic Nomenclature to provide guidelines concerning formal terminology for oceanic rocks: AAPG Bulletin, v. 62, no. 7, p. 1185-1186, 1978.
49. Caldwell, W.G.E., and N. F. Sohl, Records of Stratigraphic Commission for 1974-1976: AAPG Bulletin, v. 62, no. 7, p. 1187-1192, 1978.
50. Weiss, Malcolm P., Proposal to change name of commission: AAPG Bulletin, v. 63, no. 10, p. 1986, 1979.
51. Weiss, Malcolm P., and James D. Aitken, Records of Stratigraphic Commission, 1976-1978: AAPG Bulletin, v. 64, no. 1, p. 136-137, 1980.
52. Harrison, Jack E., and Zell E. Peterman, A preliminary proposal for a chronometric time scale for the Precambrian of the United States and Mexico: GSA Bulletin, pt. I, v. 91, no. 6, p. 377-380, 1980.

Glossary

Zounds! I was never so bethump'd
with words. . . .
***Shakespeare,* King John**

These are but wild and whirling
words, my lord.
***Shakespeare,* Hamlet**

A fine volley of words, gentleman,
and quickly shot off.
***Shakespeare,* The Two Gentlemen of Verona**

Abundance zone: A biostratigraphic unit characterized by the time range in which one or more taxa were the most abundant. Formerly called an acme zone.

Actualism: The concept that the processes that are operating today are the same as those that have operated in the geologic past.

Adhesion ripples: Small bedforms of ripplelike nature that form from the accumulation of windblown sand across a wet surface.

Aggrading, aggradation: Accumulation of sediment on a depositional surface.

Algal biscuit: Accumulation of algae in concentric layers that form a biscuit-shaped isolated mass. The modern equivalent of an oncolith.

Allostratigraphic unit: A stratigraphic unit bounded by unconformities. See Appendix A.

Angular cross laminae: Cross laminae that meet the base of the bed at an abrupt angle. See Fig. 3.3.

Angular unconformity: An erosional surface separating tilted strata below from strata parallel to the erosional surface above. See Fig. 1.13.

Anhydrite: An evaporite mineral composed of calcium sulfate.

Anorbital ripples: Ripples formed by waves whose wavelengths are not controlled by the orbital diameter of the oscillating flow that formed the ripple.

Anthracite: The highest rank of coal, recognized by a glossy luster and a concoidal fracture.

Antidune: A low-relief symmetrical bedform formed during upper regime flow. Antidunes can migrate upstream, hence the name antidune.

Argillite: A slightly metamorphosed or extremely well-indurated claystone.

Arkose: A feldspathic sandstone.

Armored mud ball: A pebble- to boulder-sized glob of mud coated by pebbles, shells, or other debris during transportation.

Assemblage zone: A biostratigraphic zone defined by the overlapping range zones of more than two taxa.

Authigenic sediment: Minerals that have grown in place in the sediment after sedimentation.

Backset bedding: Bedding inclined in the upcurrent direction.

Ball and pillow structure: A type of penecontemporaneous soft-sediment deformation in which isolated masses or lobes of sediment from an overlying bed sink into underlying sediment of a different lithology.

Barchan dune: An eolian crescentric-shaped dune that forms in areas of limited sand supply. "Horns" of the crescent point downwind.

Bed: An individual layer of sedimentary rock.

Bedded gypsum: A sedimentary rock formed from a layered accumulation of hydrous calcium sulfate.

Bedform: The shape that a bed of loose grains takes as the grains are moved during sedimentation. Bedforms produce sedimentary structures as they migrate.

Bedload: Sediment transported at the bed by traction and saltation.

Bioclasts: Fragments of the hard parts of organisms.

Bioclastic calcarenite: A bioclastic limestone composed of sand-sized grains of the hard parts of organisms.

Bioclastic limestone: A limestone composed dominantly of sedimentary grains made of the hard parts of organisms. For example, a limestone composed mostly of the whole or broken pieces of clam shells. See also coquina.

Bioclastic sediment: Grains formed by the mechanical, chemical, and biological breakdown of the skeletal parts of organisms.

Biocoenosis: An *in situ* accumulation of organisms that were living at the site of deposition.

Bioherm: An *in situ* mound-shaped structure built by sessile organisms such as corals, brachiopods, algae, etc., and enclosed in rock of a different lithologic character.

Biolithite: A carbonate rock formed by the binding together of grains by organisms. See also boundstone.

Biosparite: A limestone composed of the whole or broken hard parts of organisms and cemented together by sparry calcite.

Biostratigraphic unit: A stratigraphic unit defined or body of rock defined or characterized by its fossil contents without regard to lithologic or other physical features or relations. The basic unit is a biozone, of which there are several kinds.

Biostrome: A bedded, more-or-less tabular accumulation of organic debris from sessile organisms. In contrast to bioherms, they lack moundlike or lenslike form.

Bioturbation: The disruption of sedimentary bedding from the action of organisms, plants or animals, churning and homogenizing the sediment while it is still soft.

Biozone: The fundamental biostratigraphic unit defined by fossil organisms without regard to time or lithology.

Bird's eye structure: The infilling of small cavities, mostly resulting from bubbles in lime mud, by sparry calcite and/or anhydrite.

Bituminous: A medium-grade rank of brown/black coal characterized by a dull to bright luster and blocky fracture.

Borate: A sedimentary rock composed of evaporite minerals of the salts of boric

Boring: An epigenetic biogenic sedimentary structure formed as an organism cuts or bores into a hard substrate. Can be differentiated from burrows by truncated fossils or grains. See Fig. 3.94 and 3.95.

Bouma sequence: A characteristic sequence of sedimentary structures in a turbidity current deposit described by Arnold Bouma in the 1960s.

Boundstone: A limestone that contains structures and textures indicating that organisms bound sedimentary grains together during sedimentation.

Breccia: A coarse-grained sedimentary rock with very angular gravel-sized clasts. May also be used to describe similar accumulations of volcanic debris, the rubbly zone at the top of a lava flow, or textures in a fault zone. See sharpstone.

Bubble sand: Air-bubble cavities commonly formed in a beach where air is trapped in the sand between a rising water table from an incoming flood tide and water percolating down into the sand from waves.

Burrow: A penecontemporaneous biogenic sedimentary structure formed as organisms move through soft sediment.

Calcarenite: A limestone composed of sand-sized grains of calcium carbonate.

Calcilutite: A limestone composed of mud-sized grains of calcium carbonate.

Calcirudite: A limestone composed of gravel-sized grains of calcium carbonate.

Carbonate: A sedimentary rock composed of more than 50% calcite and/or dolomite.

Chalcedony: Cryptocrystalline quartz and chert, commonly microscopically fibrous.

Chalk: A poorly indurated, fine-grained limestone composed of the tests of microorganisms such as coccoliths and foraminiferans.

Chert: A sedimentary rock composed largely of amorphous/microcrystalline silica commonly made from the siliceous tests of diatoms and radiolarians and other microorganisms. May also form from the precipitation of silica from solution or the accumulation of fine-grained volcanic ash followed by devitrification.

Chronostratigraphic unit: A stratigraphic unit (erathem, system, series, and stage) defined as all rocks deposited during a specified interval of time.

Clast: A sedimentary grain formed from the breakdown of other solids.

Clastic dike: A roughly tabular body of clastic material, commonly sandstone, discordant to the sedimentary host rock. Formed by injection of fluidized sediment into overlying or underlying sediments or rocks.

Clay: Sedimentary particles less that 4 μm in diameter.

Clay mineral: Finely crystalline hydrous-layered silicate minerals.

Claystone: A terrigenous clastic sedimentary rock composed of clay-sized particles.

Climbing ripples: Sedimentary structures formed by the preservation of stoss and lee sides of ripple marks as they migrate downcurrent and move or climb vertically as the bed aggrades.

Coal: Compacted and metamorphosed stratified plant remains that contain more than 50% carbon and burn readily.

Compound cross laminae: Cross beds containing internal erosional surfaces resulting in cross laminae within cross beds. See Fig. 3.6.

Concretion: A penecontemporaneous or secondary sedimentary structure resulting from the nodular or irregular concentration of authigenic minerals in sedimentary rocks forming structures more resistant than the enclosing rock.

Conglomerate: A coarse-grained terrigenous sedimentary rock made of more than 25% gravel-sized clasts.

Convolute, contorted bedding: A penecontemporaneous sedimentary structure composed of bent, twisted, folded, or crumpled sedimentary beds or laminae. Individual laminae are often continuous and traceable from fold to fold. The folds tend to be bounded above and below by undisturbed beds.

Coquina: A rock made of the broken and whole transported pieces of shells.

Critical velocity: The velocity of flow that initiates movements of grains on a bed.

Cross laminae or beds: Sets of strata discordant to bedding planes that enclose them.

Cross ripples: The superposition of more than one type of ripple each with a different crest trend. Same as interference ripple.

Debris flow: A sediment-gravity flow whose movement is controlled by the strength of the matrix. Generally composed of very coarse grained clasts supported in a finer-grained matrix.

Deep-water wave: A wave in deep water where the bottom has no effect on the wave. Deep-water waves form where the ratio of depth to wavelength is greater than 4 to 1.

Degradation: The erosion or removal of sediment during transport of sediment across a bed.

Diachronic: A stratigraphic unit that comprises the unequal spans of time represented either by a specific lithostratigraphic, allostratigraphic, biostratigraphic, or pedostratigraphic unit or by an assemblage of such units.

Diamictite, Diamicton: A nonsorted conglomerate composed of large particles dispersed in a fine-grained matrix.

Diastem: Small amounts of time not recorded in the stratigraphic record.

Disconformity: An erosional surface separating two sequences of parallel strata from one another. See Fig. 1.13.

Dish structure: A penecontemporaneous water escape structure formed by the upward migration of water through the sediment. Forms nested dishlike structures seen as small changes in texture and accumulations of finer-grained sediment and mica.

Dispersive pressure: The force that keeps grains moving because of the physical interactions and collision of the grains.

Diurnal tide: A tide with one high and one low tide per day.

Dolomite: A mineral made of calcium magnesium carbonate.

Dolostone: A carbonate sedimentary rock composed of more than 50% dolomite.

Dune: A lunate and linguoid asymmetrical large ripple that has a height of several decimeters or greater and spacing of a few meters or more. Dunes are a large accumulation of sand with a gentle stoss side and steeply dipping stoss slope than form in both eolian and subaqueous environments.

Dust: Silt and clay that have been moved by suspension in an eolian environment. Makes you sneeze and creates bowls.

Epigenetic: Structures formed after the lithification of a sediment into rock.

Evaporite: A sedimentary rock composed of mineral salts precipitating from solution during evaporative concentration.

Fabric: The arrangement of the grains of a sediment or sedimentary rock in terms of orientation by form, mineral composition, sorting, and packing.

Facies: The lithologic and biological characteristics of a sedimentary rock that results from deposition in a particular environment.

Facies model: A general summary of a specific sedimentary environment and the biological and lithologic characteristics that result from processes of sedimentation within that environment.

Fanglomerate: A conglomerate deposited on an alluvial fan.

Fecal pellets: Little balls, rods, and ovoids of invertebrate excreta, generally of mud-sized grains and common in marine environments.

Feldspathic sandstone: A sandstone containing more than 50% feldspar.

Festoon cross bed: See trough cross bed.

Fissility: The property of a fine-grained sedimentary rock to split into thin layers along closely spaced bedding planes.

Flame structure: A load structure that has a flamelike shape. Generally deformed in the direction of movement of the overlying sediment.

Flaser bedding: Small thin lenses of mud in sand. Often formed from mud settling out of suspension into the troughs of ripples in sand.

Flat-pebble conglomerate: An intraformational conglomerate composed of elongate clasts of mud in a sand or gravel matrix.

Flint: Black to gray chert.

Flow lineation: The alignment of elongate grains on a bed caused by upper regime flow over the bed.

Flow regime model: The model that describes the relationship between flow depth and velocity and resulting bedforms in sediment of a particular grain size.

Fluidal flow: A sediment gravity flow that acts like a fluid.

Fluidized flow: The sediment gravity flow resulting from the upward movement of fluid through the sediment that buoys the grains.

Flume: An artificial stream used in a laboratory to study the relationship between flow parameters and sediment transport.

Flute cast: A penecontemporaneous sedimentary structure formed from the infilling of flutes scoured into mud. Flutes are scours in cohesive mud beds with the upcurrent end rounded or bulbous and the downstream end flaring out to merge with the unscoured surface of the bed.

Formation: The basic lithostratigraphic unit. Defined solely by lithologic characteristics without regard to age or fossil content and mappable at a particular scale.

Foreset: Steep lee slope of an asymmetric bedform where most deposition occurs.

Gas escape structure: A penecontemporaneous sedimentary structure that results from gas moving upward through the sediment and disrupting bedding and original textures. The gas is commonly generated by bacterial decay of organic material within the sediment.

Geochronologic: A division of geologic time traditionally distinguished on the basis of the rock record as defined by chronostratigraphic units. Geochronologic units are not material units but the time represented by a body of rocks. Subdivisions include eon, era, period, epoch, and age.

Geochronometric: A stratigraphic unit established through the direct division of geologic time expressed in years. Geochronometric units are abstractions and not material units. They are simply time divisions of convenient magnitude and were established for the development of the time scale for the Precambrian.

Geologic-time unit: See geochronologic.

Geopedal structure: The infilling of cavities in carbonate rocks by carbonate and/or terrigenous sediment in layers that indicate the original tops of beds.

Graded bed: A sedimentary bed that grades from coarse-grained particles near the base to finer-grained clasts near the top.

Grained and striped horizontal stratification: The term used by Sorby for flow lineation or streaks of coarser-grained material that form on upper flow regime plane beds.

Grain flow: A sediment flow of particles moving under completely dry conditions.

Grainstone: A limestone composed of more than 90% sand-sized grains.

Grain-supported: A texture in a sedimentary rock in which a large proportion of the grains of a particular size are in contact with one another.

Granule ridge: An asymmetric ridge of gravel formed in an eolian environment.

Group: A lithostratigraphic unit composed of two or more similar formations.

Gutter cast: A sedimentary structure formed from the infilling of grooves in underlying sediment.

Halite: A sedimentary rock formed by chemical precipitation of sodium chloride.

Halite cast, Hopper cast: A sedimentary structure formed from the infilling of cubic-shaped cavities left by the dissolution of halite crystals.

Hardground: A bed of limestone with its upper surface well indurated and bored, corroded, or eroded.

Hemipelagic: An oceanic sediment composed of a mixture of pelagic and terrigenous sediment.

Herringbone cross laminae: Tabular sets of cross laminae with opposing directions of dip.

Hiatus: The time missing or not represented at an unconformity or other break in the stratigraphic record.

High-density turbidity current: Turbidity current with a high sediment load (approximately 20–30%) in which the matrix and collision of grains are important in adding buoyant lift to the grains.

Hummocky cross-stratification (HCS): Low-angle cross-stratification with intersecting wavy laminae commonly truncated by erosional surfaces and lacking clear indication of directionality.

Hydrodynamically equivalent grain size: The grain size of a sediment determined by comparing that sediment's settling velocity in water to that of quartz spheres that have the same settling velocity.

Ichnofossil: See trace fossil.

Imbrication: A fabric in which disk- or blade-shaped particles are arranged in a overlapping shinglelike pattern dipping in one direction

Interference ripples: Intersecting sets of ripples of more than one type and/or crest orientation.

Intraclasts: Sand-sized or larger pieces of weakly consolidated carbonate sediment eroded and redeposited within the basin of deposition.

Intraformational conglomerate: A rock composed of gravel-sized clasts derived from underlying sediment within the same bed.

Inversely graded bed: A sedimentary bed in which the grain size increases from bottom to top.

Irregular bedding: A bed showing a lack of uniformity in thickness and/or shape.

Jasper: A red-, brown-, yellow-, or green-colored chert.

Lahars: A general term given to debris flows and mudflows generated on volcanoes as the result of volcanic activity.

Laminar flow: Streamline flow in a viscous or slow-moving fluid in which there is no vertical mixing. Contrast with turbulent flow.

Laminar sublayer: The layer at the base of turbulent flow in which viscous (laminar) flow dominates.

Laminite: A thinly layered (less than 1 cm) sedimentary rock.

Large ripples, large-scale ripples: Ripple-shaped bedforms that have lengths greater than 60 cm and heights greater than approximately 6 cm.

Lateral continuity: Steno's law that states that strata were deposited in laterally continuous layers across the surface of the earth.

Lensoidal bedding: A bed with an external shape approximating that of a lens.

Lenticular bedding: Small isolated lenses of sand in mud.

Lime mud: Mud-sized sediment composed of carbonate. Also called micrite.

Litharenite: A contraction of the term lithic arenite, a lithic sandstone.

Lithic sandstone: A sandstone composed of less than 10% quartz and 50% feldspar, the remainder being rock fragments.

Lithic wacke: A sandstone with 10% or more matrix and a significant portion of rock fragments.

Lithofacies: A unit of rock defined and recognized on the basis of lithologic characteristics.

Lithostratigraphic unit: A body of sedimentary, extrusive igneous, metamorphic, or metavolcanic strata that is distinguished and delimited on the basis of lithic characteristics and stratigraphic position.

Liquified flow: Flow in which grains in the flow are partially supported by escaping fluid as the grains settle and displace the fluid.

Load cast, load structure: An irregular or roll-like protrusion at the base of a bed produced by differential settling and/or compaction.

Löess: Deposits of windblown dust deposited from suspension. Generally loess contains in excess of 75% silt.

Longitudinal dune: A large linear eolian dune with its axis parallel with the dominant wind direction.

Low-density turbidity current: A sediment gravity flow with the sediment concentration less than approximately 20%. Grains in such flows are supported by turbulence and impeded from settling by the high concentration of sediment and the buoyancy of a finer-grained matrix.

Lower flow regime: That part of the flow regime where ripplelike bedforms and the lower plane beds form. Characterized by tranquil flow and water surface out-of-phase with the bedform. See Fig. 5.1.

Low-rank graywacke: A lithic sandstone with little or no feldspar.

Magnetopolarity unit: A body of rock unified by its remanant magnetic polarity and distinguished from adjacent rock that has different polarity.

Mass flow: A sediment gravity flow that flows with plastic behavior.

Massive bed: A bed with no visible internal sedimentary structures.

Matrix: Fine-grained particles that fit between larger particles in the same rock.

Matrix-supported: A texture in which large particles are supported or "float" in a finer matrix.

Maturity: The extent to which a sediment's properties approach the ultimate end product toward which they are being driven by the processes that effect it. Maturity can be measured in one of two ways. Compositionally mature sediments are composed mostly of one mineral type, usually quartz. Texturally mature sediments reflect the change in a sediment's texture as it is transported from its source area, including the percentage of clay-sized particles and the degree of rounding and sorting.

Megaripples: Equivalent to large ripples.

Member: A named entity within a formation that possesses characteristics distinguishing it from adjacent parts of the formation.

Micrite: A limestone composed of greater than 50% lime mud or carbonate cement less than mud size.

Minikarst: Small-scale secondary dissolution features on the surface of limestóne.

Mixed tide: A tide with two highs and two lows of unequal height each day.

Monomictic: A conglomerate with clasts composed of dominantly one composition.

Mudcrack: A penecontemporaneous polygonally shaped set of cracks formed from shrinkage of the sediment.

Mud drape: A layer of mud deposited from suspension over a depositional surface with some relief, for example, over bedforms.

Mudflow: A sediment flow with a large component of fine-grained material.

Mudrock, mudstone: A sedimentary rock composed of greater than 50% mud.

Nodules: Small spheroidal to irregular bodies of a different composition and harder than the surrounding sediment or rock.

Nonconformity: An erosional surface separating sedimentary strata above from nonstratified igneous or metamorphic rocks below.

Normal graded bed: A bed in which the grain size decreases from bottom to top

Oblique dunes: Large eolian dunes with their crests at an oblique angle to the dominant wind direction.

Oncolith: A spherical or subspherical mass of roughly concentric layers of carbonate. Oncoliths originate as algal biscuits from the growth of algae and/or bacteria on a particle that is periodically rolled by currents or waves.

Ooids: A smooth spherical carbonate grain less than 2 mm in diameter formed of concentric rings of carbonate surrounding a nucleus.

Opaline silica: Cryptocrystalline hydrated silicon dioxide. Commonly forms tests of microorganisms such as diatoms and radiolarians.

Orbital diameter: The difference between extreme positions within the orbit of oscillatory flow generated by waves.

Orbital ripples: Wave ripples that show a direct relationship between wavelength and orbital diameter of the oscillatory flow forming the ripples.

Orbital velocity: The circumference of the orbit of oscillatory flow divided by the time that it takes to complete one wave oscillation.

Original horizontality: One of Steno's principles that sediments are deposited in horizontal layers.

Orthoquartzite: A sandstone composed of more than 90% detrital quartz. Commonly cemented by silica.

Oxidation spot: A spheroidal to irregular area in a sedimentary rock where iron has been oxidized, leaving a reddish or orangish spot in an otherwise greenish or grayish rock.

Paraconformity: A nondepositional surface separating two parallel sequences of strata from one another. Nondeposition is inferred by evidence of missing time other than erosion, such as gaps in the fossil record. See Fig. 1.13.

Peat: A dark brown or black accumulation of partially decomposed plants, particularly those that grow in marshes.

Pedostratigraphic unit: A stratigraphic unit defined as a body of rock that consists of one or more soil horizons developed within some other lithostratigraphic unit.

Penecontemporaneous sedimentary structure: Structures formed just after sedimentation and before the sediment is lithified into a rock.

Period: The time represented by the rocks constituting a system.

Period of oscillation: The time that it takes for a complete orbit in oscillatory flow generated by waves.

Phi unit: A grain size parameter that is defined as the negative log to the base two of the diameter of the grain in millimeters divided by one millimeter. This parameter is a dimensionless number that takes into account the log normal size distribution of most sediments.

Phosphorite: A sedimentary rock containing a large percentage of calcium phosphate.

Phyllarenite: A lithic sandstone whose grains are composed largely of fine-grained metamorphic rocks.

Pillar structures: A column-shaped water escape structure.

Planar bedding: Horizontal laminae.

Plane bed: A flat smooth bedform. Generally used in context with upper flow regime plane bed or lower plane bed.

Plinian eruption: A volcanic eruption typified by a tall column of ash and gas. Named for Pliny the Younger's description of the eruption of Mount Vesuvius in A.D. 79.

Polarity-chronologic unit: Division of geologic time distinguished on the basis of magnetic polarity.

Polymictic: A conglomerate with clasts of two or more compositions.

Porcellanite: A fine-grained siliceous sedimentary rock with 50–75% nonterrigenous silica.

Primary dolostone: A sedimentary rock composed of dolomite formed during deposition.

Primary sedimentary structure: Sedimentary structures formed during the process of sedimentation.

Pseudomatrix: Fine-grained particles in a sandstone that are the pulverized remains of larger grains crushed by compaction or folding.

Puddingstone: A conglomerate.

Quartz arenite, quartz sandstone: A sandstone composed of more than 90% quartz and less than 10% matrix.

Quartzite: A metamorphic rock produced by the recrystallization of a quartz sandstone. Also used to describe quartz sandstones cemented with silica so that when broken the fracture cuts across grains.

Quartzo-feldspathic sandstone: A sandstone containing dominantly quartz and feldspar.

Reactivation surface: A broad, curving erosional surface that truncates cross-stratification that is caused by changes in flow.

Reduction spot: A spheroidal to irregular area in an otherwise oxidized rock where iron-bearing minerals have been reduced. Generally white or green in reddish or orangish surrounding rock.

Reef: A wave-resistant organic buildup.

Relaxation time: The time that it takes a bed to adjust to new conditions of flow.

Reverse grading: See inversely graded bed.

Reynolds number: A dimensionless number that describes the type of flow, turbulent or laminar, depending on the viscosity, velocity, density, and depth of the fluid.

Rill marks: Thin, stringlike erosional grooves formed on the surface of a bed by rills of water draining off an inclined surface such as a beach.

Ripple drift: The bedding formed by ripples climbing at an angle steeper than their upcurrent stoss side.

Ripple marks: See small ripples.

Ripplet: A very small ripple-shaped bedform generally a few millimeters high and a centimeter or less long.

Rip-up clast: Clasts of mud eroded from a bed by currents and redeposited after only a short distance of transport.

Rolling grain ripples: The bedform which forms during the initial stages of movement of sediment by oscillating flow.

Root structure: An organic sedimentary structure formed when plant roots disrupt bedding and other sedimentary structures.

Roundstone: A conglomerate composed of well-rounded clasts.

Runnel: A trough parallel to shore and shoreward of a breaker bar.

Runzelmarken: A set of irregular, very small, ripplelike bedforms synonymous with ripplet.

Saltation: The transport of sedimentary grains as bedload by bouncing, hopping, and skipping.

Salt polygons: Polygonal-shaped ridges formed during growth of evaporite minerals in arid environments.

Sand wave: A very large, straight-crested, asymmetric bedform with height in the meters and wavelength in the tens of meters or greater.

Sapropel: A fine-grained organic sediment or sludge originating as a product of putrification of plant remains, commonly algae. An important component of many black shales.

Scour marks: A penecontemporaneous sedimentary structure eroded in cohesive sediment by turbulent flow over the bed.

Secondary sedimentary structures: A structure formed after a sediment has lithified into a rock.

Sedimentation: The act or process of depositing sediment.

Sediment gravity flow: A flow of fluid composed of large amounts of sediment, ranging from 20 to 70%, so that the flow is controlled by the properties of the sediment–water mixture.

Sedimentology: The study of processes and mechanisms transporting and depositing sediment; the study of sedimentation.

Semidiurnal tide: A tide that has two highs and two lows of equal or subequal height per day.

Septarian concretion (nodule): A roughly spheroidal concretion divided into polyhedral blocks by radiating and intersecting cracks filled with veins of carbonate.

Set: A bed.

Shale: A fissile mudrock.

Shallow-water wave: A wave in shallow water that is strongly controlled by the bottom. A wave where the depth of water to wavelength ratio is less than 0.05 to 1.

Sharpstone: A sedimentary rock composed of more than 25% gravel-sized angular clasts.

Siliceous ooze: An unconsolidated, fine-grained siliceous sediment generally composed of accumulations of tests of organisms such as diatoms and radiolarians.

Siltite: A slightly metamorphosed or extremely well indurated siltstone.

Siltstone: A terrigenous sedimentary rock composed of more than 50% silt.

Simple cross laminae: Cross laminae with no internal erosional surfaces.

Sinter: Chemical precipitates deposited around hot springs, both calcareous and siliceous in composition.

Small ripples, small-scale ripples: Asymmetric or symmetric bedforms with a more-or-less triangular cross section with heights less than 5 cm and lengths less than 60 cm. See Fig. 3.31.

Soft-sediment folds: Penecontemporaneous folds formed by the ductile flow of sediment layers before lithification.

Sorting: A measure of the dispersion of the grain size distribution of a sediment.

Sparite: A calcarenite cemented with sparry calcite.

Sparry calcite, spar: A cement of transparent or translucent crystalline calcite with crystals larger than mud size. If composed of dolomite, it is called sparry dolomite or dolospar.

Star dune: A large eolian dune complex with a more-or-less pyramidal shape, often tens to hundreds of meters tall.

Stoke's law of settling: The relationship that defines the settling velocity of the grain to be controlled by the grain size, viscosity of the fluid, and density of the grain.

Strata: Layers of sedimentary rock.

Stratigraphy: That branch of geology that deals with the origin, composition, description, and correlation of stratified rocks through an understanding of recent sedimentary processes, modern environments, and the principles of biology, ecology, physics, and chemistry.

Stromatolite: A laminated, lithified deposit of limestone formed by algae and bacteria. Often occurs as solitary or groups of columns and mounds from a centimeter to several meters high that look like cabbages.

Structureless bed: A bed of sediment with no apparent internal structure.

Styolite: A dissolution feature, generally in carbonate rocks. Styolites are recognized by an irregular dark band defined by a concentration of nonsoluble minerals. Missing rock can be inferred by truncated grains or fossils.

Subarkose: A transitional class of sandstone with less feldspar than arkose and few or no lithic grains.

Subgraywacke: A sandstone with over 15–20% matrix and less than 10% feldspar, essentially a quartz wacke or a lithic sandstone.

Sublitharenite: A sandstone containing 5–25% rock fragments, up to 10% feldspar, and 65–95% quartz.

Suborbital ripples: Wave ripples in which the wavelength of the ripples decreases as the orbital diameter of the oscillatory flow increases.

Supergroup: A lithostratigraphic unit that is a formal assemblage of related groups, or of groups and formations.

Superposition: In a vertical stack of strata, the one at the bottom is the oldest and the one at the top is the youngest.

Surface creep: The movement of grains along a bed caused by the collisions of saltating grains.

Surface texture: The texture such as markings, pits, and roughness on the surface of a sedimentary grain.

Suspension: The state of grains when they are mixed with, but undissolved in, a fluid.

Swaley cross-stratification (SCS): Low-angle concave laminae connecting hummocky cross-stratification.

Swash marks: Overlapping hyperbola-shaped marks formed by wave swash on a sediment surface, typically a beach.

Symmetrical ripples: Small ripples with either side of the crest approximately equal in length and slope angle. See Fig. 3.31.

Syneresis crack: Shrinkage crack formed subaqueously by the dewatering of clays.

Syngenetic sedimentary structure: Sedimentary structure formed during and shortly after sedimentation but before lithification.

Synthem: The basic unconformity-bounded stratigraphic unit. They define unconformity-bounded units that stratigraphers such as Sloss (1963) identified as continent-wide transgressive and regressive sequences.

System: A chronostratigraphic unit of rank below erathem. The rocks during a period of earth history.

Tabular bed: A package of sedimentary rock with parallel or near parallel upper and lower bounding surfaces.

Taxon range zone: A biostratigraphic unit defined as the interval between the documented lowest and highest occurrence of a single organism.

Teepee structure: A penecontemporaneous structure formed by the growth of evaporite minerals and expansion of the surface causing a triangular-shaped ridge.

Terrigenous clastic sediment: Grains derived from the weathering of previous rocks.

Test: The "shell" or hard part of a unicellular organism. The tests of radiolarians and diatoms are made of opaline silica and those of foraminiferans and coccoliths of calcium carbonate. The tests of these microrganisms are an important contributor of bioclastic sedimentary particles.

Thixotropic: The property of a solid that causes it to loose strength and behave as a fluid when shaken or otherwise disturbed.

Three-dimensional dune, 3-D dune: See dune.

Tillite: A sedimentary rock composed of nonsorted sediment deposited by glaciers. Often contains grains ranging in size from boulders to clay.

Time-rock unit: See chronostratigraphic unit.

Time-stratigraphic unit: See chronostratigraphic unit.

Tool: Any particle carried by moving fluid that gouges the sediment to produce a tool mark.

Tool mark: Various marks, such as gouges, streaks, or depressions, made as tools are dragged or rolled along the bottom.

Trace fossil: The structures resulting from the disruption of sediment by the actions of organisms living and moving through or on the sediment.

Track: An individual footprint of an animal; a type of trace fossil.

Trackway: A series of tracks made by a single animal.

Traction: That portion of bedload moved by bouncing, sliding, and rolling.

Traction carpet: A sheet of sediment transported as bedload very near the bed.

Trail: The disturbances produced by an animal moving across the surface of the sediment.

Transverse dune: A large eolian dune with crest perpendicular to the dominant wind direction.

Travertine: Calcium carbonate deposited as sinter around hot springs.

Trough cross bed (laminae): A cross bed with concentric laminae within a lensoidal external bed.

Tufa: Calcium carbonate deposited around hot springs as the result of the action of bacteria and algae.

Turbidite: The sedimentary rock formed by the deposition of turbidity currents.

Turbidity current: A sediment gravity flow in which transport of grains is dependent on turbulent flow.

Turbulent flow: Fluid flow at which the velocity at a given point changes constantly in magnitude and direction. Contrasts with laminar flow.

Type section, stratotype: A designated standard sequence of sedimentary rocks at a particular locality that serves as a standard for definition and recognition of a geologic unit. Unit Stratotypes refer to the type sequence and are defined independently of the boundary stratotype that marks the type upper and lower contact of the unit.

Upper flow regime: That part of the flow regime where low amplitude antidunes and the upper plane beds form. Characterized by shooting flow and water surface in-phase with the bedform, see Fig. 5.1.

Viscosity: The resistance to flow of a fluid.

Volcaniclastic sediment: Grains generated by volcanic eruptions and subsequently reworked, transported, and deposited by sedimentary processes.

Vortex ripples: Ripples formed during oscillatory flow by separation of flow over the crests of the ripples.

Wacke: A muddy sandstone containing more than 10% matrix less than 30 μm in diameter.

Water escape structures: A penecontemporaneous structure formed by fluids being expelled from a sediment during deposition, burial, or compaction.

Wavy bedding: Intermediate between flaser and lenticular bedding. Thin irregular and lensoidal layers of mud and sand.

Wedge-shaped bed: A bed with an external shape like that of a wedge.

Wind ripples: Small asymmetric ripples formed by the transport of sediment by wind.

Wrinkle marks: See Runzelmarken.

We should have a great many fewer disputes in the world if words were taken for what they are, the signs of our ideas only, and not for things themselves.

John Locke,
An Essay Concerning Human Understanding

References Cited

Adams, F. D. 1938. *The Birth and Development of the Geological Sciences.* Dover, New York, 506 pp.

Ager, D. V. 1981. The Nature of the Stratigraphical Record, 2nd ed. Macmillan & Co., London, 122 pp.

Ager, D. V. 1984. The stratigraphic code and what it implies. In W. A. Berggern and J. A. Van Couvering (Eds.), *Catastrophies and Earth History.* Princeton Univ. Press, Princeton, N.J., 464 pp.

Alam, Mustafa M., Crook, A. W. Keith, and Taylor, G. 1985. Fluvial herring-bone cross-stratification in a modern tributary mouth bar, Coonamble, New South Wales, Australia. *Sedimentology,* **32,** 235–244.

Allen, J. R. L. 1963a. Asymmetrical ripple marks and the origin of water-laid cosets of cross-strata. *Liverpool and Manchester Geological Journal,* **3,** pt. 2, 187–236.

Allen, J. R. L. 1963b. Henry Clifton Sorby and the sedimentary structures of sands and sandstones in relation to flow conditions. *Geol. en Mijnbouw,* **42,** 223–228.

Allen, J. R. L. 1970. *Physical Processes of Sedimentation.* Allen & Unwin, London, 248 pp.

Allen, J. R. L. 1979. A model for the interpretation of wave ripple-marks using their wavelength, textural composition, and shape. *Journal of the Geological Society of London,* **136,** 673–682.

Allen, J. R. L. 1982. *Sedimentary Structures: Their Classification and Physical Basis,* Vols. 1 and 2. Elsevier, Amsterdam, 64 pp.

Allen, J. R. L. 1985a. *Principles of Physical Sedimentology.* Allen & Unwin, London, 272 pp.

Allen, J. R. L. 1985b. *Experiments in Physical Sedimentology.* Allen & Unwin, London, 63 pp.

Allen, J. R. L., and Friend, P. F. 1976. Relaxation time of dunes in decelerating aqueous flows. *Journal of the Geological Society of London,* **132,** 17–26.

Allen, P. A. 1981. Some guidelines in reconstructing ancient sea conditions from wave ripples. *Marine Geology,* **42,** M59–M67.

Anderson, R. S. 1986. Erosion profiles due to particles entrained by wind: Application of an eolian sediment-transport model. *Geological Society of America Bulletin,* **97,** 1270–1278.

Andersson, K. A. 1979. Early lithification of limestones in the Redwater Shale Member of the Sundance Formation (Jurassic) of southeastern Wyoming. *Contributions to Geology, University of Wyoming,* **18,** 1–17.

Andersson, K. A. 1981. Bernhard Lundgren's (1891) description of *Ophiomorpha. Geologiska Foreningens i Stockholm Forhandlingar,* **103,** Pt. 1, 105–107.

Bagnold, R. A. 1941. *The Physics of Blown Sand and Desert Dunes.* Methuen, London, 265 pp. (reprinted 1954).

Bagnold, R. A. 1946. Motion of waves in shallow water. Interaction of waves and sand bottoms: *Proceedings of the Royal Society of London, Series A,* **187,** 1–16.

Bagnold, R. A. 1954. Experiments on gravity-free dispersion of large solid spheres in a Newtonian fluid under shear. *Proceedings of the Roayl Society of London, Series A,* **225,** 49–63.

Bagnold, R. A. 1956. The flow of cohesionless grains in fluids. *Philosophical Transactions of the Royal Society of London, Series A,* **249,** 235–297.

Basan, P. B. (Ed.). 1978. *Trace Fossil Concepts.* Society of Economic Paleontologists and Mineralogists Short Course Notes No. 5, 181 pp.

Bathurst, R. G. C. 1971. *Carbonate Sediments and Their Diagenesis,* 2nd ed., *Developments in Sedimentology 12.* Elsevier, Amsterdam, 658 pp.

Batiza, R., Fornari, D. J., Vanko, D. A., and Lonsdale, P. 1984. Craters, calderas, and hyaloclastites on young Pacific seamounts. *Journal of Geophysical Research,* **89,** No. B10, 8371–8390.

Bell, H. S. 1940. Armored mudballs: Their origin, properties and role in sedimentation. *Journal of Geology,* **48,** 1–31.

Berg, O. R., and Wolverton, D. B. (Eds.). 1985. *Seismic Stratigraphy II, and Integrated Approach.* American Association of Petroleum Geologists Memoir 39, 276 pp.

Berggren, W. A., and Van Couvering, J. A. (Eds.). 1984. *Catastrophies and Earth History: The New Uniformitarianism.* Princeton Univ. Press, Princeton, N.J., 464 pp.

Berry, W. B. N. 1968. *Growth of a Prehistoric Time Scale Based on Organic Evolution.* Freeman, San Francisco, 158 pp.

Blackwelder, E. 1928. Mudflow as a geologic agent in semiarid mountains. *Geological Society of America Bulletin,* **39,** 465–484.

Blatt, H. 1982. *Sedimentary Petrology.* Freeman, San Francisco, 564 pp.

Blatt, H., Middleton, G. V., and Murray, R. 1980. *Origin of Sedimentary Rocks,* 2nd ed. Prentice–Hall, Englewood Cliffs, N.J., 782 pp.

Bliven, L., Huang, N. E., and Janowitz, G. S. 1977. *An Experimental Investigation of Some Combined Flow Sediment Transport Phenomena.* North Carolina State University, Center for Marine and Coastal Studies, Report No. 77-3.

Bogardi, J. 1974. *Sediment Transport in Alluvial Streams.* Akademai Kiado, Budapest, 826 pp.

Boggs, S. 1986. *Principles of Sedimentology and Stratigraphy.* Merrill, Columbus, Ohio, 784 pp.

Boltwood, B. B. 1907. On the ultimate disintegration products of the elements. *American Journal of Science,* **23,** 77–88.

Boswell, P. G. H. 1961. *Muddy Sediments: Some Geotechnical Studies for Geologists, Engineers and Soil Scientists.* Heffer, Cambridge, England, 140 pp.

Bouma, A. H. 1962. *Sedimentology of Some Flysch Deposits. A Graphic Approach to Facies Interpretation.* Elsiever, Amsterdam, 168 pp.

Bouma, A. H. 1969. *Methods for the Study of Sedimentary Structures.* Wiley–Interscience, New York, 458 pp.

Brenner, R. L. 1980. Construction of processes–response models for ancient epicontinental seaway depositional systems using partial analogs. *American Association of Petroleum Geologists Bulletin,* **64,** 1223–1244.

Brookfield, M. E., and Ahlbrandt, T. S. (Eds.). 1983. *Eolian Sediments and Processes.* Elsevier, Amsterdam, 660 pp.

Carstens, M. R., Neilsen, F. M., and Altinbilek, H. D. 1969. *Bed Forms Generated in the Laboratory under an Oscillatory Flow: Analytical and Experimental Study.* U.S. Army Corps of Engineers, Coastal Engineering Research Center, Technical Memo 28, 39 pp.

Chafetz, H. S., and Folk, R. L. 1984. Travertines: Depositional morphology and the bacterially constructed constituents. *Journal of Sedimentary Petrology,* **54,** 289–316.

Chan, M. A., and Dott, R. H., Jr. 1983. Shelf and deep-sea sedimentation in Eocene forearc basin, Western Oregon—Fan or non-fan. *American Association of Petroleum Geologists Bulletin,* **67,** 2100–2116.

Chan, M. A., and Dott, R. H., Jr. 1986. Depositional facies and progradational sequences in Eocene wave-dominated deltaic complexes, southwestern Oregon. *American Association of Petroleum Geologists Bulletin,* **70,** 415–429.

Chang, K. H. 1975. Unconformity-bounded stratigraphic units. *Geological Society of America Bulletin,* **86,** 1544–1552.

Childs, O. E. 1983. *Correlation of Stratigraphic Units of North America, COSUNA, 1977–1983.* American Association of Petroleum Geologists and U.S. Geological Survey, 49 pp.

Clifton, H. E. 1969. Beach lamination: Nature and origin. *Marine Geology,* **7,** 553–559.

Clifton, H. E. 1976. Wave-formed sedimentary structures: A conceptual model. In R. A. Davis, Jr., and R. L. Ethington (Eds.), *Beach and Nearshore Sedimentation.* Society of Economic Paleontologists and Mineralogists Special Publication No. 24, pp. 126–148.

Clifton, H. E., and Dingler, J. R. 1984. Wave-formed structures and paleoenvironmental reconstruction. *Marine Geology,* **60,** 165–198.

Clifton, H. E., Hunter, R. E., and Phillips, R. L. 1971. Depositional structures and processes in the high-energy nonbarred nearshore. *Journal of Sedimentary Petrology,* **41,** 651–670.

Cloud, P. E. 1973. Paleoecological significance of banded iron formations. *Economic Geology,* **68,** 1135–1143.

Cloud, P. E., and Glaessner, M. F. 1982. The Ediacrian Period and System: Metazoa inherit the Earth. *Science,* **217,** 783–792.

Cohee, G. V., Glaessner, M. F., and Hedberg, H. D. 1978. *Contributions to the Geologic Time Scale.* American Association of Petroleum Geologists, Tulsa, 388 pp.

Collinson, J. D., and Thompson, D. B. 1982. *Sedimentary Structures.* Allen & Unwin, London, 194 pp.

Committee on Stratigraphic Nomenclature. 1933. Classification and nomenclature of rock units. *Geological Society of America Bulletin,* **44,** 423–459, and *American Association of Petroleum Geologists Bulletin,* **17,** 843–868.

Compton, R. R. 1985. *Geology in the Field.* Wiley, New York.

Conkin, B. M., and Conkin, J. E. (Eds.). 1984. *Stratigraphy: Foundations and Concepts.* Van Nostrand–Reinhold, New York, 365 pp.

Costello, W. R., and Southard, J. B. 1981. Flume experiments on lower-flow-regime bed forms in coarse sand. *Journal of Sedimentary Petrology.* **51,** 849–864.

Craig, G. Y., McIntyre, D. B., and Watterson, C. D. 1978. *James Hutton's Theory of the Earth: The Lost Drawings.* Scottish Academic Press, Edinburgh.

Crimes, T. P., and Harper, J. C. (Eds.). 1970. *Trace Fossils.* Seel House Press, Liverpool, 547 pp.

Crimes, T. P., and Harper, J. C. (Eds.). 1977. *Trace Fossils 2.* Seel House Press, Liverpool, 351 pp.

Crowley, K. D. 1983. Large-scale bed configurations (macroforms), Platte River Basin, Colorado and Nebraska: Primary structures and formative processes. *Geological Society of America Bulletin,* **94,** 117–133.

Cubitt, J. M., and Reyment, R. A. (Eds.). 1982. *Quantitative Stratigraphic Correlation.* Wiley, New York, 301 pp.

Cummins, W. A. 1962. The greywacke problem: *Liverpool Manchester Geological Society Journal,* **3,** 51–72.

Curry, R. R. 1966. Observations of alpine mudflows in the Ten Mile Range, central Colorado. *Geological Society of America Bulletin,* **77,** 771–776.

Dabrio, C. J. 1982. Sedimentary structures generated on the foreshore by migrating ridge and runnel systems on microtidal and mesotidal coasts of S. Spain. *Sedimentary Geology,* **32,** 141–151.

Dalrymple, R. W., Knight, R. J., and Lambiase, J. J. 1978. Bedforms and their hydraulic stability relationships in a tidal environment, Bay of Fundy, Canada. *Nature,* **275,** 100–104.

Davidson-Arnott, R. G. D., and Greenwood, B. 1976. Facies relationships on a barred coast, Kouchibouguac Bay, New Brunswick, Canada. In R. A. Davis, Jr., and R. L. Ethington (Eds.), *Beach and Nearshore Sedimentation.* Society of Economic Paleontologists and Mineralogists Special Publication No. 24, pp. 149–168.

Davis, R. A., Jr. 1983. *Depositional Systems: A Genetic Approach to Sedimentary Geology.* Prentice–Hall, Englewood Cliffs, N.J., 669 pp.

Davis, R. A., Jr. (Ed.). 1985. *Coastal Sedimentary Environments,* 2nd ed. Springer-Verlag, New York, 716 pp.

Davis, R. A., Jr., and Ethington, R. L. (Eds.). 1976. *Beach and Nearshore Sedimentation.* Society of Economic Paleontologists and Mineralogists Special Publication No. 25, 187 pp.

Davis, R. A., Fox, W. T., Hayes, M. O., and Boothroyd, J. C. 1972. Comparison of ridge and runnel systems in tidal and nontidal environments. *Journal of Sedimentary Petrology,* **42,** 413–421.

DeCelles, P. G., Langford, R. P., and Schwartz, R. K. 1983. Two new methods of paleocurrent determination from trough cross-stratification. *Journal of Sedimentary Petrology,* **53,** 629–642.

Degens, E. T. 1967. Diagenesis of organic matter. In G. Larsen and G. V. Chilingar (Eds.), *Diagenesis in Sediments,* pp. 343–390. Elsevier, New York.

Dickinson, W. R. (Ed.). 1974. *Tectonics and Sedimentation.* Society of Economic Paleontologists and Mineralogists Special Publication No. 22, 204 pp.

Dingler, J. R. 1974. *Wave-Formed Ripples in Nearshore Sands.* Thesis, University of California, San Diego, Calif., 136 pp.

Dingler, J. R., and Inman, D. L. 1977. Wave-formed ripples in nearshore sands. *Proceedings of the 15th Conference on Coastal Engineering, Honolulu, Hawaii.* pp. 2109–2126.

Donovan, D. T. 1966. *Stratigraphy—An Introduction to Principles.* Wiley, New York, 199 pp.

Dott, R. H. 1964. Wacke, graywacke and matrix—What approach to immature sandstone classification. *Journal of Sedimentary Petrology,* **34,** 625–632.

Dott, R. H., Jr. 1983. 1982 SEPM Presidential Address: Episodic sedimentation—How normal is average? How rare is rare? Does it matter? *Journal of Sedimentary Petrology,* **53,** 5–23.

Dott, R. H., Jr., and Batten, R. L. 1981. *Evolution of the Earth,* 3rd ed. McGraw–Hill, New York, 573 pp.

Dott, R. H., Jr., and Bourgeois, J. 1982. Hummocky stratification: Significance of its variable bedding sequences. *Geological Society of America Bulletin,* **93,** 663–680.

Dunbar, C. O., and Rodgers, J. 1957. *Principles of Stratigraphy.* Wiley, New York, 356 pp.

Dunbar, C. O., and Waage, K. M. 1969, *Historical Geology,* 3rd ed. Wiley, New York, 356 pp.

Dunham, R. J. 1962. Classification of carbonate rocks according to depositional textures. In W. E. Ham (Ed.), *Classification of Carbonate Rocks.* American Association of Petroleum Geologists Memoir 1, pp. 108–121.

Edwards, M. R. 1979. Late Precambrian glacial loessite from North Norway and Svalbard. *Journal of Sedimentary Petrology,* **49,** 85–92.

Eicher, D. L. 1976. *Geologic Time.* Prentice–Hall, Englewood Cliffs, N.J., 150 pp.

Ekdale, A. A., Bromley, R. G., and Pemberton, S. G. 1984. *Ichnology, Trace Fossils in Sedimentology and Stratigraphy.* Society of Economic Paleontologists and Mineralogists Short Course Notes No. 15, 317 pp.

Embry, A. F., and Klovan, J. E. 1971. A late Devonian reef tract on the northeastern Banks Island, N.W.T. *Canadian Petroleum Geology Bulletin,* **19,** 730–781.

Esteban, M. 1976. Vadose pisolite and caliche. *American Association of Petroleum Geologists Bulletin,* **60,** 2048–2057.

Evans, G. 1965. Intertidal flat sediments and their environments of deposition in The Wash. *Geological Society of London Quarterly Journal,* **121,** 209–241.

Fairbridge, R. W., and Bourgeois, J. W. (Eds.). 1978. *The Encyclopedia of Sedimentology.* Dowden, Hutchinson and Ross, Stroudsburg, Penn., 901 pp.

Faul, H. and Faul, C. 1983. *It Began with a Stone.* Wiley, New York, 270 pp.

Fenton, C. L. and Fenton, M. A. 1952. *Giants of Geology.* Doubleday, Garden City, N.Y., 333 pp.

Fisher, R. V. 1961. Proposed classification of volcaniclastic sediments and rocks. *Geological Society of America Bulletin,* **72,** 1409–1414.

Fisher, R. V. 1966. Rocks composed of volcanic fragments and their classification. *Earth-Science Review,* **1,** 287–298.

Fisher, R. V., and Schmincke, H.-U. 1984. *Pyroclastic Rocks.* Springer-Verlag, New York, 472 pp.

Fisk, L. H. 1974. Inverse grading as stratigraphic evidence of large floods: Comment. *Geology,* **2,** 613–615.

Flint, R. F., and Skinner, B. J. 1974. *Physical Geology.* Wiley, New York, 497 pp.

Folk, R. L. 1951. Stages of textural maturity in sedimentary rocks. *Journal of Sedimentary Petrology,* **21,** 127–130.

Folk, R. L. 1954. The distinction between grain size and mineral composition in sedimentary rock nomenclature. *Journal of Geology,* **62,** 344–359.

Folk, R. L. 1959. Practical petrographic classification of limestones. *American Association of Petroleum Geologists Bulletin,* **43,** 1–38.

Folk, R. L. 1962. Spectral subdivision of limestone types. In W. E. Ham (Ed.), *Classification of Carbonate Rocks.* American Association of Petroleum Geologists Memoir 1, pp. 62–84.

Folk, R. L. 1974. *Petrology of Sedimentary Rocks.* Hemphill, Austin, Texas, 182 pp.

Folk, R. L., and Weaver, C. E. 1952. A study of the texture and composition of chert. *American Journal of Science,* **250,** 498–510.

Frey, R. W. 1973. Concepts in the study of biogenic sedimentary structures. *Journal of Sedimentary Petrology,* **43,** 6–19.

Frey, R. W. (Ed.). 1975. *The Study of Trace Fossils.* Springer-Verlag, New York.

Frey, R. W., and Pemberton, S. G. 1984. Trace fossil facies models. In R. G. Walker (Ed.), *Facies Models,* 2nd ed. Geoscience Canada, Reprint Series 1, pp. 189–207.

Frey, R. W., and Pemberton, S. G. 1986. Vertebrate lebensspuren in intertidal and supratidal environments, Holocene barrier islands, Georgia. *Senckenbergiana Marit.,* **18,** 45–95.

Friedman, G. M., and Sanders, J. E. 1978. *Principles of Sedimentology.* Wiley, New York, 792 pp.

Fritz, W. J. 1980. Reinterpretation of the depositional environments of the Yellowstone "fossil forests." *Geology,* **8,** 309–313.

Fritz, W. J. 1982. Geology of the Lamar River Formation, northeast Yellowstone National Park. In S. G. Reid and D. J. Foote (Eds.), *Geology of Yellowstone Park Area.* Wyoming Geological Association Guidebook, 33rd Annual Field Conference, pp. 73–101.

Fritz, W. J., and Harrison, S. 1983. Giant armored mud boulder from the 1982 Mount St. Helens mudflows. *Journal of Sedimentary Petrology,* **53,** 131–133.

Fritz, W. J., and Harrison, S. 1985a. Early Tertiary volcaniclastic deposits of the northern Rocky Mountains. In R. M. Flores and S. S. Kaplan (Eds.), *Cenozoic Paleogeography of West-Central United States.* Rocky Mountain Section of SEPM Symposium 3, pp. 383–402.

Fritz, W. J., and Harrison, S. 1985b. Transported trees from the 1982 Mount St. Helens sediment flows: Their use as paleocurrent indicators. *Sedimentary Geology,* **42,** 49–64.

Fritz, W. J., and Ogren, D. O. 1984. Clast orientations in the Mount St. Helens sediment flows. *Geological Society of America Abstracts with Programs,* **16,** No. 6, 513.

Fryberger, S. G. 1986. Stratigraphic traps for petroleum in wind-laid rocks. *American Association of Petroleum Geologists Bulletin,* **70,** 1765–1776.

Galloway, W. E., and Hobday, D. K. 1983. *Terrigenous Clastic Depositional Systems—Applications to Petroleum, Coal, and Uranium Exploration.* Springer-Verlag, New York, 423 pp.

Geikie, A. 1905. *The Founders of Geology.* Dover, New York, 486 pp. (1962 reprint).

Gibbs, R. J., Matthews, M. D., and Link, D. A. 1971. The relationship betwen sphere size and settling velocity. *Journal of Sedimentary Petrology,* **41,** 7–18.

Ginsburg, R. N. (Ed.). 1975. *Tidal Deposits.* Springer-Verlag, New York, 428 pp.

Greenwood, B., and Sherman, D. J. 1986. Hummocky cross-stratification in the surf zone: Flow parameters and bedding genesis. *Sedimentology,* **33,** 33–45.

Grim, R. E. 1968. *Clay Mineralogy,* 2nd ed. McGraw–Hill, New York, 596 pp.

Grim, R. E., and Guven, N. 1978. Bentonites. In *Developments in Sedimentology,* Vol. 24. Elsevier, Amsterdam, 256 pp.

Guy, H. P., Simons, D. B., and Richardson, E. V. 1966. *Summary of Alluvial Channel Data from Flume Experiments, 1956–1961.* U. S. Geological Survey Professional Paper 462-I, 96 pp.

Hall, A. M., and Fritz, W. J. 1984. Armored mud balls from Cabretta and Sapelo barrier islands, Georgia. *Journal of Sedimentary Petrology,* **54,** 831–835.

Hallam, A. 1983. *Great Geological Controversies.* Oxford Univ. Press, New York, 182 pp.

Hampton, M. A. 1970. *Subaqueous Debris Flow and Generation of Turbidity Currents.* Ph.D. dissertation, Stanford University, Palo Alto, Calif., 80 pp.

Hampton, M. A. 1972. The role of subaqueous debris flows in generating turbidity currents. *Journal of Sedimentary Petrology,* **42,** 775–793.

Hampton, M. A. 1975. Competence of fine-grained debris flows. *Journal of Sedimentary Petrology,* **45,** 834–844.

Hampton, M. A. 1979. Buoyancy in debris flows. *Journal of Sedimentary Petrology,* **49,** 753–758.

Harbaugh, J. W. 1968. *Stratigraphy and Geologic Time.* W. C. Brown, Dubuque, Iowa, 113 pp.

Harland, W. B., Cox, A. V., Llewellyn, P. G., Picton, C. A. G., Smith, A. G., and Walters, R. 1982. *A Geologic Time Scale.* Cambridge Univ. Press, Cambridge, England, 131 pp.

Harland, W. B., Herod, K. N., and Krinsley, D. H. 1966. The definition and identification of tills and tillites. *Earth Science Review,* **2,** 225–256.

Harms, J. C. 1969. Hydraulic significance of some sand ripples. *Geological Society of America Bulletin,* **80,** 363–396.

Harms, J. C., Southard, J. B., Spearing, D. R., and Walker, R. G. 1975. *Depositional Environments as Interpreted from Primary Sedimentary Structures and Stratification Sequences.* Society of Economic Paleontologists and Mineralogists Short Course No. 2, 161 pp.

Harms, J. C., Southard, J. B., and Walker, R. G. 1982. *Structures and Sequences in Clastic Rocks.* Society of Economic Paleontologists and Mineralogists Lecture Notes for Short Course No. 9.

Harrison, S., and Fritz, W. J. 1982. Depositional features of March 1982 Mount St. Helens sediment flows. *Nature,* **299,** 720–722.

Hawkins, J. W., and Whetten, J. T. 1969. Graywacke matrix minerals. Hydrothermal reactions with Columbia River sediments. *Science,* **166,** 868–870.

Hjulstrom, F. 1935. Studies of the morphological activities of rivers as illustrated by the River Fyris. *Geol. Inst. Bull., University of Uppsala,* **25**, 221–527.

Hjulstrom, F. 1939. Transportation of detritus by moving water. In P. D. Trask (Ed.), *Recent Marine Sediments: A Symposium,* Pt. 1, *Transportation,* pp. 5–31, American Association of Petroleum Geologists, Tulsa, Okla. Murby, London, 736 pp. (reprinted 1968 by Dover, New York).

Holmes, A. 1911. The association of lead with uranium in rock minerals and its application to the measurement of geologic time. *Proceedings of the Royal Society of London, Series A,* **85**, 248–256.

Honnorez, J., and Kirst, P. 1976. Submarine basaltic volcanism: Metamorphic parameters for discriminating hyaloclastites from hyalotuffs. *Bulletin of Volcanology,* **39**, 441–465.

Hooke, R. LeB. 1967. Processes on arid-region alluvial fans. *Journal of Geology,* **75**, 438–460.

Hoyt, J. H. and Henry, V. J. Jr., 1963. Rhomboid ripple mark, indicator of current direction and environment. *Journal of Sedimentary Petrology,* **33**, 604–608.

Hoyt, J. H. and Henry, V. J. Jr., 1964. Development and geologic significance of soft beach sand. *Sedimentology,* **3**, 44–51.

Hunter, R. E. 1977. Basic types of stratification in small eolian dunes. *Sedimentology,* **24**, 361–387.

Hunter, R. E. 1981. Stratification styles in eolian sandstones: Some Pennsylvanian to Jurassic examples from the western interior U.S.A. In F. G. Ethridge and R. M. Flores (Eds.), *Recent and Ancient Nonmarine Depositional Environments: Models for Exploration.* Society of Economic Paleontologists and Mineralogists Special Publication No. 31, pp. 315–329.

Hutton, J. 1788. Theory of the earth; or an investigation of the laws observable in the composition, dissolution, and restoration of land upon the globe. *Transactions of the Royal Society of Edinburgh,* **1**, 209–304.

Hutton, J. 1795. *Theory of the Earth with Proof and Illustrations.* Cadell and Davies, London.

Hutton, J. 1899. *Theory of the Earth with Proofs and Illustrations,* A. Geike (Ed.), Vol. 3. Burlington House, London.

Imbrie, J., and Purdy, E. G. 1962. *Classification of Modern Bahamian Carbonate Sediments.* American Association of Petroleum Geologists Memoir 1, pp. 253–272.

Inman, D. L. 1957. *Wave Generated Ripples in Nearshore Sands.* Technical Memorandum No. 100, Beach Erosion Board, U.S. Army Corps of Engineers.

Inman, D. L. and Bowen, A. J. 1963. Flume experiments on sand transport by waves and currents. *Proceedings of the 8th Conference on Coastal Engineering, Berkeley, California,* pp. 137–150.

International Subcommission on Stratigraphic Classification (H. D. Hedberg, Ed.). 1976. *International Stratigraphic Guide: A Guide to Stratigraphical Classification, Terminology and Procedure.* International Subcommission on Stratigraphic Classification of IUGS Commission on Stratigraphy. Wiley, New York, 200 pp.

International Subcommission on Stratigraphic Classification (Amos Salvador, Chairman). 1987. Unconformity-bounded stratigraphic units. *Geological Society of America Bulletin,* **98,** 232–237.

Jameson, R. 1808. *Elements of Geognosy,* being Vol. III and Part II of the *System of Mineralogy,* Edinburgh.

Jameson, R. 1976. *The Wernerian Theory of the Neptunian Origin of Rocks.* A facsimile reprint of *Elements of Geognosy* (1808) edited by R. Jameson and G. W. White. Hafner, New York, 368 pp.

Janda, R. J., Scott, K. M., Nolan, K. M., and Martinson, H. A. 1981. Lahar movement, effects, and deposits. In P. W. Lipman and D. R. Mullineaux (Eds.), *The 1980 Eruptions of Mount St. Helens, Washington.* U.S. Geological Survey Professional Paper 1250, pp. 461–478.

Johnson, A. M. 1965. *A Model for Debris Flow.* Unpublished Ph. D. dissertation. The Pennsylvania State University, University Park, Penn., 205 pp.

Johnson, A. M. 1970. *Physical Processes in Geology.* Freeman, San Francisco, 571 pp.

Jopling, A. V., and Richardson, E. V. 1966. Backset bedding developed in shooting flow in laboratory experiments. *Journal of Sedimentary Petrology,* **36,** 821–825.

Kauffman, E. G., and Hazel, J. E. (Eds.). 1977. *Concepts and Methods of Biostratigraphy.* Dowden, Hutchinson and Ross, Stroudsberg, Penn., 658 pp.

Kennett, J. P. 1980. *Magnetic Stratigraphy of Sediments.* Benchmark Papers in Geology 54. Dowden, Hutchinson and Ross, Stroudsberg, Penn., 438 pp.

Klein, G. deV. 1963. Analysis and review of sandstone classifications in North American geological literature, 1940–1960. *Geological Society of America Bulletin,* **74,** 555–576.

Klein, G. deV. 1977. *Clastic Tidal Facies.* CEPCO: Continuing Education Publication Co., Champaign, Ill., 147 pp.

Knight, R. J., and Dalrymple, R. W. 1975. Intertidal sediments from the south shore of Coequid Bay, Bay of Fundy, Nova Scotia, Canada. In R. N. Ginsburg (Ed.), *Tidal Deposits. A Casebook of Recent Examples and Fossil Counterparts,* Chap. 6. Springer-Verlag, New York.

Kocurek, G., and Dott, R. H., Jr. 1981. Distinctions and uses of stratification types in the interpretation of eolian sand. *Journal of Sedimentary Petrology,* **51,** 579–595.

Komar, P. D. 1974. Oscillatory ripple marks and the evaluation of ancient wave conditions and environments. *Journal of Sedimentary Petrology,* **44,** 169–180.

Komar, P. D. 1976. *Beach Processes and Sedimentation.* Prentice–Hall, Englewood Cliffs, N.J., 429 pp.

Koster, E. H., and Steel, R. J. (Eds.). 1984. *Sedimentology of Gravels and Conglomerates.* Canadian Society of Petroleum Geologists Memoir 10, 441 pp.

Krumbein, W. C. 1934. Size frequency distributions of sediments. *Journal of Sedimentary Petrology,* **4,** 65–77.

Krumbein, W. C. 1936. The application of logarithmic moments to size frequency distributions of sediments. *Journal of Sedimentary Petrology,* **6,** 35–47.

Krumbein, W. C., and Sloss, L. L. 1963. *Stratigraphy and Sedimentation.* Freeman, San Francisco, 660 pp.

Krynine, P. D. 1948. The megascopic study and field classification of sedimentary rocks. *Journal of Geology,* **56,** 130–165.

Leeder, M. R. 1982. *Sedimentology: Process and Product.* Allen & Unwin, London, 344 pp.

Lees, A. 1975. Possible influence of salinity and temperature on modern shelf carbonate sedimentation. *Marine Geology,* **19,** 159–198.

Lewan, M. D. 1978. Laboratory classification of very fine-grained sedimentary rocks. *Geology,* **6,** 745–748.

Link, M. H. 1975. Matilija Sandstone: A transition from deep-water turbidite to shallow-marine deposition in Eocene of California. *Journal of Sedimentary Petrology,* **45,** 63–78.

Little, R. R. 1982. Lithified armored mud balls of Lower Jurassic Turner Falls sandstone, north-central Massachusetts. *Journal of Geology,* **90,** 203–207.

Lofquist, K. E. G. 1978. *Sand Ripple Growth in an Oscillatory-Flow Water Tunnel.* Technical Paper 75-8, Coastal Engineering Research Center, U.S. Army Corps of Engineers.

Logan, B. W., Rezak, R., and Ginsburg, R. N. 1964. Classification and environmental significance of algal stromatolites. *Journal of Geology,* **72,** 68–83.

Lowe, D. R. 1976a. Grain flow and grain flow deposits. *Journal of Sedimentary Petrology,* **46,** 188–199.

Lowe, D. R. 1976b. Subaqueous liquified and fluidized sediment flows and their deposits. *Sedimentology,* **23,** 285–308.

Lowe, D. R. 1979a. Stratigraphy and sedimentology of the Pigeon Point Formation, San Mateo County, California. In T. H. Nilsen and E. E. Brabb (Eds.), *Geology of the Santa Cruz Mountains, California.* Geological Society of America, Cordilleran Section, Guidebook, pp. 17–29, 56–60.

Lowe, D. R. 1979b. Sediment gravity flows: Their classification and some problems of application to natural flows and deposits. In L. J. Doyle and O. H. Pilkey, Jr. (Eds.), *Geology of Continental Slopes.* Society of Economic Paleontologists and Mineralogists Special Publication 27, pp. 75–82.

Lowe, D. R. 1982. Sediment gravity flows. II. Depositional models with special reference to the deposits of high-density turbidity currents. *Journal of Sedimentary Petrology,* **52,** 279–297.

Lyell, C. 1842. *Principles of Geology.* Reprinted from the 6th English ed., Hilliard, Gray, Boston.

Mathews, R. K. 1984. *Dynamic Stratigraphy: An Introduction to Sedimentation and Stratigraphy.* Prentice–Hall, Englewood Cliffs, N.J., 489 pp.

McBride, E. F. 1963. A classification of common sandstones. *Journal of Sedimentary Petrology,* **33,** 664–669.

McKee, E. D. 1965. Experiments on ripple lamination. In G. V. Middleton, (Ed.), *Primary Sedimentary Structures and Their Hydrodynamic Interpretation:* Society of Economic Paleontologists and Mineralogists Special Publication 12, pp. 66–83.

McKee, E. D. 1966. Structures of dunes at White Sands National Monument, New Mexico (and comparison with structures of dunes from other selected areas). *Sedimentology,* **7,** 1–69.

McKee, E. D. (Ed.). 1979. *A Study of Global Sand Seas.* U.S. Geological Survey Professional Paper 1052, 429 pp.

Meissner, F. 1972. Cyclic sedimentation in Middle Permian strata of the Permian Basin, West Texas and New Mexico. In J. C. Elam and S. Chuber (Eds.), *Cyclic Sedimentation in the Permian Basin,* 2nd ed., pp. 203–232. West Texas Geological Society.

Miall, A. D., 1978a. Lithofacies types and vertical profile models in braided river deposits: a summary. In A. D. Miall (Ed.), *Fluvial Sedimentology,* pp. 597–604. Canadian Society of Petroleum Geologists Memoir 5.

Miall, A. D. (Ed.). 1978b. *Fluvial Sedimentology.* Canadian Soc. Pet. Geol. Memoir 5, 859 pp.

Miall, A. D. 1984. *Principles of Sedimentary Basin Analysis.* Springer-Verlag, New York, 490 pp.

Middleton, G. V. (Ed.). 1965a. *Primary Sedimentary Structures and Their Hydrodynamic Interpretation.* Society of Economic Paleontologists and Mineralogists Special Publication 12, 265, pp.

Middleton, G. V. 1965b. Antidune cross-bedding in a large flume. *Journal of Sedimentary Petrology,* **35,** 922–927.

Middleton, G. V. 1969. Turbidity currents and grain flows and other mass movements down slopes. In D. J. Stanley (Ed.), *The New Concepts of Continental*

Margin Sedimentation, pp. GM-A-1 to GM-B-14. American Geological Institute Short Course Lecture Notes.

Middleton, G. V. 1973. Johannes Walther's law of the correlation of facies. *Geological Society of America Bulletin,* **84,** 979–988.

Middleton, G. V. (Ed.). 1977. *Sedimentary Processes: Hydraulic Interpretation of Primary Sedimentary Structures.* Society of Economic Paleontologists and Mineralogists Reprint Series No. 3, 285 pp.

Middleton, G. V., and Bouma, A. H. (Eds.). 1973. *Turbidites and Deep-Water Sedimentation.* Society of Economic Paleontologists and Mineralogists, Pacific Section, Short Course Lecture Notes, 158 pp.

Middleton, G. V., and Hampton, M. A. 1973. Sediment gravity flows: Mechanics of flow and deposition. In *Turbidites and Deep Water Sedimentation,* pp. 1–38. Society of Economic Paleontologists and Mineralogists, Pacific Section, Short Course Lecture Notes.

Middleton, G. V., and Hampton, M. A. 1976. Subaqueous sediment transport and deposition by sediment gravity flows. In D. J. Stanley and D. J. P. Swift (Eds.), *Marine Sediment Transport and Environmental Management,* pp. 197–218. Wiley, New York.

Miller, H. 1860. *The Old Red Sandstone: Or, New Walks in an Old Field: Boston.* Gould and Lincoln, New York, 427 pp.

Miller, M. C., and Komar, P. D. 1980a. Oscillation sand ripples generated by laboratory apparatus. *Journal of Sedimentary Petrology,* **50,** 173–182.

Miller, M. C., and Komar, P. D. 1980b. A field investigation of the relationship between oscillation ripple spacing and the near-bottom water orbital motions. *Journal of Sedimentary Petrology,* **50,** 183–191.

Miller, M. F., Ekdale, A. A., and Picard, M. D. (Eds.). 1984. Trace fossils and paleoenvironments: Marine carbonate, marginal marine terrigenous and continental terrigenous settings. *Journal of Paleontology,* **58,** 283–597.

Mills, H. H. 1984. Clast orientation in Mount St. Helens debris-flow deposits, North Fork Toutle River, Washington. *Journal of Sedimentary Petrology,* **2,** 626–634.

Mitchum, R. M., Jr., Vail, P. R., and Thompson, S., III. 1977. The depositional sequence as a basic unit for stratigraphic analysis. In C. E. Payton (Ed.), *Seismic Stratigraphy—Applications to Hydrocarbon Exploration.* American Association of Petroleum Geologists Memoir 26, pp. 53–62.

Moore, J. N., Fritz, W. J., and Futch, R. S. 1984. Occurrence of megaripples in a ridge and runnel system, Sapelo Island, Georgia: Morphology and processes. *Journal of Sedimentary Petrology,* **54,** 615–625.

Moore, L. R. 1968. Cannel coals, bogheads and oil shales. In D. Murchinson and T. S. Westoll (Eds.), *Coal and Coal Bearing Strata,* pp. 19–29. Oliver & Boyd, Edinburgh.

Mount, J. 1984. Mixing of siliclastic and carbonate sediments in shallow shelf environments. *Geology,* **12,** 432–435.

Mount, J. 1985. Mixed siliciclastic and carbonate sediments: A proposed first-order textural and compositional classification. *Sedimentology,* **32,** 435–442.

Murray, J. W. 1981. *A Guide to Classification in Geology.* Ellis Horwood, London, and Halstead, New York, 112 pp.

Mustoe, G. E. 1982. The origin of honeycomb weathering. *Geological Society of America Bulletin,* **93,** 108–115.

Mutti, E., and Ricci-Lucchi, F. 1972. Le Torbiditi dell' Appennino settrentrionale: introduzione all' analisi di facies. *Soc. Geol. Italiana, Mem.,* **11,** 161–199; reprinted and translated: by T. H. Nielson: Turbidites of the Northern Apennines: Introduction to facies analysis. *International Geology Review,* **20,** (1978), 125–166.

Naranjo, J. L., Sigurdsson, H., Carey, S. N., and Fritz, W. 1986. Eruption of the Nevado del Ruiz volcano, Colombia, on 13 November 1985: Tephra fall and lahars. *Science,* **233,** 961–963.

Nardin, T. R., Hein, F. J., Gorsline, D. S., and Edwards, B. D. 1979. *A Review of Mass Movement Processes, Sediment and Acustic Characteristics, and Contrasts in Slope and Base-of-Slope Systems versus Canyon–Fan–Basin Floor Systems.* Society of Economic Paleontologists and Mineralogists Special Publication 27, pp. 61–73.

Naylor, M. A. 1980. The origin of inverse grading in muddy debris flow deposits—A review. *Journal of Sedimentary Petrology* **50,** 1111–1116.

Neidell, N. S. 1979. *Stratigraphic Modeling and Interpretation: Geophysical Principles and Techniques.* American Association of Petroleum Geologists Education Course Notes 13, 141 pp.

Nelson, C. H., and Nielson, T. H. 1984. *Modern and Ancient Deep-Sea Fan Sedimentation.* Society of Economic Paleontologists and Mineralogists Short Course 14, 404 pp.

Nelson, C. H., Johnson, K. R., and Barber, J. H., Jr. 1987. Gray whale and walrus feeding excavation on the Bering Shelf, Alaska. *Journal of Sedimentary Petrology,* **50,** 419–430.

Nielson, T. H., and Abbott, P. L. 1981. Paleogeography and sedimentology of upper Cretaceous turbidites, San Diego, California. *American Association of Petroleum Geologists Bulletin,* **65,** 1256–1284.

North American Commission on Stratigraphic Nomenclature. 1983. North American Stratigraphic Code. *American Association of Petroleum Geologists Bulletin,* **67,** 841–875 (reprinted as Appendix A in this text).

Nøttvedt, A., and Kreisa, R. D. 1987. Model for the combined-flow origin of hummocky cross stratification. *Geology,* **15,** 357–361.

Odin, G. S. (Ed.). 1982. *Numerical Dating in Stratigraphy,* Pt. I, pp. 1–630, and Pt. II, pp. 631–1040. Wiley, New York.

Okada, H. 1971. Classification of sandstone: Analysis and proposal. *Journal of Geology,* **79,** 509–525.

Orford, J. D., and Wright, P. 1978. What's in a name?—Descriptive or genetic implications of ridge and runnel topography. *Marine Geology,* **28,** M1–M8.

Owen, D. E. 1978. Usage of stratigraphic nomenclature and concepts in the *Journal of Sedimentary Petrology* or time, place, and rocks—How to keep them separate. *Journal of Sedimentary Petrology,* **48,** 355–358.

Owen, D. E. 1987. Commentary: Usage of stratigraphic terminology in papers, illustrations, and talks. *Journal of Sedimentary Petrology,* **57,** 363–372.

Palmer, A. R. (comp.). 1983. The decade of North American geology 1983 geologic time scale. *Geology,* **11,** 503–504.

Peltz, S. 1971. Quelques considerations sur la nomenclature et classification des pyroclastites. *Bulletin of Volcanology,* **35,** 295–302.

Pettijohn, F. J. 1975. *Sedimentary Rocks,* 3rd ed. Harper & Row, New York, 628 pp.

Pettijohn, F. J., and Potter, P. E. 1964. *Atlas and Glossary of Primary Sedimentary Structures.* Springer-Verlag, New York, 370 pp.

Pettijohn, F. J., Potter, P. E., and Siever, R. 1972. *Sand and Sandstone.* Springer-Verlag, New York, 618 pp.

Picard, M. D. 1971. Classification of fine-grained sedimentary rocks. *Journal of Sedimentary Petrology,* **41,** 179–195.

Pierson, T. C. 1982. Flow behavior of two major lahars triggered by the May 18, 1980 eruption of Mount St. Helens, Washington. *Proceedings, Symposium on Erosion Control in Volcanic Areas,* Public Works Research Institute, Japan, Technical Memorandum No. 1908, pp. 99–129.

Pierson, T. C. 1985. Initiation and flow behavior of the 1980 Pine Creek and Muddy River lahars, Mount St. Helens, Washington. *Geological Society of America Bulletin,* **96,** 1056–1069.

Pierson, T. C., and Costa, J. E. 1984. A rheologic classification of subaerial sediment-water flows (abstract). *Geological Society of America Abstracts with Programs,* **16,** No. 6, 623.

Pierson, T. C., and Scott, K. M. 1985. Downstream dilution of a lahar: Transition from debris flow to hyperconcentrated streamflow. *Water Resource Research,* **21,** 1511–1524.

Playfair, J. 1805. Biographical account of the late Dr. James Hutton. *Transactions of the Royal Society of Edinburgh,* **5,** 39–99.

Poag, C. W., and Ward, L. W. 1987. Cenozoic unconformities and depositional sequences of North Atlantic continental margins: Testing the Vail model. *Geology,* **15,** 159–162.

Potter, P. E. 1984. South African (sic) modern beach sand and plate tectonics. *Nature,* **311,** 645–648.

Potter, P. E., Maynard, J. B., and Pryor, W. A. 1980. *Sedimentology of shale.* Springer-Verlag, New York, 306 pp.

Potter, P. E., and Pettijohn, F. J. 1975. *Paleocurrents and Basin Analysis,* 2nd ed. Springer-Verlag, New York, 460 pp.

Power, W. R., Jr. 1961. Backset beds in the Coso Formation, Inyo County, California. *Journal of Sedimentary Petrology,* **31,** 603–607.

Powers, M. C. 1953. A new roundness scale for sedimentary particles. *Journal of Sedimentary Petrology,* **23,** 117–119.

Prandtl, L. 1952. *Essentials of Fluid Dynamics.* Blackie, London.

Raiswell, R. 1971. The growth of Cambrian and Liassic concretions. *Sedimentology,* **17,** 147–171.

Reading, H. G. (ed.). 1986. *Sedimentary Environments and Facies,* 2nd ed. Blackwell, Oxford, 615 pp.

Reeves, C. C. 1970. Origin, classification, and geologic history of caliche on the southern High Plains, Texas and eastern New Mexico. *Journal of Geology,* **78,** 352–362.

Reineck, H. E. 1972. Tidal flats. In J. K. Rigby and W. K. Hamblin (Eds.), *Recognition of Ancient Depositional Environments.* Society of Economic Paleontologists and Mineralogists Special Publication No. 16, pp. 146–159.

Reineck, H. E., and Singh, I. B. 1980. *Depositional Sedimentary Environments,* 2nd ed. Springer-Verlag, Berlin and New York, 549 pp.

Reineck, H. E., and Wunderlich, F. 1969. Die Eentstehung von Schichten und Schichtbanken im Watt. *Senckenbergiana Maritima,* **1,** 85–106.

Rouse, H. 1937. *Nomogram for Settling Velocity of Spheres,* App. 1, pp. 136–137. Report of the Commission of Sedimentation.

Rubey, W. W. 1933. Settling velocities of gravel, sand, and silt particles. *American Journal of Science, 5th Series,* **25,** 325–338.

Rubin, D. M., and Hunter, R. E. 1982. Bedform climbing in theory and nature. *Sedimentology,* **29,** 121–138.

Rubin, D. M., and Hunter, R. E. 1983. Reconstructing bedform assemblages from compound crossbedding. In M. E. Brookfield and T. S. Ahlbrandt (Eds.), *Eolian Sediments and Processes.* Developments in Sedimentology 38, Elsevier, Amsterdam.

Rubin, D. M., and McCulloch, D. S. 1980. Single and superimposed bedforms: A synthesis of San Francisco Bay and flume observations. *Sedimentary Geology,* **26,** 207–231.

Saddler, P. M. 1981. Sediment accumulation rates and the completeness of stratigraphic sections. *Journal of Geology,* **89,** 569–584.

Sanders, J. E. 1965. Primary sedimentary structures formed by turbidity currents and related resedimentation mechanisms. In G. V. Middleton (Ed.), *Primary Sedimentary Structures and Their Hydrodynamic Interpretation.* Society of Economic Paleontologists and Mineralogists Special Publication 12, pp. 192–219.

Sarjeant, W. A. S. (Ed.). 1983. *Terrestrial Trace Fossils.* Dowden, Hutchinson and Ross, Stroudsburg, Penn., 415 pp.

Schmid, R. 1981. Descriptive nomenclature and classification of pyroclastic deposits and fragments: Recommendations of the IUGS Subcommission on the Systematics of Igneous Rocks. *Geology,* **9,** 41–43.

Schneck, W. M., and Fritz, W. J. 1985. An amphibian trackway (*Cincosaurus cobbi*) from the lower Pennsylvanian ("Pottsville") of Lookout Mountain, Georgia: A first occurrence. *Journal of Paleontology,* **59,** 1243–1250.

Schneer, C. J. (Ed.). 1969. *Toward a History of Geology.* M.I.T. Press, Cambridge, Mass.

Scholle, P. A., Bebout, D. G., and Moore, C. H. (Eds.). 1983. *Carbonate Depositional Environments.* American Association of Petroleum Geologists Memoir 33, 708 pp.

Scholle, P. A., and Spearing, P. R. (Eds.). 1982. *Sandstone Depositional Environments.* American Association of Petroleum Geologists Memoir 31, 410 pp.

Scoffin, T. P. 1987. *An Introduction to Carbonate Sediments and Rocks.* Blackie, Glasgow, 274 pp.

Selley, R. C. 1982. *An Introduction to Sedimentology,* 2nd ed. Academic Press, London, 417 pp.

Selley, R. C. 1985. *Ancient Sedimentary Environments,* 3rd ed. Cornell Univ. Press, Ithaca, N.Y.

Sharp, R. P. 1963. Windripples: *Journal of Geology,* **71,** 617–636.

Shaw, D. B., and Weaver, C. E. 1965. The mineralogical composition of shales. *Journal of Sedimentary Petrology,* **35,** 213–222.

Shea, J. H. 1982. Twelve fallacies of uniformitarianism. *Geology,* **10,** 455–460.

Simons, D. B., Richardson, E. V., and Nordin, C. F., Jr. 1965. *Sedimentary Structures Generated by Flow in Alluvial Channels.* Society of Economic Paleontologists and Mineralogists Special Publication 12, pp. 34–52.

Simonson, B. E. 1985. Sedimentological constraints on the origins of Precambrian iron-formations. *Geological Society of America Bulletin,* **96,** 244–252.

Skipper, K. 1971. Antidune cross-stratification in a turbidite sequence, Cloridrome Formation, Gaspe, Quebec. *Sedimentology,* **17,** 51–68.

Sloss, L. L. 1963. Sequences in the cratonic interior of North America. *Geological Society of America Bulletin,* **74,** 93–114.

Smalley, I. J. (Ed.). 1975. *Loess, Lithology and Genesis.* Dowden, Hutchinson and Ross, Stroudsburg, Penn.

Smith, G. A. 1986. Coarse-grained nomarine volcaniclastic sediment: Terminology and depositional process. *Geological Society of America Bulletin,* **97.** p. 1–10.

Smosna, R. 1987. Compositional maturity of limestones—A review. *Sedimentary Geology,* **51,** 137–146.

Sneed, E. D., and Folk, R. L. 1958. Pebbles in the lower Colorado River, Texas, a study in particle morphogenesis. *Journal of Geology,* **66,** 114–150.

Sorby, H. C. 1859. On the structures produced by the current present during the deposition of stratified rocks. *The Geologist,* **2,** 137–147.

Southard, J. B. 1971. Representation of bed configurations in depth–velocity–size diagrams. *Journal of Sedimentary Petrology,* **41,** 903–915.

Sparks, R. S. J., and Walker, G. P. L. 1973. The ground surge deposit: A third type of pyroclastic rock. *Nature, Phys. Sci.,* **241,** 62–64.

Stanley, D. J., and Kelling, G. (Eds.). 1978. Sedimentation in Submarine Canyons, Fans, and Trenches. Dowden, Hutchinson and Ross, Stroudsburg, Penn., 395 pp.

Stauffer, P. H. 1967. Grain-flow deposits and their implications, Santa Ynez Mountains, California. *Journal of Sedimentary Petrology,* **37,** 487–508.

Steno, Nicolaus. 1667a. *Elementorvm myologiae specimen sev descripto geometrica: Cvi Accedvnt Canis carchariae dissectvm Capvt, et Dissectvs piscis ex canvm genere.* Florence.

Steno, Nicolaus. 1667b. *De solido intra solidvm naturaliter contento dissertation is prodromvs.* Florence.

Steno, Nicolaus. 1916. *The Prodromus of Nicolaus Steno's Dissertation Concerning a Solid Body Enclosed by Processes of Nature within Solid.* John Garrett Winter, Transactions, University of Michigan Studies, Humanistic Series, Vol. 9, Part 2. Reprinted by Hafner, New York, 1968.

Steno, Nicolaus. 1969. *Steno Geological Papers,* Gustav Scherz (Ed.) and Alex J. Pollock (trans.). University Press, Odense, Denmark.

Stokes, G. G. 1851. On the effect of the internal friction of fluids on the motion of pendulums. *Cambridge Philosophical Society,* **9,** No. 2, 8–108.

Stokes, G. G. 1891. *Natural Theology.* Adam and Charles Black, London, 272 pp. (reprinted by AMS Press, New York, 1979).

Sundborg, Å. 1956. The river Klaralven, a study of fluvial processes. *Geog. Ann.,* **38,** 127–316.

Suthern, R. J. 1985. Facies analysis of volcaniclastic sediments: A review. In P. J. Brenchley and B. P. J. Williams (Eds.), *Sedimentology—Recent Developments and Applied Aspects,* pp. 123–146. The Geological Society and Blackwell, Oxford.

Tarling, D. H. 1983. *Paleomagnetism: Principles and Applications in Geology, Geophysics and Archaeology.* Chapman & Hall, London, 379 pp.

Teichmuller, M., and Teichmuller, R. 1968. Geological aspects of coal metamorphism. In D. Murchinson and T. S. Wetoll, (Eds.)., *Coal and Coal Bearing Strata,* pp. 233–267. Oliver & Boyd, Edinburgh.

Tisote, B. P., and Welte, D. H. 1978. *Petroleum Formation and Occurrence.* Springer-Verlag, New York, 538 pp.

Tucker, M. E. 1981. *Sedimentary Petrology: An Introduction.* Halstead, New York, 252 pp.

Tucker, M. E. 1982. *The Field Description of Sedimentary Rocks.* Halstead, New York, 112 pp.

Vail, P. R., and Mitchum, R. M., Jr. 1977. Seismic stratigraphy and global changes of sea level. Part 1. Overview. In C. E. Peyton (Ed.), *Seismic Stratigraphy—Application to Hydrocarbon Exploration.* American Association of Petroleum Geologists Memoir 26, pp. 51–52.

Vail, P. R., and Mitchum, R. M., Jr., and Thompson, S., III. 1977a. Seismic stratigraphy and global changes of sea level. Part 3. Relative changes of sea level from coastal onlap. In C. E. Peyton (Ed.), *Seismic Stratigraphy—Application to Hydrocarbon Exploration.* American Association of Petroleum Geologists Memoir 26, pp. 63–81.

Vail, P. R., and Mitchum, R. M., Jr., and Thompson, S., III. 1977b. Seismic stratigraphy and global changes of sea level. Part 4. Global cycles of relative changes in sea level. In C. E. Peyton (Ed.), *Seismic Stratigraphy—Application to Hydrocarbon Exploration.* American Association of Petroleum Geologists Memoir 26, pp. 83–97.

Van der Kamp, P. C., Harper, J. D., Conniff, J. J., and Moris, D. A. 1973. Facies relations in the Eocene–Oligocene in the Santa Ynez Mountains, California. *Journal of the Geological Society of London,* **130,** 545–565.

van Straaten, L. M. J. U. 1954. Sedimentology of recent tidal flat deposits and the Psammites du Condroz. *Geol. en Mijnbouw* **16,** 197–216.

Visher, G. S. 1984. *Exploration Stratigraphy.* Penwell, Tulsa, Okla., 334 pp.

Waag, C., and Ogren, D. O. 1984. Shape evolution and fabric in a boulder beach, Monument Cove, Maine. *Journal of Sedimentary Petrology,* **54,** 98–102.

Wadell, H. 1932. Volume, shape and roundness of rock particles. *Journal of Geology,* **40,** 443–451.

Waitt, R. B., Jr., Pierson, T. C., MacLeod, N. S., Janda, R. J., Voight, B., and Holcomb, R. T. 1983. Eruption-triggered avalanche, flood and lahar at Mount St. Helens—Effects of winter snowpack. *Science,* **221,** 1394–1397.

Walker, R. G. 1975a. *Conglomerate: Sedimentary Structures and facies models.* Society of Economic Paleontologists and Mineralogists Short Course 2, pp. 133–161.

Walker, R. G. 1975b. Generalized facies models for resedimented conglomerates of turbidite association. *Geological Society of America Bulletin,* **86,** 737–748.

Walker, R. G. (Ed.). 1984. *Facies Models,* 2nd ed. Geological Association of Canada, 317 pp.

Walther, J. 1893–1894. *Einleitung in die Geologie als historische Wissenschaft.* Von Gustav Fisher, Jena, East Germany, 3 vols., 1055 pp.

Weaver, C. E. 1958. Geological interpretation of argillaceous sediments. *American Association of Petroleum Geologists Bulletin,* **42,** 254–271.

Weaver, C. E. 1959. *The Clay Petrology of Sediments.* Proceedings, 6th National Conference on Clays and Clay Minerals, National Research Council Publication 566, pp. 154–187.

Weller, J. M. 1960. *Stratigraphic Principles and Practices.* Harper, New York, 725 pp.

Wentworth, C. K. 1922. A scale of grade and class terms for clastic sediments. *Journal of Geology,* **30,** 377–392.

Wentworth, C. K., and Williams, H. 1932. The classification and terminology of the pyroclastic rocks. *Bulletin of the National Research Council,* **89,** 19–53.

Werner, Abraham Gottlob. 1774. *On the External Characters of Minerals,* Albert Carozzi (trans.). Univ. of Chicago Press, Chicago, 1962.

Werner, Abraham Gottlob. 1786. *Short Classification and Description of the Various Rocks,* A. Ospovat (trans.). Hafner, New York, 1971.

Whetten, J. T. 1966. Sediments from the lower Columbia River and origin of graywacke. *Science,* **152,** 1057–1058.

Whetten, J. T. 1972. *Matrix-Rich Pleistocene Sediments from Western Washington: Incipient Graywacke-Type Sedimentary Rocks?* Geological Society of America Memoir 132, pp. 573–584.

Whetten, J. T., and Hawkins, J. W. 1970. Diagenetic origin of graywacke matrix. *Sedimentology,* **15,** 347–361 (see also discussion by J. P. B. Lovell and reply, **19,** (1972), 141–146).

White, G. W. (Ed.). 1970. James Hutton's *System of the Earth* (1785), *Theory of the Earth* (1788), and *Observations on Granite* (1794). Hafner, Darien, Conn., 203 pp.

Wickman, F. E. 1954. The total amount of sediments and the composition of the average igneous rock. *Geochim. Cosmochim. Acta,* **5,** 97–100.

Wilson, J. L. 1975. *Carbonate Facies and Geologic History.* Springer-Verlag, Berlin, 471 pp.

Wright, A. E., and Bowes, D. R. 1963. Classification of volcanic breccias: A discussion. *Geological Society of America Bulletin,* **74,** 79–86.

Zarillo, G. A. 1982. Stability of bedforms in a tidal environment. *Marine Geology,* **48,** 337–351.

Zingg, Th. 1935. Beitrage zur Schotteranalyse: Schweiz. *Mineralog. Petrog. Mitt.,* **15,** 39–140.

Index

"index . . . a list of items treated in a printed work that gives for each item the page number where it may be found"

Webster's New Collegiate Dictionary